PHYSICOCHEMICAL PRINCIPLES
OF PHARMACY

PHYSICOCHEMICAL PRINCIPLES OF PHARMACY

A. T. FLORENCE

School of Pharmacy
University of London

and

D. ATTWOOD

Department of Pharmacy
University of Manchester

SECOND EDITION

MACMILLAN

First edition 1981
Reprinted 1985, 1986
Second edition 1988
Reprinted 1989, 1990, 1991

Published by
THE MACMILLAN PRESS LTD
Houndmills, Basingstoke, Hampshire RG21 2XS
and London
Companies and representatives
throughout the world

Typeset by TecSet Ltd, Wallington, Surrey
Printed in Hong Kong

ISBN 0–333–44995–9 (hard cover)
ISBN 0–333–44996–7 (paper cover)

Contents

Preface to Second Edition

The text of the first edition has been updated and new material added, but we have endeavoured to make the book no longer than before. Some material has been discarded and some of the sections rearranged to provide a more coherent flow to the text. However, our aim has remained the same: to provide the physicochemical background to drug formulation and delivery. Some of the basic physical chemistry has been removed from the text, not because it has diminished in importance but perhaps because it gave an undue emphasis, for example to thermodynamics, which could not be justified in pharmaceutical applications. The purists might query this decision, but there are two defences. First, there is available an abundance of good straight physical chemistry text-books. Second, there is a limit to the depth and rigour of the basic sciences to which pharmacy undergraduates can be exposed, without deflecting them from the main goal, namely applying their knowledge to pharmacy.

A. T. Florence
Glasgow

D. Attwood
Manchester

Preface to First Edition

This book sets out to provide the physicochemical background to the design and use of pharmaceutical products. It does not cover processing technology as such as this is dealt with adequately elsewhere. Rather an attempt is made to relate the physical chemistry of the drug or drug system to clinical usage. The book deals with the basic situations encountered in the progress of a drug from the dosage form to its site of action and how this can be controlled. Adhesion, de-aggregation, solution, rates of solution, stability, diffusion, partition, aggregation, ionisation, interaction with water, and interaction with other molecules are topics which have been tackled. The special problems of the various routes of administration of particular dosage forms are considered from a physicochemical viewpoint, bearing in mind the physiological constraints. Where relevant, the physical chemistry of adjuvant substances such as surface active agents and polymers has been included as these substances are becoming more widely used to effect changes in the extent or duration of drug activity; often the basic mechanism of their effect is a physical and not a biological one.

Above all, however, an effort is made to unite the physical and biological aspects of pharmaceutics. Students sometimes forget that the same forces operate in inanimate and animate systems and early on in their training cannot see the relevance of the physical chemistry that is taught. It is hoped that this book will go some way towards bridging the gap between the fundamental and applied aspects of physical chemistry, pharmaceutical chemistry and biopharmaceutics. It does not purport to be a complete physical chemistry textbook but should be useful as a textbook which follows on from the standard physical chemistry texts, for use in all years of the undergraduate course. As it is aimed at undergraduates the reference lists at the end of each chapter have been kept to a minimum size. It has frequently been difficult to decide which facts to reference and which not to, but we feel that our approach has been correct. We hope that the book will be of use to undergraduate students of pharmacy and other life sciences and to postgraduate students and practising pharmacists wishing to refresh their memories.

We would be pleased to hear from readers of any errors in our treatment of subjects. It has not been possible to acknowledge by citation of references the contribution of many pharmaceutical scientists who have made this book possible but we nevertheless would have found it impossible to write without recourse to the literature. This is, indeed, one of the reasons why we felt the book was necessary — undergraduate students in pharmacy have had to rely too much on seeking out facts in the original literature. Perhaps this book will ease the way somewhat.

A. T. Florence
Glasgow

D. Attwood
Manchester

1 Gases and Volatile Agents

Gases and volatile substances are encountered in pharmacy mainly as anaesthetic gases or as propellants in aerosols. This chapter deals briefly with the properties of both gases and vapours and examines the application of physico-chemical principles in the formulation of aerosols. The factors governing the solubility of gases in liquids are reviewed and related to the solubility of anaesthetic gases in the complex solvent systems of blood and tissues.

1.1 Ideal and non-ideal gases

Ideal gases obey the combined gas law

$$PV = nRT \tag{1.1}$$

where P is the pressure in $N\,m^{-2}$, V is the volume in m^3, n is the number of moles of gas, T is the temperature in kelvins and R is the gas constant ($8.314\ J\ mol^{-1}\ K^{-1}$).

Equation 1.1 may be derived from the kinetic theory of gases assuming the gas molecules to behave as perfectly elastic spheres having negligible volume with no intermolecular attraction or repulsion.

For a given number of moles of gas the quantity PV/RT should, according to equation 1.1, be independent of changes in P, V or T providing such changes do not involve a change of state. A convenient means of expressing departure from ideality is by plots such as that shown in figure 1.1 in which PV/RT is plotted as a function of pressure for 1 mole of each gas. It is important to note the magnitude of the pressures involved in figure 1.1. The narrow shaded area represents the pressure normally met in pharmaceutical systems and it is clear that the ideal gas laws are sufficient for most purposes.

Where it is clear that equation 1.1 is inadequate in describing the behaviour of a gaseous system, a better approximation to real behaviour may be achieved using the van der Waals' equation

$$\left(P + \frac{an^2}{V^2}\right)(V - nb) = nRT \tag{1.2}$$

where a and b are constants for a particular gas. At the moment of impact of a molecule with the container wall the molecule is subjected to an imbalance of forces which tend to pull it back into the bulk of the gas and so lessen the force of impact. Since pressure is a consequence of collisions of molecules with the walls there is a resultant reduction of pressure which may be corrected by addition of the a/V^2 term. Around each molecule of a gas is a particular volume from which other molecules are excluded for purely physical reasons. The bulk molar volume, V, of the gas is consequently an over-estimation of the true molar volume. In the van der Waals' equation allowance is made for the excluded volume by subtraction of the constant b. Table 1.1 gives values of a and b for some common gases.

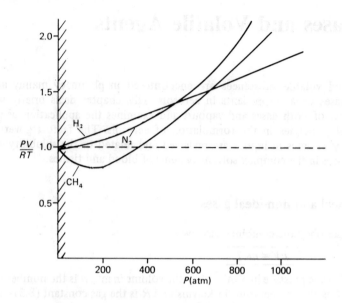

figure 1.1   Departure of gases from ideal behaviour

table 1.1   **Van der Waals' constants for some gases**

	a $(N\,m^4\,mol^{-2})$	b $(m^3\,mol^{-1} \times 10^5)$
H_2	0.0248	2.661
H_2O	0.5537	3.047
O_2	0.1378	3.183
N_2	0.1408	3.913
CO_2	0.3640	4.267
N_2O*	0.3832	4.415
$CH_2\!=\!CH_2$*	0.4530	5.714
C_2H_5OH	1.2180	8.407
C_2H_5Cl*	1.1055	8.651
$CHCl_3$*	1.5372	10.220
$CCl_2\!=\!CHCl$*	1.7206	11.280
$(C_2H_5)_2O$*	1.7611	13.440

*Gases producing anaesthesia

Direct correlation between the anaesthetic potency for a series of gases and the values of the van der Waals' constants has been reported.[1]

1.2 Vapour pressure

1.2.1 Vapour pressure and solution composition – Raoult's law

In pharmaceutical systems in which an equilibrium exists between a liquid and
its vapour, such as in certain types of aerosols, it is important to be able to
calculate the vapour pressure from a knowledge of the composition of the solution. In an ideal solution, the relationship between the partial vapour pressure,
p_i, of a component i in the vapour phase and the mole fraction of that component in solution, x_i, is expressed by Raoult's law as

$$p_i = p_i^{\ominus} x_i \qquad (1.3)$$

where $p_i^{\ominus}$ is the vapour pressure of the pure component.

Binary mixtures of the fluorinated hydrocarbon aerosol propellants show
behaviour which approaches ideality. Figure 1.2*a* shows the vapour pressure-
composition plots for a mixture of the propellants 12 and 114. Deviation from
the Raoult's law plot does not exceed 5 per cent.

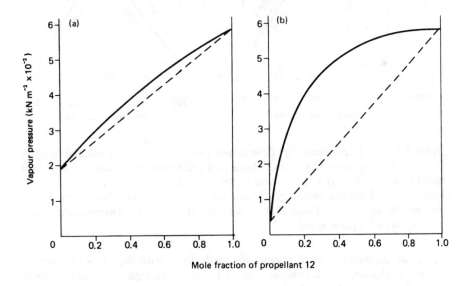

figure 1.2 Vapour pressure–composition curves for (a) propellant 12–propellant
114 mixtures, and (b) propellant 12–ethanol mixtures at 21 °C. Broken line
shows ideal, Raoult's law, line

In aerosol systems consisting of a binary mixture of a propellant and a solvent
such as acetone or alcohol, there is a pronounced departure from ideal behaviour.
Figure 1.2*b* shows a large positive deviation from the Raoult's law plot for mixtures of propellant 12 and ethanol. Such positive deviations usually arise when
the attraction between molecules of one component is greater than that between
the molecules of the two components. This form of interaction is referred to as
association.

Vapour pressure–concentration curves for mixtures of two anaesthetic agents also show positive deviations from ideality (figure 1.3). These are greatest for enflurane and least for halothane. Such curves are of value in assessing errors which may arise through the incorrect usage of agent-specific anaesthetic vaporisers. As the name suggests, these vaporisers are specifically calibrated for a

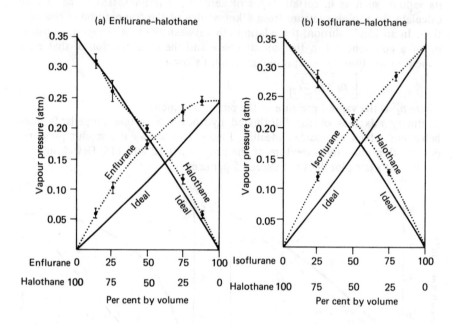

figure 1.3 (a) Experimentally determined (·················) and calculated ideal (———) vapour pressures of halothane and enflurane when combined at 760 mmHg and 22 °C. (b) Experimentally determined (·················) and calculated ideal (———) vapour pressures of halothane and isoflurane when combined at 760 mmHg and 22 °C. From D. L. Bruce and H. W. Linde. *Anesthesiology*, 60, 342 (1984) with permission

particular anaesthetic gas. If a vaporiser partly filled with the correct gas is mistakenly replenished with another, then it is clear from figure 1.3 that, because of the facilitation of vaporisation in the gas mixtures, more of each agent will be delivered than would be the case if ideal mixtures were formed. The clinical consequences of this error will of course depend upon the potencies of each agent as well as the delivered vapour concentrations.

Another system which shows positive deviation is the binary mixture of the volatile drug, methylamphetamine, and eucalyptol. Combinations of these volatile compounds are used in nasal inhalations, and an examination of the vapour pressure of such systems is of interest since the partial pressure exerted by the drug in the presence of other volatile constituents is a major factor governing the dose delivered to the patient. The slight positive deviation from Raoult's law exhibited by this system (figure 1.4) may be explained by hydrogen bonding between the amine groups of methylamphetamine molecules.

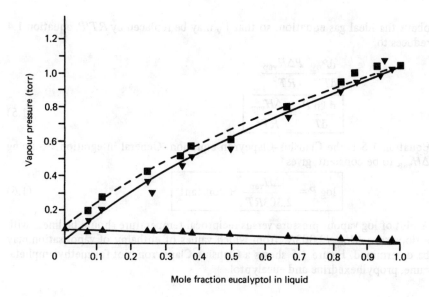

figure 1.4 Effect of mole fraction of eucalyptol in methylamphetamine on total and partial vapour pressures of the mixtures at 25.25 °C. ▲, Partial pressure methylamphetamine; ▼, partial pressure eucalyptol; ■, total vapour pressure of mixture. (Note: 1 torr = 133.3 N m^{-2}.) From P. A. M. Armstrong, J. E. Carless and R. B. Enever. *J. Pharm. Pharmacol.*, 23, 473 (1971) with permission

Negative deviations from Raoult's law may arise when the specific attractions between the component molecules exceed the normal attractions which exist between the molecules of each pure component.

1.2.2 Variation of vapour pressure with temperature – Clausius–Clapeyron equation

The increased motion of the molecules of the liquid following an increase of temperature leads to a greater tendency for escape of molecules into the vapour phase with a consequent increase of vapour pressure. The variation of vapour pressure with temperature may be expressed in terms of the molar enthalpy of vaporisation of the liquid, ΔH_{vap}, using the Clapeyron equation

$$\frac{dP}{dT} = \frac{\Delta H_{vap}}{T \Delta V} \tag{1.4}$$

In this equation ΔV is the difference in molar volumes of the two phases. Since the molar volume of the vapour is very much larger than that of the liquid, ΔV may be approximately equated with V_v. If it is also assumed that the vapour obeys the ideal gas equation, so that V_v may be replaced by RT/P, equation 1.4 reduces to

$$\frac{dP}{dT} = \frac{P\Delta H_{vap}}{RT^2}$$

or

$$\frac{d\ln P}{dT} = \frac{\Delta H_{vap}}{RT^2} \qquad (1.5)$$

Equation 1.5 is the Clausius-Clapeyron equation. General integration, assuming ΔH_{vap} to be constant, gives

$$\log P = \frac{-\Delta H_{vap}}{2.303RT} + \text{constant} \qquad (1.6)$$

A plot of log vapour pressure versus reciprocal temperature should be linear with a slope of $-\Delta H_{vap}/2.303R$, from which values of enthalpy of vaporisation may be determined. Figure 1.5 shows a Clausius-Clapeyron plot for methylamphetamine, propylhexedrine and eucalyptol.

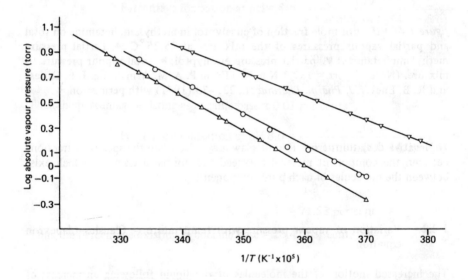

figure 1.5 Logarithm of vapour pressure against reciprocal of temperature. $\triangle$, Methylamphetamine, $\bigcirc$, propylhexedrine; $\triangledown$, eucalyptol. (Note: 1 torr = 133.3 N m^{-2}.) From P. A. M. Armstrong, J. E. Carless and R. P. Enever. *J. Pharm. Pharmacol.*, 23, 473 (1971) with permission

   The Clausius-Clapeyron equation is useful in the calculation of the enthalpy of vaporisation, and also in the study of phase transitions, for example, the melting of a solid or vaporisation of a liquid.

Example 1.1

The slope of a plot of log P against $1/T$ for methylamphetamine (see figure 1.5) is -2.727×10^3 K. Calculate the molar enthalpy of vaporisation of this compound over the given temperature range.

From the form of the Clausius–Clapeyron equation given by equation 1.6, the slope of a plot of log P against $1/T$ is

$$\text{slope} = -\Delta H_{vap}/2.303R = -2.727 \times 10^3$$

$$\therefore -\Delta H_{vap} = -2.727 \times 10^3 \times 2.303 \times 8.314$$

$$= -52.2 \times 10^3 \text{ J mol}^{-1}$$

The molar enthalpy of vaporisation of methylamphetamine is 52.2 kJ mol^{-1}. ∎

An interesting application of the Clausius–Clapeyron equation to a biological system was reported by Cammarata[2]. Gaseous anaesthesia in mice is an equilibrium process between the anaesthetic gas and the phase in which the gas exerts its effect (the biophase). As such, it should be amenable to treatment by the Clausius–Clapeyron equation. Modification of this equation is required when the distribution of a series of gases is to be compared. So that the same equation of state will apply to each gas, it is necessary to use 'reduced' thermodynamic variables. The reduced physiological temperature T_r may be obtained by dividing the physiological temperature for mice (310 K) by the critical temperature of each gas, or, as an approximation, by the boiling point of the gas. Equation 1.6 now becomes

$$\log P = \frac{-\Delta H_{vap}}{2.303RT_r} + \text{constant} \tag{1.7}$$

Figure 1.6 shows the logarithm of the partial pressure of each anaesthetic gas required to produce a given level of anaesthesia plotted against the reciprocal of its reduced physiological temperature. Apart from the fluorinated compounds (which possess unique solubility properties), a close adherence to equation 1.7 is noted.

1.2.3 Vapour pressure lowering

The change of vapour pressure following the addition of a non-volatile solute to a solvent may be determined by application of Raoult's law. Since the solute is non-volatile the total vapour pressure, P, above the dilute solution is due entirely to the solvent and may be equated with p_1, the vapour pressure of the solvent. From Raoult's law we have

$$P = p_1 = p_1^{\ominus} x_1 = p_1^{\ominus} (1 - x_2) \tag{1.8}$$

where x_2 is the mole fraction of the added solute. Rearranging gives

$$\boxed{\frac{p_1^{\ominus} - p_1}{p_1^{\ominus}} = x_2} \tag{1.9}$$

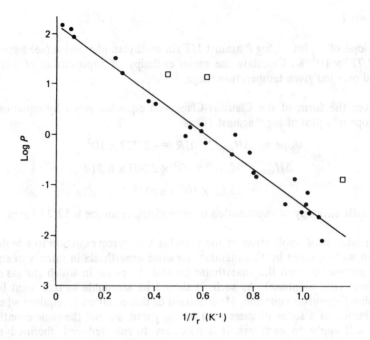

figure 1.6 Graph of anaesthetic pressure P against the reciprocal of the physiological reduced temperature. Compounds not following the relationship (fluoromethane, perfluoromethane, and perfluoroethane) are identified by the symbol □. From reference 2 with permission

That is, the relative lowering of the vapour pressure is equal to the mole fraction of the solute.

A direct consequence of the reduction of vapour pressure by the added solute is that the temperature at which the vapour pressure of the solution attains atmospheric pressure (that is, the boiling point) must be higher than that of the pure solvent.

In figure 1.7 points A and B represent the boiling points of pure solvent and solution respectively. The boiling point is thus raised by an amount $T - T_0 = \Delta T_b$. AE represents the lowering of vapour pressure, $p^{\ominus} - p$, by the solution.

An expression for the boiling point elevation may readily be derived using the Clausius–Clapeyron equation. The vapour pressure of the solution is p at temperature T_0 and $p^{\ominus}$ (equal to that of pure solvent) at temperature T. According to the Clausius–Clapeyron equation we may write

$$\ln \frac{p}{p^{\ominus}} = - \frac{\Delta H_{vap}}{R} \left[\frac{T - T_0}{T T_0} \right] \tag{1.10}$$

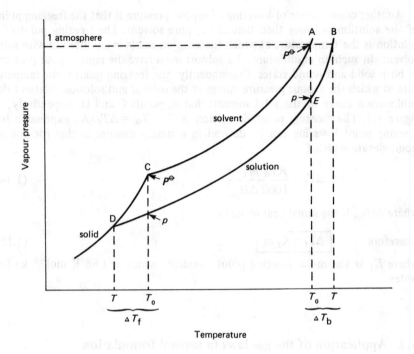

figure 1.7 Freezing point depression and boiling point elevation for a binary system with a non-volatile solute

Assuming that the magnitude of the elevation is small, T may be replaced by T_0 in the denominator and hence

$$\ln \frac{p}{p^{\ominus}} = - \frac{\Delta H_{vap} \, \Delta T_b}{RT_0^2} \tag{1.11}$$

According to Raoult's law, the relative lowering of the vapour pressure is equal to the mole fraction of solute; that is $(p^{\ominus} - p)/p^{\ominus} = x_2$, or $p/p^{\ominus} = 1 - x_2$.

Thus $\ln (p/p^{\ominus}) = \ln (1 - x_2)$

For small values of x_2 (that is, dilute solutions), $\ln(1 - x_2) \simeq -x_2$, and equation 1.11 becomes

$$\Delta T_b = x_2 \, T_0^2 \, R/\Delta H_{vap} \tag{1.12}$$

Mole fraction may be replaced by molality, m, using equation 3.29.

Therefore $\boxed{\Delta T_b = \frac{RT_0^2 \, M_1 \, m}{1000 \, \Delta H_{vap}} = K_b m}$ \tag{1.13}

where K_b is the molal elevation constant which has the value 0.511 K mol^{-1} kg for water.

Another consequence of lowering of vapour pressure is that the freezing point of the solution is lower than that of the pure solvent. The freezing point of a solution is the temperature at which the solution exists in equilibrium with solid solvent. In such an equilibrium, the solvent must have the same vapour pressure in both solid and liquid states. Consequently, the freezing point is the temperature at which the vapour pressure curves of the solvent and solution intersect the sublimation curve of the solid solvent; that is, points C and D, respectively, in figure 1.7. The freezing point depression is $T - T_0 = \Delta T_f$. An expression for freezing point lowering can be derived in a similar manner to that for boiling point elevation giving

$$\Delta T_f = \frac{RT_0^2\, M_1\, m}{1000\, \Delta H_{fus}} \qquad (1.14)$$

where ΔH_{fus} is the molal heat of fusion.

Therefore $\boxed{\Delta T_f = K_f\, m}$ (1.15)

where K_f is the molal freezing point constant, which is 1.86 K mol^{-1} kg for water.

1.3 Application of the gas laws in aerosol formulation

Pharmaceutical aerosols are often pressurised packages which contain the therapeutically active ingredient dissolved, suspended or emulsified in a propellant which is capable of expelling that product through an opened valve. There are two basic types of pressurised aerosols: the liquefied gas and the compressed gas systems (figure 1.8).

Liquefied gas systems

In the simplest type of liquefied gas aerosol an equilibrium is established between the liquefied gas and its vapour. On opening the aerosol valve, the pressure forces some of the liquid up the dip tube and out through the valve. When the expelled liquid comes in contact with the surrounding air at atmospheric pressure, the propellant gas vaporises and so disperses the active ingredients as a fine spray. Since some of the liquid has now been lost from the container the volume of free space above the liquid is increased, causing a temporary drop in pressure due to expansion of the gas in the vapour phase. Equilibrium conditions are rapidly established by the vaporisation of more propellant and the original pressure is restored. Not only does the vapour pressure remain constant in this type of aerosol at a constant temperature, but it is also possible, by careful blending of propellants, to formulate aerosols with a given vapour pressure. This is important since the vapour pressure is a key factor in the control of the nature of the extrusion, that is whether as a fine spray or as a foam, from the aerosol container. The calculation of the proportion of propellant gases of differing vapour pressures required to produce a desired total pressure may be achieved using Raoult's

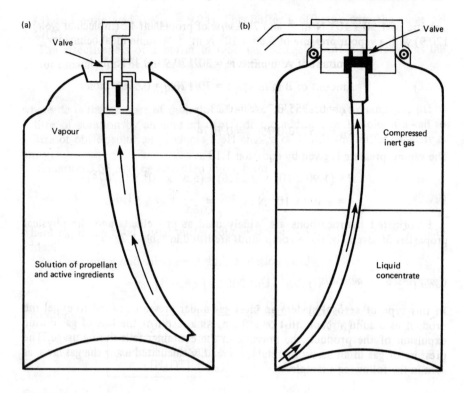

figure 1.8 Cross-sections of a typical liquefied gas aerosol spray (a) and a compressed gas aerosol foam (b)

law. For two components A and B with respective vapour pressures $p_A^\ominus$ and $p_B^\ominus$ we have

$$p_A = p_A^\ominus x_A \tag{1.16}$$

$$p_B = p_B^\ominus x_B \tag{1.17}$$

where p_A and p_B are the partial pressures of components A and B, respectively, and x_A and x_B are the mole fractions of these components in the liquid phase.

From Dalton's law of partial pressures, the total vapour pressure P is the sum of the partial pressures of the component gases, assuming ideal behaviour. Thus

$$\boxed{P = p_A^\ominus x_A + p_B^\ominus x_B} \tag{1.18}$$

Example 1.2

Calculate the vapour pressure at 298 K above an aerosol mixture consisting of 30% w/w of aerosol propellant 114 (molecular weight 170.9) with a vapour pressure of 1.90×10^5 N m^{-2} and 70% w/w of propellant 12 (molecular weight 120.9) with a vapour pressure of 5.85×10^5 N m^{-2}. Assume ideal behaviour.

$$\text{Amount of A in mixture} = 30/170.9 = 0.1755 \text{ moles}$$

$$\text{Amount of B in mixture} = 70/120.9 = 0.5790 \text{ moles}$$

$$x_A = 0.1755/0.7545 = 0.2326$$

$$x_B = 0.5790/0.7545 = 0.7674$$

The vapour pressure is given by equation 1.18 as

$$P = (1.90 \times 10^5 \times 0.2326) + (5.85 \times 10^5 \times 0.7674)$$

$$= 4.492 \times 10^5 \text{ N m}^{-2}. \quad \blacksquare$$

Fluorinated hydrocarbons are widely used as propellants and the physical properties of several of these compounds are listed in table 1.2.

Compressed gas aerosols

In this type of aerosol system an inert gas under pressure is used to expel the product as a solid stream, mist or a foam. As a result of the loss of gas during expulsion of the product, the pressure in the container falls during usage. The pressure of gas in an aerosol container is readily calculated using the gas laws, as seen in the following example.

Example 1.3

Calculate the pressure at 25°C within an aerosol container of internal volume 250 cm^3 containing 160 cm^3 of concentrate above which has been introduced 0.04 mol of nitrogen gas. Assume ideal behaviour.

$$PV = nRT$$

$$P = \frac{0.04 \times 8.314 \times 298}{(250 - 160) \times 10^{-6}}$$

$$P = 1.01 \times 10^6 \text{ N m}^{-2}$$

That is, the pressure within the container is 1.01×10^6 N m^{-2}. $\quad \blacksquare$

1.4 Solubility of gases in liquids

The amount of gas which can be dissolved by a particular liquid depends on the temperature, pressure and the nature of both the gas and the liquid solvent. The solubility may be expressed by Bunsen's absorption coefficient, a, which is the volume of gas reduced to 273 K and a pressure of 1 atmosphere which dissolves in a unit volume of the liquid at the given temperature when the partial pressure of the gas is 1 atmosphere.

table 1.2 Physicochemical properties of some aerosol propellants

Identification number	Chemical name	Formula	Mol. wt.	Vapour pressure (N m^{-2} × 10^{-5})	Boiling point (1 atm.) K	Boiling point (1 atm.) °C	Liquid density (kg m^{-3} × 10^{-3} at 21°C)
11	Trichloromonofluoromethane	CCl_3F	137.4	0.924	296.9	23.7	1.485
12	Dichlorodifluoromethane	CCl_2F_2	120.9	5.854	243.4	−29.8	1.325
114	Dichlorotetrafluoroethane	$CClF_2C\ ClF_2$	170.9	1.903	276.8	3.6	1.468
115	Chloropentafluoroethane	$CClF_2CF_3$	154.5	8.102	234.5	−38.7	1.290
142b	Monochlorodifluoroethane	CH_3CClF_2	100.5	3.020	263.8	−9.4	1.119
152a	Difluoroethane	CH_3CHF_2	66.1	5.268	249.2	−24.0	0.911
C318	Octafluorocyclobutane	$CF_2CF_2CF_2CF_2$	200.0	2.765	267.1	−6.1	1.513

Example 1.4

If the solubility of N_2 in water at 25°C and a nitrogen pressure of 450 torr is 0.378 mol m^{-3}, calculate the Bunsen coefficient.

The volume, V, of dissolved nitrogen at 0°C and a pressure of 1 atm (1.013×10^5 N m^{-2}), assuming ideality, is given by equation 1.1 as

$$V = \frac{0.378 \times 8.314 \times 273.16}{1.013 \times 10^5}$$

$$= 8.474 \times 10^{-3} \text{ m}^3$$

The volume of N_2 that would dissolve at a nitrogen pressure of 1 atm (760 torr) is

$$V = 8.474 \times 10^{-3} \times \frac{760}{450}$$

$$= 0.0143 \text{ m}^3$$

That is, the Bunsen absorption coefficient for N_2 at 25°C is 0.0143. ∎

1.4.1 Effect of temperature on solubility

When gases dissolve in water without chemical reaction there is generally an evolution of heat. Hence by Le Chatelier's principle an increase in temperature usually leads to a decreased solubility. The effect of temperature on the absorption coefficient may be determined from an equation analogous to the van't Hoff equation

$$\log \frac{a_2}{a_1} = \frac{\Delta H}{2.303R} \left(\frac{T_2 - T_1}{T_1 T_2} \right) \tag{1.19}$$

where a_1 and a_2 are the absorption coefficients at temperatures T_1 and T_2, respectively, and ΔH is the change in enthalpy accompanying the solution of 1 mole of gas.

1.4.2 Effect of pressure on solubility

The influence of pressure on solubility is expressed by Henry's law, which states that the mass of gas dissolved by a given volume of solvent at a constant temperature is proportional to the pressure of the gas in equilibrium with the solution. If w is the mass of gas dissolved by unit volume of solvent at an equilibrium pressure, p, then from Henry's law

$$w = kp \tag{1.20}$$

where k is a proportionality constant. Most gases obey Henry's law under normal conditions of temperature and at reasonable pressures, providing the solubility is not too high.

If a mixture of gases is equilibrated with a liquid, the solubility of each component gas is proportional to its own partial pressure; that is, Henry's law may be applied independently to each gas.

Rather than considering equation 1.20 as a means of expressing the solubility of a gas in terms of vapour pressure, we could also view it as a way of expressing the vapour pressure developed by a given concentration of dissolved gas. In so doing we invoke an analogy with Raoult's law which gives the vapour pressure p_1 of the solvent in equilibrium with a solution in which the solvent mole fraction is x_1, as

$$p_1 = x_1 p_1^{\ominus} \tag{1.21}$$

where $p_1^{\ominus}$ is the vapour pressure of pure solvent. Assuming the solute rather than the solvent to be the volatile component, we may write

$$p_2 = x_2 p_2^{\ominus} \tag{1.22}$$

For a dilute solution of a gas we may express the concentration of gas in terms of the mole fraction and thus Henry's law may be written

$$x_2 = k' p_2$$

or $$\tag{1.23}$$

$$p_2 = x_2 / k'$$

Comparing equations 1.22 and 1.23 it is clear that the Henry's law and Raoult's law expressions would become identical if k' could be equated with $1/p_2^{\ominus}$. Such an equating of terms is valid in the case of ideal solutions only, and in most solutions of gases in liquids, although k' is constant, it is not equal to $1/p_2^{\ominus}$.

Figure 1.9 shows the plot of percentage of chloroform in an oleyl alcohol-

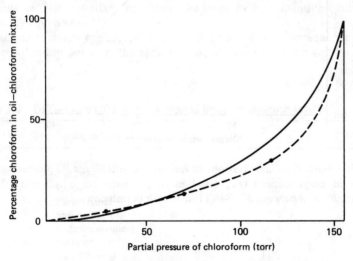

figure 1.9 Percentage of chloroform in an oleyl alcohol–chloroform mixture as a function of the partial pressure of chloroform in gas phase at 20 °C. Solid circles represent experimental results. Full curve is the Raoult's law plot. (Note: 1 torr = 133.3 N m^{-2}.) From J. F. Nunn. *Br. J. Anaesth.*, 32, 346 (1960) with permission

chloroform mixture as a function of the partial pressure of chloroform in the gas phase. Significant departures from Raoult's law (full curve) are apparent when the amount of dissolved chloroform exceeds about 20 per cent.

1.4.3 The solubility of volatile anaesthetics in oil

The oil solubility of an anaesthetic is of interest, not only because it governs the passage of the anaesthetic into and out of the fat depots of the body, but also because there is a well established correlation between anaesthetic potency and oil solubility. Figure 1.10 shows a linear inverse relationship between log narcotic concentration and log solubility in oleyl alcohol (expressed in terms of the oil/ gas partition coefficient) for a series of common anaesthetic gases.

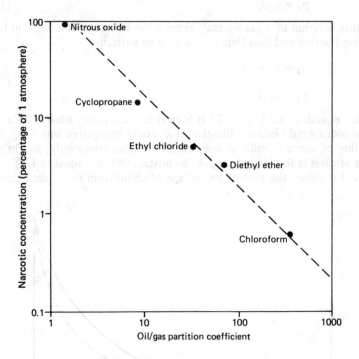

figure 1.10 Narcotic concentrations of various anaesthetic agents plotted against solubility in oleyl alcohol (expressed as oil/gas partition coefficient). From J. F. Nunn. *Br. J. Anaesth.*, 32, 346 (1960) with permission

1.5 The solubility of gases in blood and tissues

The application of physicochemical principles in the consideration of the solubility of gases in blood and tissues is complicated by the complex nature of these solvent systems.

1.5.1 The solubility of oxygen in the blood

The major respiratory function of the lungs is to add oxygen to the blood and to remove carbon dioxide. Thus the measurement of the concentration of these gases in the arterial blood leaving the lungs, combined with a knowledge of the partial pressure of oxygen in the inspired air (approximately 147 mmHg at $37\,^{\circ}C$), allows an assessment of the gas exchanging function of the lungs[3].

The solubility of oxygen in the blood is dependent upon the concentration of haemoglobin, each gram of which can combine with 1.34 ml of oxygen at $37\,^{\circ}C$, and upon the presence of other ligands which combine with haemoglobin and affect oxygen binding. The oxygen saturation, S_{O_2}, of a particular blood sample, which determines the colour of the blood, is defined by the ratio of the oxygen concentration in the blood sample to the oxygen concentration when that blood is fully saturated (i.e. the oxygen capacity of that blood). Defined in this manner, it is clear that S_{O_2} for an anaemic patient, where there is a low haemoglobin content, may be the same as that for a patient with polycythaemia, but the oxygen concentration of the blood would be much less in the anaemic patient.

The partial pressure, P_{O_2}, of the oxygen in the blood (oxygen tension) is related to S_{O_2} by the oxygen dissociation curve (figure 1.11). The shape and position of this sigmoidal curve depend on the temperature, the hydrogen ion concentration and the concentration within the red cells of other ligands of haemoglobin which may also bind to this molecule in addition to oxygen. The position of the curve is defined by the P_{O_2} at 50 per cent saturation, which is

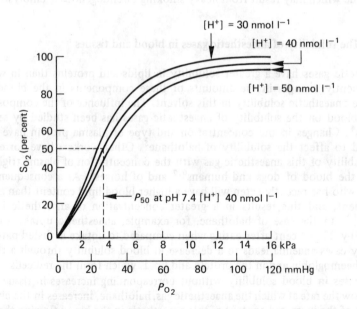

figure 1.11 The oxygen dissociation curve relating blood oxygen saturation, S_{O_2}, to oxygen tension, P_{O_2}, at three different levels of hydrogen ion activity (H^+). From reference 3 with permission

denoted as P_{50}. An alternative method of plotting the data uses the logarithmic equation

$$\log\left(\frac{S_{O_2}}{1 - S_{O_2}}\right) = n \log P_{O_2} \qquad (1.24)$$

Plots of $\log [S_{O_2}/(1 - S_{O_2})]$ against $\log P_{O_2}$ are linear over most of the range with a gradient n.

As seen from figure 1.11, the P_{50} value is affected by pH change. Oxygenation of the haemoglobin molecule releases hydrogen ions, i.e. oxygenated haemoglobin behaves as a stronger acid (proton donor) than reduced haemoglobin. The ratio of ΔP_{50} to ΔpH is referred to as the Bohr effect and normally has a value of 0.5. It is usual to correct the P_{50} value to a plasma pH of 7.4 (although the pH in the red blood cell is about 7.18).

Normal haemoglobin has a P_{50} of 3.4 kPa and n of 2.6–3.0 at pH 7.4. Values of both P_{50} and n are affected by genetic abnormalities in haemoglobin synthesis which alter the amino acid sequence. Over 190 such variants are known with a wide range of P_{50} and n values. Other ligands of the haemoglobin molecule apart from oxygen which can affect these values include 2,3-diphosphoglycerate, a by-product within the Embden–Meyerhof glycolytic pathway in the red cell. This is normally present in equimolar concentrations to haemoglobin. Transfused blood stored in acid–citrate dextrose, however, contains very little of this compound and a lowering of P_{50} is noted over several hours in patients receiving massive blood transfusions. P_{50} is also affected by the presence of carbon monoxide which may result from heavy smoking or endogenous haemolysis.

1.5.2   The solubility of anaesthetic gases in blood and tissues

Anaesthetic gases have a greater solubility in lipids and proteins than in water. Consequently changes in the amounts of these components in the blood can alter the anaesthetic solubility in this solvent. The influence of the composition of the blood on the solubility of anaesthetic gases has been studied by several workers[4]. Changes in the concentration and type of plasma protein have been reported to affect the solubility of halothane[5]. Other workers have correlated the solubility of this anaesthetic gas with the concentration of plasma triglycerides in the blood of dogs and humans[6,7] and of horses[8]. As a consequence, a patient who has recently eaten will have a higher blood lipid content than a fasting patient, and this results in a greater concentration of anaesthetic in the blood[9,10]. In the case of halothane, for example, anaesthetic uptake was increased by 22 per cent after eating when compared to uptake in fasted patients. In many cases anaemia leads to a decrease in blood solubility through a reduction in haemoglobin and in the protein and lipids which form the red cells.

Increases in blood solubility without corresponding increases in tissue solubility slow the rate at which the anaesthetic gas, halothane, increases in the alveoli. Because of the increased content of this anaesthetic in the blood flowing through the tissues, however, the halothane partial pressure in the tissues approaches more rapidly that in the alveoli. The net consequence is that the time for induction with halothane is not greatly affected by changes in blood solubility although the depth of anaesthesia achieved after 10-30 minutes may be considerably affected.

The high lipid solubility of anaesthetics is an important factor in determining the extent of tissue solution. A successful prediction of the partition coefficients of anaesthetic gases into the white and grey matter of the brain from a knowledge of their water and lipid content has been reported[11]. The solubility of xenon and krypton in human liver tissue has been found to be proportional to its triglyceride content[12]. Halothane solubility has been correlated with the fat content of the muscles of horses[8]. This relationship between solubility and fat content implies a greater muscle solubility in adults than in children because of a greater infiltration of fat in adult muscles. The consequence of increased tissue solubility on the depth and rate of onset of anaesthesia is different from that caused by increased blood solubility. Although a similar slowing of the rate of rise of anaesthetic in the alveoli is observed, the increased capacity of the tissues for the anaesthetic leads to an increase in the time required for the partial pressure in the tissues to approach that in the alveoli. The resultant effect of increased tissue solubility is a delayed onset of anaesthesia and also a decreased eventual depth of anaesthesia produced by a given inspired concentration.

It is perhaps not surprising, bearing in mind the complexity of blood and tissue fluids, that Henry's law is frequently disobeyed for these. Departures of the solubility versus pressure relationships from Henry's law have been reported, for example, for cyclopropane in blood[11,13] and have been attributed to the binding of the cyclopropane by the haemoglobin molecule.

An increase in pressure at low partial pressures of cyclopropane simply results in an increase in the proportion of cyclopropane binding sites on the haemoglobin molecule that are occupied. At higher pressures, however, nearly all the sites become occupied and further pressure increases cannot further increase the extent of cyclopropane binding and a deviation from Henry's law becomes apparent. Similar deviations from Henry's law have been reported for xenon in the presence of myoglobin[14].

The effect of temperature on anaesthetic solubility is of interest because of possible wide variation of body temperature in the surgical patient. Body temperature may be lowered as a result of preoperative sedation, by cutaneous vasodilation, the infusion of cold fluids and reduced metabolism under operating conditions. Temperature decrease leads to an increased solubility of the anaesthetic as shown by increased blood/gas and tissue/gas partition coefficients. The general effect of the increased solubility in conditions of hyperthermia is a decreased rate of onset of (and recovery from) anaesthesia.

References

1. R. J. Wolf and R. M. Featherstone. *Anesthesiology*, 18, 97 (1957)
2. A. Cammarata. *J. pharm. Sci.*, 64, 2025 (1975)
3. D. C. Flenley. *Br. J. clin. Pharm.*, 9, 129 (1980)
4. E. I. Eger. *Anaesthetic Uptake and Action*, Williams and Wilkins, Baltimore, 1974, chapter 9
5. L. J. Laasberg and J. Hedley-Whyte. *Anesthesiology*, 32, 351 (1970)

6. P. D. Wagner, P. F. Naumann and R. B. Laravuso. *J. appl. Physiol.*, 36, 600 (1974)
7. R. A. Saraiva, B. A. Willis, A. Steward, J. N. Lunn and W. W. Mapleson. *Br. J. Anaesth.*, 49, 115 (1977)
8. B. M. Q. Weaver and A. I. Webb. *Br. J. Anaesth.*, 53, 487 (1981)
9. R. L. Kozam, S. M. Landau, J. M. Cubina and D. S. Lukas. *J. appl. Physiol.*, 29, 593 (1970)
10. E. S. Munson, E. I. Eger, M. K. Tham and W. J. Embro. *Anesth. Analg. (Cleve.)*, 57, 224 (1978)
11. H. J. Lowe and K. Hagler. In *Gas Chromatography in Biology and Medicine* (ed. R. Porter), Churchill, London, 1969, pp. 86-112
12. K. Kitani and K. Winkler. *Scand. J. clin. lab. Invest.*, 29, 173 (1972)
13. G. A. Gregory and E. I. Eger. *Fedn. Proc.*, 27, 705 (1968)
14. B. P. Schoenborn. *Nature*, 214, 1120 (1967)

2 Properties of the Solid State

The physical properties of the solid state of both drugs and adjuvants are of considerable interest as they can affect both the production of dosage forms and the biological behaviour of the finished form. Powders, as Pilpel[1] reminds us, 'can float like a gas, flow like a solid or support a weight in the same way as hard-packed snow'. We deal with aerosolised powders in chapter 9, and some properties of compacted solids in chapter 6, but here our main interest is in the effect of crystallinity and particle size on drug behaviour especially dissolution and bioavailability.

The nature of the crystalline form of a drug substance may affect its stability in the solid state, its flow properties and its biological availability, the last mainly through the effect of crystal properties on dissolution rate. It is with this latter subject that we shall deal first, and then proceed to other properties of the solid state of importance in production and formulation.

2.1 Crystal form and polymorphism

An *ideal crystal* is a regular polyhedral solid bounded by plane faces. The basic component of the crystal is the unit cell or unit structure, repetition of which in three dimensions produces the crystal. Each unit cell is the same size for a specific crystal and contains the same number of molecules or ions similarly arranged. Even though the size and shape of crystals of a given compound may vary, the angles between the crystal faces remain constant. The crystals of a given substance may vary in size, relative development of the given faces and the number and kind of the faces (that is, forms) present; that is, they may have different crystal habits. Such habits include needle, tabular, equant, columnar and lamellar types. Figure 2.1 shows the seven crystal systems, and crystal habits of one of these, a hexagonal crystal.

Crystal form is described by two terms: *habit* and the *combination of crystallographic forms*. The habit bears upon the over-all shape of the crystal in rather general terms: acicular, prismatic, pyramidal, etc.; the combination of crystallographic forms refers to the faces of the crystal. Two crystals with the same habit may have a different combination of faces; the most obvious is with the orthorhombic form shown in figure 2.2, where the same combination of forms leads to prismatic, isometric and tabular forms.

Many substances can exist in more than one crystalline form, the different forms being termed *polymorphs* (or modifications) and the condition *polymorphism*. The various polymorphic forms arise through packing of the molecules in different arrays within the crystal or by differences in the orientation or conformation of molecules at lattice sites. The resulting crystalline material may, depending on the mode of preparation or its subsequent treatment, have different physical properties amongst the most important of which is the aqueous solubility of the substance. Hence one polymorphic form may have a higher biological activity than another if the rate of solution is the rate-limiting step in its absorp-

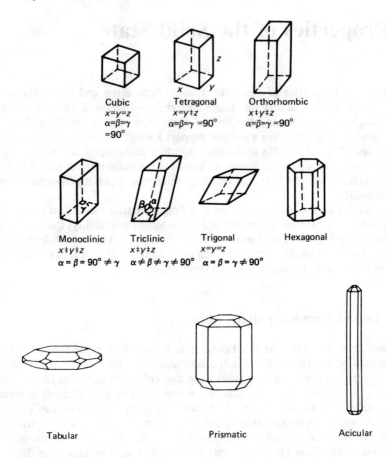

figure 2.1 In the upper part of the figure the seven possible crystal systems are shown, namely cubic, tetragonal, orthorhombic, monoclinic, triclinic, trigonal and hexagonal. In the lower part three crystal habits of a hexagonal crystal are illustrated. Shown are tabular, prismatic and acicular habits

tion across the gastro-intestinal barrier. The more soluble forms of crystalline substances, having less internal cohesion, are less stable and transformation into stable modifications can occur by a process which may be rapid or slow. Use of more soluble but metastable forms of a drug thus presents special pharmaceutical problems.

We are interested in the solubility, stability, and rate of solution of polymorphic forms and the factors which result in their formation. The awareness of polymorphism dates back to 1821 when Mitscherlich discovered two forms of sodium phosphate. In 1832 the same phenomenon was discovered in an organic compound, benzamide, and by 1942, 1200 organic compounds were known to exist in two or more crystalline forms. Many drugs come into this category, the steroids producing many examples. The various 'forms' usually exhibit different physical properties such as density, X-ray diffraction pattern, infrared spectrum,

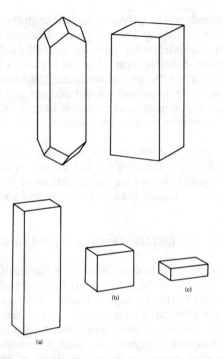

figure 2.2 Above, two orthorhombic crystals are shown with the same habit but a different combination of forms. Below, three orthorhombic crystals are illustrated with the same combination of forms but different habits: prismatic (a), isometric (b) and tabular (c)

and frequently have different melting points, although this is not always a useful diagnostic feature (table 2.1).

In the liquid and gaseous states the molecules which show polymorphism in the solid state are of course indistinguishable. The question as to why one substance is polymorphic and another is not cannot be answered yet; one cannot

table 2.1 Characteristics of the three polymorphs of sulphathiazole*

	Form		
	I	II	III
Melting point (°C)	200-202	200-202	200-202
Transition temperature (to form I) (°C)	–	173-175	173-175
Habit	Rods	Prisms	Plates
Molecules per asymmetric unit	8	4	4
Density (g m ℓ^{-1})	1.5	1.55	1.57

*From S. R. Byrn. *J. pharm. Sci.*, 65, 1 (1976)

predict which compounds will be polymorphic, although there are classes of sub-stances which frequently display polymorphism. Unstable modifications may occur with any type of crystallisation from the melt (the solid above its melting point) or from solvents. Solvates form when the solvent is incorporated in the crystal lattice and thus alters the crystal structure; this results in what is some-times known as *pseudopolymorphism*. Solvent adducts and polymorphic forms should be distinguished; polymorphs and solvates have different pharmaceutical properties. Where the solvent is toxic its presence is, of course, of concern in drug solvates.

Different habits can appear without there being polymorphism. The habit, however, can affect flow and adhesion and the properties of suspensions of the crystal. The habits acquired depend on the conditions of crystallisation such as solvent used, the temperature, and the concentration and presence of impurities.

2.2 Crystallisation and factors affecting crystal form[2]

Crystallisation from solution can be considered to be the result of three successive processes, (i) supersaturation of the solution, (ii) formation of crystal nuclei, and (iii) crystal growth round the nuclei. Supersaturation can be achieved by cooling, by evaporation, by the addition of a precipitant or by a chemical reaction which changes the nature of the solute. Supersaturation itself is insufficient to cause crystals to form; the crystal embryos must form by collision of molecules of solute in the solution, or sometimes by the addition of seed crystals, or dust particles, or even particles from the container walls. Deliberate seeding is often carried out in industrial processes; seed crystals do not necessarily have to be of the substance concerned but may be isomorphous substances. As soon as stable nuclei are formed they begin to grow into visible crystals. Crystal growth can be considered to be a reverse dissolution process and the diffusion theories of Noyes and Whitney, and of Nernst, consider that matter is deposited continuously on a crystal face at a rate proportional to the difference of concentration between the surface and the bulk solution. So an equation for crystallisation can be proposed in the form

$$\frac{dm}{dt} = Ak_m(c_{ss} - c_s) \tag{2.1}$$

where m is the mass of solid deposited in time t, A is the surface area of the crystal, c_s is the solute concentration at saturation and c_{ss} is the concentration at supersaturation. As $k_m = D/\delta$ (D being the diffusion coefficient of the solute and δ the diffusion layer thickness (see figure 2.6)), the degree of agitation of the system which affects δ also influences crystal growth. As crystals generally dis-solve faster than they grow, growth is not simply the reverse of dissolution. It has been suggested that there are two steps involved in growth, namely transport of the molecules to the surface and their arrangement in an ordered fashion in the lattice. Equation 2.1 turns out to be better written in a modified form

$$\frac{dm}{dt} = Ak_g(c_{ss} - c_s)^n \tag{2.2}$$

k_g being the over-all crystal growth coefficient and n the 'order' of the crystal growth process. For more details reference 2 should be consulted.

Precipitation may occur by altering the pH of the solution so that the saturation solubility is exceeded. Precipitation may occur from a homogeneous solution by slowly generating the precipitating agent by means of a chemical reaction, a process likely to occur, for example, in intravenous infusion fluids and liquid pharmaceuticals. Precipitation by direct mixing of two reacting solutions sometimes does not bring about immediate nucleation and, as a result, the mixing stage may be followed by an appreciable lag time. The rate of precipitation is an important factor in determining habit, as might be imagined with a dynamic process such as crystallisation, involving nucleation and subsequent crystal growth.

The crystal form of phenylsalicylate, for example, depends on the rate of growth. Transition to an acicular shape occurs when the rate of growth increases. At low rates of growth crystals of a more regular shape are obtained. In a study of the effect of solvents on habit, less viscous media favoured the growth of coarse and more equidimensional forms of some minerals.

Modification of crystal habit can be obtained by the addition of impurities or 'poisons'; for example, sulphonic acid dyes alter the crystal habit of ammonium, sodium and potassium nitrates.

Surfactants in the solvent medium used for crystal growth (or, for example, in stabilisation or wetting of suspensions) can alter crystal form by adsorbing onto growing faces during crystal growth. This is best illustrated by the effect of anionic and cationic surfactants on the habit of adipic acid crystals[3]. In the description of crystal form Miller indices (010), (001), etc.* are used to identify the crystal faces. X-ray analysis showed that the linear six-carbon dicarboxylic acid molecules were aligned end to end in a parallel array in the crystal with their long axis parallel to the (010) faces so that the (001) face is made up entirely of —COOH groups while the (010) and (110) faces contain both —COOH and hydrocarbon portions of the molecule (figure 2.3). The cationic surfactant trimethyldodecylammonium chloride was twice as effective in hindering the growth of the (001) face as that of the (110) and (010) faces. In high concentrations it caused the formation of very thin plates or flakes. Conversely, the anionic surfactant sodium dodecylbenzene sulphonate at 50 ppm was three times as effective in reducing the growth rates of the (110) and (010) faces as that of the (001) face. Higher levels of sodium dodecylbenzene sulphonate caused extreme habit modification, producing not hexagonal plates but long thin rods or needles. The crystallographic faces whose growth rates were depressed most were those upon which surfactant adsorption was the greatest. Cationic additives adsorb on the face composed of carboxylic groups (001), and anionic additives on the (110) and (200) faces which are hydrophobic. A coulombic interaction of the cationic head groups and the —COO⁻ groups on the (001) faces has been suggested. The adsorption of the anionic surfactant, repelled from the anionic (001) faces, takes place amphipathically on the hydrophilic (110) faces and (100) faces (figure 2.3).

Miller indices: Miller (1839) designated each face of a crystal by three integers which give the relation of the face to axes through the crystal.

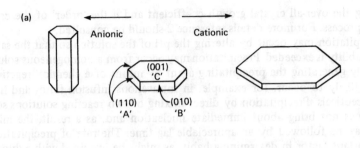

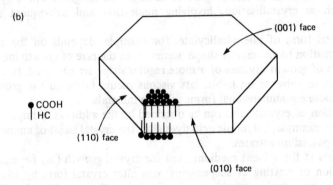

figure 2.3 (a) Effect of anionic and cationic surfactants on the habit of adipic acid crystals. (b) A diagrammatic representation of the arrangement of molecules at the crystal surface

2.2.1 Solvates

Modification of the solvent of crystallisation may result in different solvated forms. This is of particular relevance because the hydrated and anhydrous forms of a drug can have melting points and solubilities sufficiently different to affect their pharmaceutical behaviour. For example, glutethimide exists in both an anhydrous form (m.p. 83 °C, solubility 0.042 per cent at 25 °C) and a hydrated form (m.p. 68 °C, solubility 0.026 per cent at 25 °C). Other anhydrous forms show similar higher solubilities than the hydrated materials and, as expected, the anhydrous forms of caffeine, theophylline, glutethimide and cholesterol show correspondingly higher dissolution rates than their hydrates. One can assume that as the hydrate has already interacted intimately with water (the solvent), then the energy released for crystal break-up, on interaction of the hydrate with solvent, is less than for the anhydrous material. The non-aqueous solvates, on the other hand, tend to be more soluble in water than the non-solvates. Shefter and Higuchi[4] found that the n-amyl alcohol solvate of fludrocortisone acetate is at least five times as soluble as the parent compound, while the ethyl acetate solvate was twice as soluble.

The stoichiometry of some of the solvates is unusual. Fludrocortisone pentanol solvate, for example, contains 1.1 molecules of pentanol for each steroid molecule, and its ethyl acetate solvate 0.5 molecules per steroid. A succinyl-sulphathiazole solvate appears to have 0.9 moles of pentanol per mole of drug.

Infrared measurements show that cephaloridine exists in α, β, δ, ϵ, ζ, and μ forms (that is, six forms after recrystallisation from different solvents)[5]. Proton magnetic resonance spectroscopy showed that although the μ form contained about 1 mole of methanol and the ϵ form about 1 mole of dimethylsulphoxide, ethylene glycol or diethylene glycol (depending on the solvent), the α, β, anhydrous δ and ϵ forms contained less than 0.1 mole, that is non-stoichiometric amounts of solvent. The α form is characterised by containing about 0.05 mole of N,N-dimethylacetamide. This small amount of 'impurity', which cannot be removed by prolonged treatment under vacuum at 10^{-5}–10^{-6} torr, is apparently able to 'lock' the cephaloridine molecule in a particular crystal lattice.

The equilibrium solubility of the non-solvated form of a crystalline organic compound which does not dissociate in the solvent (for example, water) can be represented as

$$A_{(c)} \underset{\longleftarrow}{\overset{K_s}{\rightleftharpoons}} A_{(aq)}$$

where K_s is the equilibrium constant. This equilibrium will, of course, be influenced by the crystal form as we have seen, as well as by temperature and pressure. For a hydrate $A.xH_2O$, we can write

$$A.xH_2O_{(c)} \underset{\longleftarrow}{\overset{K_{sh}}{\rightleftharpoons}} A_{(aq)} + xH_2O$$

K_{sh} is then the solubility of the hydrate. The process of hydration of an anhydrous crystal in water is represented by an equation of the type

$$A_{(c)} + xH_2O_{liquid} \underset{K_{sh}}{\overset{K_s}{\rightleftharpoons}} A.xH_2O_{(c)}$$

(anhydrate)   (hydrate)

and the free energy of the process is written

$$\Delta G_{trans} = RT \ln \frac{(K_{sh})}{(K_s)} \tag{2.3}$$

ΔG_{trans} can be obtained from the solubility data of the two forms at a particular temperature, as for theophylline and glutethimide in table 2.2.

2.3 The extent of the problem of polymorphism

Eight crystal modifications of phenobarbitone have been isolated[6] but eleven have been identified with melting points ranging from 176 to 112 °C. Examples of the differing solubility and melting points of polymorphic barbiturates and steroids are given in table 2.3. Of the barbiturates used medicinally, about 70 per cent exhibit polymorphism. A fairly large number of barbiturates possess four or five modifications. The steroids frequently possess polymorphic modifications, testosterone having four. These are cases of true polymorphism and not pseudo-polymorphism where solvent is the cause. Of the commercial sulphonamides about 65 per cent are found to exist in several polymorphic forms.

table 2.2 Solubility of theophylline and glutethimide forms at various temperatures*

Temperature (°C)	Solubility	
	Hydrate	Anhydrate
Theophylline	(mg mℓ^{-1})	(mg mℓ^{-1})
25	6.25	12.5
35	10.4	18.5
45	17.6	27.0
55	30	38.0
Glutethimide	(per cent w/v)	(per cent w/v)
25	0.0263	0.042
32	0.0421	0.0604
40	0.07	0.094

*From Erikson. *Am. J. pharm. Educ.*, 28, 47 (1964)

table 2.3 Melting points of some polymorphic forms of barbiturates, steroids, sulphonamides and riboflavine*

	Form and/or melting point (°C)				
Polymorphic barbiturates					
Cyclobarbitone	(I) 173	(II) 161			
5,5-Dipropylbarbituric acid	(I) 148	(II) 146	(III) 126	(IV) 120	(V) 110
Pentobarbitone	(I) 129	(II) 114	(III) 108		
Polymorphic steroids	(I)	(II)	(III)	(IV)	
Corticosterone	180–186	175–179	163–168	155–160	
β-Oestradiol	178	169			
Oestradiol	225	223			
Testosterone	155	148	144	143	
Methylprednisolone	I (205, aqueous solubility 0.075 mg mℓ^{-1})				
	II (230, aqueous solubility 0.16 mg mℓ^{-1})				
Polymorphic sulphonamides					
Sulphafurazole	190–195	131–133			
Acetazolamide	258–260	248–250			
Tolbutamide	127	117	106		
Others					
Riboflavine	I (291, aqueous solubility 60 mg mℓ^{-1})				
	II (278, aqueous solubility 80 mg mℓ^{-1})				
	III (183, aqueous solubility 1200 mg mℓ^{-1})				

*From reference 7

Predictability of the phenomenon is difficult except by reference to past experience. The importance of the phenomenon pharmaceutically depends very much on the stability and solubility of the forms concerned. It is difficult therefore to generalise, except to say that where polymorphs of insoluble compounds occur there are likely to be biopharmaceutical implications. Table 2.4 is a fairly

table 2.4 **Polymorphic and pseudopolymorphic drugs***

| | | Number of forms | |
	Polymorphs	Amorphous	Pseudopolymorphs
Allobarbitone	2	–	–
Amobarbitone	2	–	–
Ampicillin	1	–	1
Androsterone	1	–	1
Barbitone	6	–	–
Betamethasone	1	1	–
Bethamethasone 21-acetate	1	1	–
Betamethasone 17-valerate	1	1	–
Brucine	–	–	1
Caffeine	1	–	1
Cephaloridine	4	–	2
Chloramphenicol palmitate	3	1	–
Chlordiazepoxide HCl	2	–	1
Chlorthalidone	2	–	–
Cortisone	2	–	1
Cortisone acetate	8	–	–
Cyclobarbitone	2	–	–
Dehydropregnenolone	1	–	7
Dexamethasone acetate	3	–	1
Dexamethasone pivalate	4	–	7
Digoxin	–	1	–
Erythromycin	2	–	–
Fludrocortisone acetate	3	1	–
Fluocortilone	2	–	19
Fluprednisolone	2	–	2
Glutethimide	1	–	1
Hydrocortisone TBA‡	1	–	3
Indomethacin	3	–	–
Menadione	2	–	–
Mefenamic acid	2	–	–
Meprobamate	2	–	–
Methyl *p*-hydroxybenzoate	6	–	–
Methylprednisolone	2	–	–
Novobiocin	1	1	–
Ouabain	1	–	6
Pentobarbitone	3	–	–
Prednisolone	2	–	–
Prednisolone TBA‡	2	–	2
Prednisolone TMA§	3	–	–
Prednisolone acetate	2	–	–
Prednisone	1	–	1
Progesterone	2	–	–
Sorbitol	3	–	–
Succinylsulphathiazole	1	–	3
Sulphaguanidine	4	–	1
Sulphamethazine	2	–	–
Sulphamethoxypyridazine	2	–	–
Sulphanilimide	4	–	–
Sulphapyridine	6	–	–
Testosterone	4	–	–
Theophylline	1	–	1
Tolbutamide	3	–	–
Triamcinolone	2	–	–

*After R. Bouché and M. Draguet-Brughmans, *J. Pharm. belg.*, 32, 23 (1977)
‡Tertiary butyl acetate (tebutate)
§Trimethyl acetate

comprehensive listing of the drugs for which polymorphic and pseudopolymorphic states have been identified or for which an amorphous state has been reported.

2.4   Pharmaceutical implications of crystal habit modification and solvate formation

For analytical work it is sometimes necessary to establish conditions whereby different forms of a substance, where they exist, might be converted to a single form to eliminate differences in the solid state infrared spectra which result from the different internal structures of the crystal forms. As different crystal forms arise through different arrangements of the molecules or ions in a three dimensional array, this implies different interaction energies in the solid state. Hence one would expect different melting points and different solubilities (and of course different infrared spectra). Changes in infrared spectra of steroids due to grinding with KBr have been reported; in some substances changes in the spectrum have been ascribed to conversion of a crystalline form into an amorphous form (as in the case of digoxin)[8], or into a second crystal form. Changes in crystal form can also be induced by solvent extraction methods used for isolation of drugs from formulations prior to examination by infrared. Difficulties in identification arise when samples that are thought to be the same substance give different spectra in the solid state, as, for example, cortisone acetate which exists in at least seven forms, or dexamethasone acetate which exists in four. As an example of possible confusion the 'authentic' cortisone acetate of the BP is form II*, but USP and WHO authentic samples are form III; therefore, where there is a likelihood of polymorphism it is best where possible to record the solution spectra if chemical identification only is required. The normal way to overcome the effects of polymorphism is to convert both samples into the same form by recrystallisation from the same solvent, although obviously this technique should not be used to hide the presence of polymorphs.

2.5   Relative stability of polymorphs and their bioavailability

Since polymorphism involves differences in crystal structure, it is evident that different polymorphs may have different energy contents. Since two polymorphs differ only in the arrangement of the molecules within them the energy difference is largely concerned with the forces between the molecules in their structure and is essentially the difference between their binding energies. Under a given set of conditions the polymorphic form that has the lowest free energy is the most stable and the other polymorphs will tend to transform into it.

It has been proposed[9] that when the free energy differences between the polymorphs are small there may be no significant differences in their biopharmaceutical behaviour as measured by the blood levels they achieve. Only when the differences are large may they affect the extent of absorption. $\Delta G_{B \to A}$ for the transition of chloramphenicol palmitate form B to form A is -3.24 kJ mol^{-1};

*BP: British Pharmacopoeia
 USP: United States Pharmacopeia
 WHO: World Health Organisation

ΔH is -27.32 kJ mol^{-1}. For mefenamic acid $\Delta G_{\text{II}\to\text{I}}$ is -1.05 kJ mol^{-1} and ΔH is -4.18 kJ mol^{-1}. Whereas differences are shown by the palmitate polymorphs, no such differences in biological activity are observed with the mefenamic acid polymorphs. When little energy is required to convert one polymorph into another it is likely that the forms will interconvert *in vivo* and that the administration of one in place of the other form will be clinically unimportant.

Particle size reduction may lead to fundamental changes in the properties of the solid. Grinding of crystalline substances such as digoxin can lead to the formation of amorphous material which has an intrinsically higher rate of solution and therefore apparently greater activity. Such is the importance of the polymorphic form of poorly soluble drugs that it has to be controlled. There is a limit on the inactive polymorph of chloramphenicol palmitate. Of the three polymorphic forms of chloramphenicol palmitate one is virtually without biological activity because it is so slowly hydrolysed *in vivo* to free chloramphenicol (figure 2.4).

Other examples of biological differences may be found. The anhydrous butyl acetate ester of prednisolone has an absorption rate *in vivo* of 1.84×10^{-3} mg h^{-1} cm^{-2}, whereas the monoethanol solvate is absorbed at a rate nearly five times greater. Differences in the absorption of ampicillin and its trihydrate can be observed (figure 2.5) but the extent of the difference is of doubtful clinical significance. The more soluble anhydrous form appears at a faster rate in the serum and produces higher peak serum levels.

During formulation development it is vital that sufficient care is taken to determine polymorphic tendencies of poorly soluble drugs. It is important that

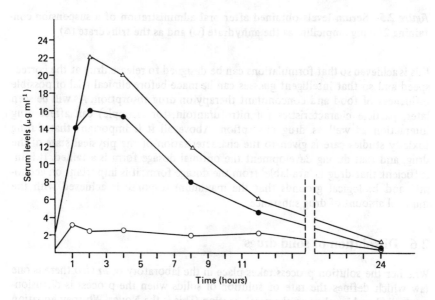

figure 2.4   Comparison of serum levels (μg mℓ^{-1}) obtained with suspensions of chloramphenicol palmitate after oral administration of a dose equivalent to 1.5 g of chloramphenicol. $\triangle$, 100 per cent form B; $\bullet$, 50 per cent form A and 50 per cent form B; and $\circ$, 100 per cent form A

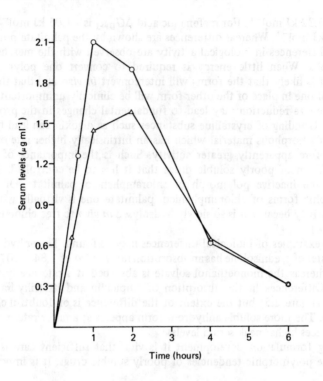

figure 2.5 Serum levels obtained after oral administration of a suspension containing 250 mg ampicillin as the anhydrate (○) and as the trihydrate (△)

this is achieved so that formulations can be designed to release drug at the correct speed and so that intelligent guesses can be made before clinical trial of possible influences of food and concomitant therapy on drug absorption. As will be seen later, particle characteristics (of nitrofurantoin, for example) can affect drug interaction as well as drug absorption. Above all it is important that during toxicity studies care is given to the characterisation of the physical state of the drug, and that during development the optimal dosage form is attained. It is insufficient that drug is 'available' from the dosage form; it is important on economic and biological grounds that the maximum response is achieved with the minimal amount of drug substance.

2.6 Dissolution of solid drugs

Whether the solution process takes place in the laboratory or *in vivo* there is one law which defines the rate of solution of solids when the process is diffusion-controlled and involves no chemical reaction. This is the Noyes-Whitney equation which may be written

$$\frac{dw}{dt} = k \, (c_s - c) \tag{2.4}$$

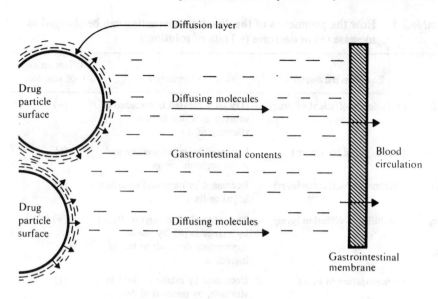

figure 2.6   Schematic diagram of dissolution from a solid surface

where $k = DA/\delta$.

The equation is the analogue of equation 2.1 discussed previously. Figure 2.6 shows the model on which this equation is based. The terms of the equation are: dw/dt, the rate of increase of the amount of material in solution dissolving from a solid; k, the rate constant of dissolution (time^{-1}); c_s, the saturation solubility of the solvate in bulk. A is the area of the solvate particles exposed to the solvent, δ the thickness of the diffusion layer, and D the diffusion coefficient of the dissolved solute. The relevance of polymorphism and solid state properties to this equation lies in the fact that A is determined by particle size. Particle size reduction if it leads to a change in polymorph results in a change in c_s, and if dissolution is the rate-limiting step in absorption then bioavailability is affected. In more general terms, one can use the equation to predict the effect of solvent change or other parameters on the dissolution rate of solid drugs. These factors are listed in table 2.5.

2.7 Biopharmaceutical importance of particle size

It has generally been believed that only substances in the molecularly dispersed form (that is, in solution) are transported across the intestinal wall and absorbed into the systemic circulation. This is the premise on which much thinking on bioavailability from pharmaceutical dosage forms is based. However it has been shown that very small particles in the nanometre size range can also be transported through enterocytes by way of pinocytosis. It has more recently been shown that absorption of solid particles of starch, for example, in the micrometre size range is a fact, the mechanism being passage of particles between the entero-

table 2.5   **How the parameters of the dissolution equation can be changed to increase (+) or decrease (−) rate of solution**

	Equation parameter	Comments	Effect on rate of solution
D	(diffusion coefficient of drug)	May be decreased in presence of substances which increase viscosity of the medium	(−)
A	(area exposed to solvent)	Increased by micronisation and in 'amorphous' drugs	(+)
δ	(thickness of diffusion layer)	Decreased by increased agitation in gut or flask	(+)
c_s	(solubility in diffusion layer)	That of weak electrolytes altered by change in pH, by use of appropriate drug salt or buffer ingredient	(−) (+)
c	(concentration in bulk)	Decreased by intake of fluid in stomach, by removal of drug by partition or absorption	(+)

cytes[10]. These normally form an impenetrable barrier to the passage of particles but there are places where (in the neighbourhood of goblet cells) and times when (as the result of cell desquamation) particles may penetrate. However, because of the much greater absorptive area available to molecules the opportunity for molecules to penetrate the cell membrane is obviously higher than that for particles.

It has been observed that the rate of absorption of many slightly soluble drugs from the gastro-intestinal tract and other sites is limited by the rate of dissolution of the drug substance. The particle size of a drug is therefore of importance if the substance in question has a low solubility.

The Noyes-Whitney equation demonstrates that solubility is one of the main factors determining rate of solution. When the rate of solution is less than the rate of absorption the solution process becomes rate limiting. Generally speaking it should become so only when the drug is of low solubility at the pH of the stomach and intestinal contents. Both rate of absorption, speed of onset of effect and the duration of therapeutic response can be determined by particle size for most routes of administration. Figure 2.7 shows the effect of particle size of phenobarbitone suspensions on its bioavailability after intramuscular injection, compared with a solution of the drug, which probably precipitates in fine crystal form at the site of injection. The rate of solution of the drug crystals controls the extent of absorption from the intramuscular site.

The vital influence (for mainly aerodynamic reasons) of particle size in the activity of drug particles which have been inhaled is discussed in chapter 9.

Particle size control has to be exerted sometimes for other than therapeutic reasons; the range of substances over which there is pharmacopoeial control of particle size is shown in table 2.6. The control exercised over the particle size of cortisone acetate, griseofulvin and prednisolone is due to their very low solu-

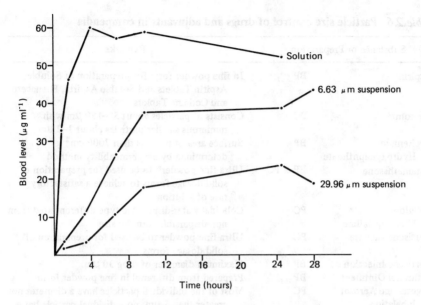

figure 2.7 Blood levels of phenobarbitone versus time after intramuscular injection of three dosage forms. Redrawn from Miller and Fincher, *J. pharm. Sci.*, 60, 1733 (1971)

bility; the experience is that if the solubility of a drug substance is about 0.3 per cent or less, then the dissolution rate *in vivo* may be the rate-controlling step in absorption[11].

The effect of particle size reduction on dissolution rate is one primarily of exposure of increasing amounts of surface of the drug to the solvent. It is only when comminution reduces particle size below 0.1 μm that there is an effect on the intrinsic solubility of the substance (see chapter 5), and thus on its intrinsic dissolution rate. Very small particles have a very high surface/bulk ratio. If the surface layer has a higher energy than the bulk, as is the case with these small particles, they will interact more readily with solvent to produce higher degrees of solubility.

It was with the action of phenothiazine that the importance of particle size was first recognised, in 1939, in relation to its toxicity to codling moth larvae, and in 1940 in relation to its anthelmintic effect, where in both it was shown that reduction in particle size increased activity. The improvement in biological response to griseofulvin on micronisation is well known; similar blood levels of the drug were obtained with half the dose of micronised drug compared to those of non-micronised griseofulvin[12]. More recently the influence of particle size on the bioavailability of digoxin[13] and dicoumarol (bishydroxycoumarin)[14] has been investigated. In both cases plasma levels of drug are of high significance in clinical and toxic responses.

In the case of digoxin there is evidence that milling to reduce particle size can produce an amorphous modification of the drug with enhanced solubility and hence increased bioavailability. The possibility of changing the crystal structure

table 2.6   **Particle size control of drugs and adjuvants in compendia**

Substance or Preparation		Remarks
Aspirin	BP	In fine powder form for preparation of Soluble Aspirin Tablets and Soluble Aspirin, Phenacetin and Codeine Tablets
Bentonite	PC	Consists of particles about 50–150 μm with numerous smaller particles about 1–2 μm
Bephenium Hydroxynaphthoate	BP	Surface area of not less than 7000 cm^2 g^{-1} determined by air permeability method
Betamethasone	EP & PC	Ultra-fine powder* to be used for preparation of solid dosage forms to achieve a satisfactory rate of solution
Cellulose Microcrystalline	PC	Colloidal water-dispersible type differentiated from non-dispersible form by size
Cortisone Acetate	PC	Ultra-fine powder to be used for preparation of solid dosage forms
Cortisone Injection	BP	Maximum diameter of drug 30 μm
Dithranol Ointment	BP	Prepared from dithranol in fine powder form
Ergotamine Aerosol Inhalation	PC	Most of the individual particles have a diameter not greater than 5 μm; no individual particle has a length greater than 20 μm
Fusidic Acid Mixture	BPC	95 per cent of particles have a maximum diameter of not more than 5 μm
Gold (^{198}Au) Injection	PC	About 80 per cent of radioactivity is present in particles between 5 and 50 nm in diameter
Griseofulvin Tablets	PC & EP	Particle size determined from disintegrated tablet generally up to 5 μm in maximum dimension although larger particles may occasionally be greater than 30 μm
Hydrocortisone preparations	BP & PC	All subject to limit on particle size of Hydrocortisone or Hydrocortisone Acetate. See Hydrocortisone Acetate Ointment BP, Hydrocortisone Cream BPC, Hydrocortisone and Neomycin Cream BPC, Hydrocortisone and Neomycin Ear drops and Eye drops BPC, Hydrocortisone Eye Ointment BPC, Hydrocortisone Lotion BPC and Hydrocortisone Suppositories BPC
Insulin preparations	BP	See Insulin Zinc Suspension (Crystalline) BP, Insulin Zinc Suspension (Amorphous) BP, Biphasic Insulin Injection BP
Isoprenaline Inhalation Aerosol		As for Ergotamine Aerosol Inhalation BPC
Kaolin, Light	EP & PC	Size and shape of particles stated
Kaolin, Heavy	EP & PC	Size and shape of particles stated
Macrisalb (^{131}I) Injection	BP	Sterile suspension of denatured albumin iodinated with ^{131}I: particle size in range 10–100 μm
Magnesium Carbonate, Heavy	PC	Subspherical particles, 10–20 μm in diameter, usually arranged in clumps of 4 to 20 particles
Magnesium Carbonate, Light	PC	Small acicular crystals 7 μm long, 1–2 μm thick in clumps of about 10–100 crystals

Substance or Preparation		Remarks
Magnesium Trisilicate	PC	Rounded particles and thin flat lamellae, plus smaller particles
Methisazone Mixture†	BP	Weight median diameter of particles not greater than 15 μm
Novobiocin Mixture	PC	75 per cent of particles less than 10 μm; not less than 99 per cent less than 20 μm
Nystatin Ointment	BPC	No particle of nystatin has a maximum diameter greater than 75 μm
Orciprenaline Aerosol Inhalation	BPC	As for Ergotamine Aerosol Inhalation BPC
Phenolphthalein	PC	Microcrystalline phenolphthalein** to be used in Liquid Paraffin Emulsion with Phenolphthalein BPC to prevent sedimentation of the phenolphthalein
Prednisolone	PC	Ultra-fine particles to be used in solid dosage form to achieve satisfactory solution rate
Prednisolone Pivalate Injection	BP	Particles rarely exceed 20 μm in diameter
Prednisone	PC	As Prednisolone
Salbutamol Aerosol Inhalation	BPC	As for Ergotamine Aerosol Inhalation BPC
Sodium Cromoglycate Cartridges	BP	When examined immediately after dispersion in pentan-1-ol by exposure for 20 sec to low intensity ultrasonic waves exhibit 2 types of particles: small rounded particles of maximum diameter 10 μm, together with larger angular particles of length 10–150 μm but mostly within the range 20–80 μm
Sulphur Precipitated	BP & PC	Grouped amorphous subglobular particles free from crystals (BP); spherules 1.5–11 μm in diameter
Talc purified	BP & PC	Irregularly shaped angular particles, either as flakes about 3–5 μm long or fragments about 14 μm long with jagged and laminated ends

*The following terms are used *inter alia* in the description of powders in the British Pharmacopoeia and the Pharmaceutical Codex, 1979.

Coarse powder: a powder of which all the particles pass through a No. 170 sieve and not more than 40 per cent pass through a No. 355 sieve.

Fine powder: a powder of which all the particles pass through a No. 180 sieve.

Ultra-fine powder: a powder of which the maximum diameter of 90 per cent of the particles is not greater than 5 μm and of which the diameter of none is greater than 50 μm.

**BPC 1973, p. 679.   †BP Addendum 1975.

BP: British Pharmacopoeia   EP: European Pharmacopoeia

PC: Pharmaceutical Codex, 11th edn 1979   BPC: British Pharmaceutical Codex, 1973

From E. G. Salole, in A. H. Beckett and J. B. Stenlake. *Practical Pharmaceutical Chemistry*, vol. 2, Athlone Press, London, 1987

during processing is therefore important: comminution, recrystallisation and drying can all affect crystal properties.

During the pharmacological and toxicological testing of drugs before formal formulation exercises have been carried out, insoluble drugs are frequently administered in suspension form, often routinely in a vehicle containing gum arabic or methylcellulose. Without adequate control of particle size or adequate monitoring the results of these tests must sometimes be in doubt, as both pharmacological activities and toxicity generally result from absorption of the drug. In a few cases particle size influences side-effects such as gastric bleeding or nausea. Gastric bleeding may in part be the direct result of contact of acidic particles of aspirin or non-steroidal anti-inflammatory agents with the mucosal wall. Ritschel[11] has examined the influence of drug form on the LD_{50} of pentobarbitone in mice; the results are shown in table 2.7. A two-fold range of LD_{50} values is obtained by the use of different, simple formulations of the barbiturate. Even in solution form sodium carboxymethylcellulose affects the LD_{50} by mechanisms which are not confirmed. Adsorption of the polymer at the intestinal surface may retard absorption or a portion of the drug may be adsorbed onto the polymer.

table 2.7   **Influence of formulation on the potency ratios of pentobarbitone in the form of the sodium salt and the free acid**

Pentobarbitone form	Dosage form	Vehicle	Particle size (μm)	LD_{50}	Potency ratio**
Sodium salt	Solution	Water	–	132	1
Sodium salt	Solution	1% NaCMC*	–	170	0.78
Free acid	Suspension	1% NaCMC*	<44	189	0.7
Free acid	Suspension	1% NaCMC*	297–420	288	0.46

*Aqueous solution of sodium carboxymethylcellulose
**Relative to aqueous solution of the sodium salt
From reference 11

The deliberate manipulation of particle size leads to a measure of control of activity and side-effects. Rapid solution of nitrofurantoin from tablets of fine particulate material led to a high incidence of nausea in patients, as local high concentrations of the drug produce a centrally mediated nausea. Development of macrocrystalline nitrofurantoin (as in Macrodantin) has led to the introduction of a form of therapy in which the incidence of nausea is reduced. Capsules are used to avoid compression of the crystals during manufacture. Although the urinary levels of the antibacterial are also lowered by the use of a more slowly dissolving form of the drug, levels are still adequate to produce efficient antibacterial effects[15].

2.8 Wetting of powders

Penetration of water into tablets or into granules precedes dissolution. The wettability of the powders, as measured by the contact angle (θ) (figure 2.8) of the substance with water, therefore determines the contact of solvent with the particulate mass. The measurement of the contact angle gives an indication of the nature of the surface. The behaviour of crystalline materials can be related to the chemical structure of the materials concerned, as is shown by the results in table 2.8 on a series of substituted barbiturates. The more hydrophobic the individual barbiturate molecules, the more hydrophobic the crystal which forms, although this would not be necessarily a universal finding, but one dependent on the orientation of the drug molecules in the crystal and the composition of the faces, as we have already seen with adipic acid. Thus hydrophobic drugs have dual problems: they are not readily wetted, and even when wetted they have low solubility. On the other hand, being lipophilic, absorption across lipid membranes is facilitated.

table 2.8 **Relationship between chemical structure and contact angle (θ) with water***

$$R_1 \diagdown \hspace{-0.5em} \underset{\displaystyle R_2}{\diagup} \hspace{-1em} \begin{array}{c} CO-NH \\ \\ CO-NH \end{array} \hspace{-1em} \diagdown \hspace{-0.5em} \underset{\diagup}{} C=O$$

R_1	R_2	θ (deg)
Et	Et	70
Et	Bu	78
Et	$CH_2 CH_2 CH(CH)_2$	102
$-CH-CH_3$ $\quad\vert$ $\quad CH_3$	$CH_2-CH=CH_2$	75
$-CH_2 CHCH_3$ $\qquad\vert$ $\qquad CH_3$	$CH_2-CH=CH_2$	87

*From C. F. Lerk *et al. J. pharm. Sci.*, 66, 1480 (1977)

2.8.1 Contact angle and wettability of solid surfaces

A representation of the several forces acting on a drop of liquid placed on a flat, solid surface is shown in figure 2.8*a*. The surface tension of the solid $\gamma_{S/A}$, will favour spreading of the liquid, but this is opposed by the solid–liquid interfacial tension, $\gamma_{S/L}$, and the horizontal component of the surface tension of the liquid, $\gamma_{L/A}$ in the plane of the solid surface, that is, $\gamma_{L/A} \cos \theta$. Equating these forces gives

$$\boxed{\gamma_{S/A} = \gamma_{S/L} + \gamma_{L/A} \cos \theta} \tag{2.5}$$

Equation 2.5 is generally referred to as Young's equation. The angle θ is termed the *contact angle*. The condition for complete wetting of a solid surface is that the contact angle should be zero. This condition is fulfilled when the forces of attraction between the liquid and solid are equal to or greater than those between liquid and liquid.

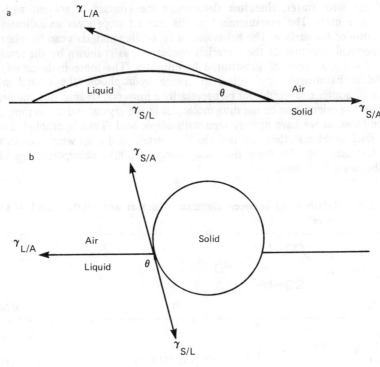

figure 2.8 Equilibrium between forces acting on (a) a drop of liquid on a solid surface, and (b) a partially immersed solid

The type of wetting in which a liquid spreads over the surface of the solid is referred to as *spreading wetting*. The tendency for spreading may be quantified in terms of the spreading coefficient

$$S = \gamma_{L/A} (\cos \theta - 1) \qquad (2.6)$$

If the contact angle is larger than $0°$, the term $(\cos \theta - 1)$ will be negative, as will the value of S. The condition for complete, spontaneous wetting is thus a zero value for the contact angle.

2.8.2 Wettability of powders

When a solid is immersed in a liquid, the initial wetting process is referred to as *immersional wetting*. The effectiveness of immersional wetting may be related to the contact angle which the solid makes with the liquid–air interface (see figure 2.8b). The condition for complete immersion of the solid in the liquid is that there should be a decrease in surface free energy as a result of the immersion process. Once the solid is submerged in the liquid, the process of spreading wetting (see previous section) becomes important.

Table 2.9 gives the contact angles of a series of pharmaceutical powders. These values were determined using compacts of the powder (produced by compressing the powder in a large-diameter tablet die) and a saturated aqueous solution of each compound as the test liquid. Many of the powders are slightly hydrophobic (for example, indomethacin and stearic acid), or even strongly hydrophobic (for example, magnesium stearate, phenylbutazone and chloramphenicol palmitate). Formulation of these drugs as suspensions (for example, Chloramphenicol Palmitate Oral Suspension USP) presents wetting problems. Table 2.9 shows that θ can be affected by the crystallographic structure (see chloramphenicol palmitate). Surface modification or changes in crystal structure are clearly not feasible methods of lowering the contact angle and the normal method of improving wettability is by the inclusion of surfactants in the formulation. The surfactants not only reduce $\gamma_{L/A}$ but also adsorb onto the surface of the powder, thus reducing $\gamma_{S/A}$. Both of these effects reduce the contact angle and improve the dispersibility of the powder.

table 2.9 **Contact angles of some pharmaceutical powders***

Material	Contact angle, θ (deg)	Material	Contact angle, θ (deg)
Acetylsalicylic acid	74	Lactose	30
Aluminium stearate	120	Magnesium stearate	121
Aminophylline	47	Nitrofurantoin	69
Aminopyrine	60	Pentobarbitone	86
Aminosalicylic acid	57	Phenacetin	78
Ampicillin (anhydrous)	35	Phenobarbitone	70
Ampicillin (trihydrate)	21	Phenylbutazone	109
Amylobarbitone	102	Prednisolone	43
Barbitone	70	Prednisone	63
Boric acid	74	Quinalbarbitone	82
Caffeine	43	Salicylamide	70
Calcium carbonate	58	Salicylic acid	103
Calcium stearate	115	Stearic acid	98
Chloramphenicol	59	Succinylsulphathiazole	64
Chloramphenicol palmitate (α form)	122	Sulphacetamide	57
		Sulphadiazine	71
Chloramphenicol palmitate (β form)	108	Sulphamerazine	58
Diazepam	83	Sulphamethazine	48
Digoxin	49	Sulphanilamide	64
Indomethacin	90	Sulphathiazole	53
Isoniazid	49	Theophylline	48
		Tolbutamide	72

*Selected values from C. F. Lerk *et al. J. pharm. Sci.*, 65, 843 (1976); *ibid.*, 66, 1481 (1977)

2.9 Solid dispersions

Over the past few years interest has been shown in solid solutions of drugs in attempts to change the biopharmaceutical properties of drugs which are poorly soluble or difficult to wet. The object is usually to provide a system in which the crystallinity of the drug is so altered as to change its solubility and solution rate, and to surround the drug intimately with water-soluble material. A *solid solution* comprises solute and solvent — a solid solute molecularly dispersed in a solid solvent. These systems are sometimes termed *mixed crystals* because the two components crystallise together in a homogeneous one-phase system. To understand the systems and their potential use an arbitrary system might be considered. In figure 2.9 the melting temperature of mixtures of A and B is plotted against mixture composition. On addition of B to A or of A to B melting points are reduced. At a particular composition the eutectic point is reached, the eutectic mixture, the composition at that point, having the lowest melting point of any mixture of A and B. Below the eutectic temperature no liquid phase exists. The phenomenon is important because of the change in the crystallinity at this point. If we cool a solution of A and B which is richer in A than the eutectic mixture (see M in figure 2.9), crystals of pure A will appear. As the solution is cooled further more and more A crystallises out and the solution becomes richer in B. When the eutectic temperature is reached, however, the remaining solution crystallises out forming a microcrystalline mixture of pure A and pure B, differing markedly at least in superficial characteristics from either of the pure solids. This has obvious pharmaceutical possibilities. This method of obtaining micro-crystalline dispersions for administration of drugs was first suggested by Sekiguchi and co-workers[16,17]. This involved the administration of a eutectic mixture composed of drug and a substance readily soluble in water. The soluble 'carrier' dissolves leaving the drug in a fine state of solution *in vivo*, often, that is, in a state which predisposes to rapid solution.

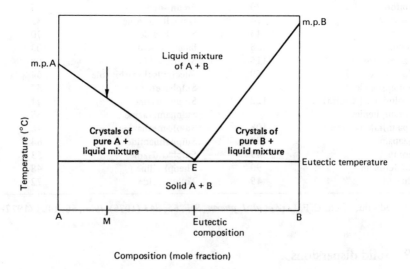

figure 2.9 Phase diagram (temperature versus composition) showing boundaries between liquid and solid phases, and the eutectic point, E

The technique has been applied to several poorly soluble drugs such as griseo-fulvin. A griseofulvin–succinic acid (soluble carrier) system has a eutectic point at 0.29 mole fraction of drug (55 per cent griseofulvin) (figure 2.10*a*). The eutectic mixture consists here of two physically separate phases; one is almost pure griseofulvin, while the other is a saturated solid solution of griseofulvin in succinic

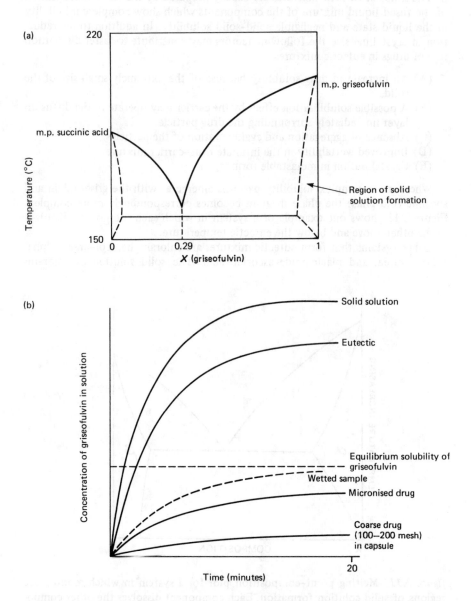

figure 2.10 (a) Griseofulvin–succinic acid phase diagram. (b) Rate of solution of griseofulvin solid solutions, eutectic and crystalline material

acid. The solid solution contains about 25 per cent griseofulvin; the eutectic mixture, which has a fixed ratio of drug to carrier, thus comprises 60 per cent solid solution and 40 per cent almost pure griseofulvin. As can be seen from figure 2.10*b*, which shows the solution profiles of the different forms, the solid solution dissolves six to seven times faster than pure griseofulvin.

The simplest eutectic mixtures are usually prepared by the rapid solidification of the fused liquid mixture of the components which show complete miscibility in the liquid state and negligible solid-solid solubility. In addition to the reduction in crystalline size the following factors may contribute to faster dissolution rate of drugs in eutectic mixtures:

(A) An increase in drug solubility because of the extremely small size of the solid.
(B) A possible solubilisation effect by the carrier may operate in the diffusion layer immediately surrounding the drug particle.
(C) Absence of aggregation and agglomeration of the particles.
(D) Improved wettability in the intimate drug–carrier mixture.
(E) Crystallisation in metastable forms.

Where more complex solubility patterns emerge as with the griseofulvin and succinic acid phase the phase diagram becomes correspondingly more complex. Figure 2.11 shows one example of a system in which each component dissolves in the other above and below the eutectic temperature.

Other systems that form eutectic mixtures are chloramphenicol–urea, sulphathiazole–urea, and niacinamide–ascorbic acid. The solid solution of chloram-

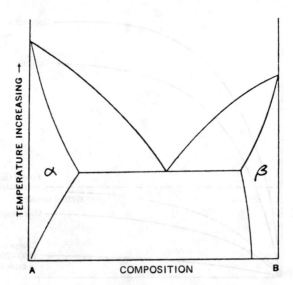

figure 2.11   Melting point–composition plot for a system in which α and β are regions of solid solution formation. Each component dissolves the other component to a degree above the eutectic temperature. As the temperature is lowered, the solid solution regions become narrower

phenicol in urea was found to dissolve twice as rapidly as a physical mixture of the same composition and about four times as rapidly as the pure drug dissolves. However, *in vivo* the system failed to display improved bioavailability although the eutectic mixture of sulphathiazole–urea studied by Sekiguchi did give higher blood levels than pure sulphonamide.

A formulation containing a eutectic

A topical preparation for intradermal anaesthesia to reduce the pain of vene-puncture is now available. The cream, Emla (eutectic mixture of local anaesthetics) (from Astra), contains a eutectic of procaine and lignocaine[18]. The eutectic mixture (50:50 mixture) is an oil, which is then formulated as an oil-in-water emulsion, allowing much higher concentrations than would have been possible by using the individual drugs dissolved in an oil.

2.9.1 Eutectics and drug identification

As the eutectic temperatures of a substance with various other compounds are, as a rule, different even when the other substances have the same melting point, this parameter can be used for identification purposes. Benzanilide (m.p. 163 °C), phenacetin (m.p. 134.5 °C) and salophen (m.p. 191 °C) are often used as test substances[7]. The eutectic temperatures of mixtures of benzanilide with various drugs are shown in table 2.10. Substances of identical melting points can be distinguished by measurement of the eutectic temperature with another suitable compound.

table 2.10 **Eutectic temperatures of drugs with benzanilide***

Compound	Melting point (°C)	Eutectic temperature (°C)
Allobarbitone	173	144
Ergotamine	172–174	135
Imipramine HCl	172–174	109

*From reference 7

Ternary eutectics are also possible. The binary eutectic points of three mixtures are, for aminophenazone–phenacetin 82 °C, aminophenazone–caffeine 103.5 °C, and for phenacetin–caffeine 125 °C; the ternary eutectic temperature of aminophenazone–phenacetin–caffeine is 81 °C. In this mixture the presence of aminophenazone and phenacetin can be tested by the mixed melting point, but the caffeine causes little depression of the eutectic given by the other two components. The possibility of determining eutectics of multicomponent mixtures has practical value in another respect. During tabletting, for example, heat is generated in the punch and die and in the powder compact; measurement of the eutectic can give information on whether this rise in temperature is likely to cause problems of melting and fusion.

References

1. N. Pilpel. *Endeavour* (NS) 6, 1983 (1982)
2. J. W. Mullin. *Crystallization*, 2nd Edition, Butterworths, London, 1972
3. A. S. Michaels and A. R. Colville. *J. phys. Chem.*, 64, 13 (1960)
4. E. Shefter and T. Higuchi. *J. pharm. Sci.*, 52, 781 (1963)
5. J. H. Chapman, *et al. J. Pharm. Pharmacol.*, 20, 418 (1968)
6. M. Kuhnert-Brandstätter, *et al. Mikrochim. Acta*, 1055 (1961)
7. M. Kuhnert-Brandstätter. *Thermomicroscopy in the Analysis of Pharmaceuticals*, Pergamon Press, New York, 1971
8. A. T. Florence and E. G. Salole. *J. Pharm. Pharmacol.*, 28, 637 (1976)
9. A. J. Aguiar and J. E. Zelmer. *J. pharm Sci.*, 58, 983 (1969)
10. G. Volkheimer. *Adv. Pharmacol. Chemother.*, 14, 163 (1977)
11. W. A. Ritschel. *Arzneim. Forsch.*, 25, 853 (1975)
12. R. M. Atkinson, A. Bedford, K. J. Child and E. G. Tomich. *Nature*, 193, 588 (1962)
13. T. R. D. Shaw, J. E. Carless, M. R. Howard and K. Raymond, *Lancet*, ii, 209 (1973)
14. J. F. Nash, *et al. Drug Dev. Commun.*, 1, 443 (1974/75)
15. J. H. Fincher. *J. pharm Sci.*, 57, 1825 (1968)
16. K. Sekiguchi and N. Obi. *Chem. pharm. Bull.*, 9, 866 (1961)
17. K. Sekiguchi, N. Obi and Y. Ueda. *ibid.*, 12, 134 (1964)
18. B. F. J. Broberg and H. C. A. Evers. Eur. Pat. 0 002 425 (1981)

3 Physicochemical Properties of Drugs in Solution

This chapter will deal with some of the physicochemical properties of drugs in aqueous solution which are of relevance to such liquid dosage forms as parenteral solutions, eye-drops, infusions and irrigations. The concept of thermodynamic activity, which is an important parameter in determining drug potency, will be discussed, as will the clinical relevance of osmotic effects of both intravenous and oral preparations. Most drugs are at least partially ionised at physiological pH values and many studies have suggested that the charged group is essential for biological activity. In this chapter the influence of pH on the ionisation of several types of drugs in solution will be dealt with in detail.

3.1 Concentration units

A wide range of units is commonly used to express solution concentration and confusion often arises in the interconversion of one set of units to another. Wherever possible throughout this book we have used the SI system of units. Although this is the currently recommended system of units in Great Britain, other more traditional systems are still widely used and these will be described in this section.

(A) Weight concentration
Concentration is often expressed as a weight of solute in a unit volume of solution; for example, $g \, \ell^{-1}$, or % w/v which is the number of grams of solute in 100 mℓ of solution. This is not an exact method when working at a range of temperatures since the volume of the solution is temperature dependent and hence the weight concentration also changes with temperature. Whenever a hydrated compound is used it is important to use the correct state of hydration in the calculation of weight concentration. Thus 10% w/v $CaCl_2$ anhydrous is approximately equivalent to 20% w/v $CaCl_2.6H_2O$ and consequently the use of the vague statement '10% calcium chloride' could result in gross error. The SI unit is kg m^{-3} which is numerically equal to $g \, \ell^{-1}$.

(B) Molarity and molality
It is important to distinguish between these two similarly sounding terms. The *molarity* of a solution is the number of moles (gram molecular weights) of solute in *1 litre of solution*. The *molality* is the number of moles of solute in *1 kg of solvent*. Molality has the unit mol kg^{-1} and is an accepted SI unit. Molarity may be converted to SI units using the relationship 1 mol ℓ^{-1} = 10^3 mol m^{-3}. Interconversion between molarity and molality requires a knowledge of the density of the solution. Of the two units molality is preferable for a precise expression of concentration because it does not depend on the solution temperature as does molarity; also, the molality of a component in a solution remains unaltered by the addition of a second solute, whereas the molarity of this component de-

creases due to an increase in the total volume of solution following the addition of the second solute.

(C) Milliequivalents
These units are commonly used clinically in expressing the concentration of an ion in solution. The term 'equivalent' or gram equivalent weight is analogous to the mole or gram molecular weight. When monovalent ions are considered, these two terms are identical; thus a 1 molar solution of sodium bicarbonate, $NaHCO_3$, contains 1 mol or 1 Eq of Na^+ and 1 mol or 1 Eq of HCO_3^- per litre of solution. With multivalent ions attention must be paid to the valency of each ion; for example, 10% w/v $CaCl_2.2H_2O$ contains 6.8 mmol or 13.6 mEq of Ca^{2+} in 10 mℓ. In analytical chemistry a solution which contains 1 Eq $ℓ^{-1}$ is referred to as a *normal* solution. Unfortunately the term 'normal' is also used to mean physiologically normal with reference to saline solution. In this usage, a physiologically normal saline solution contains 0.9 g NaCl in 100 mℓ aqueous solution and not 1 equivalent (58.44 g) per litre.

The *Pharmaceutical Codex*[1] gives a table of milliequivalents for various ions and also simple formulae for the interconversion of percentage strength and milliequivalents per litre.

(D) Mole fraction
The mole fraction of a component of a solution is the number of moles of that component divided by the total number of moles present in solution. In a two-component (binary) solution, the mole fraction of solvent, x_1, is given by $x_1 = n_1/(n_1 + n_2)$, where n_1 and n_2 are the number of moles of solvent and solute present in solution. Similarly the mole fraction of solute, x_2, is given by $x_2 = n_2/(n_1 + n_2)$. The sum of the mole fractions of all components is, of course, unity.

3.2 Activity and chemical potential

3.2.1 Activity and standard states

The *activity* is a term which has been introduced to describe departure of the behaviour of a solution from ideality. In any real solution interactions occur between the components which reduce the *effective concentration* of the solution. The activity is a way of describing this effective concentration. In an ideal solution or in a real solution at infinite dilution there are no interactions between components and the activity equals the concentration. Non-ideality in real solutions at higher concentrations causes a divergence between the values of activity and concentration. The ratio of the activity to the concentration is called the activity coefficient, γ

$$\gamma = \frac{\text{activity}}{\text{concentration}} \qquad (3.1)$$

Depending on the units used to express concentration we can have either a molal activity coefficient, γ_m, a molar activity coefficient, γ_c, or, if mole fractions are used, a rational activity coefficient, γ_x.

In order to be able to express the activity of a particular component numerically, it is necessary to define a reference state in which the activity is arbitrarily unity. The activity of a particular component is then the ratio of its value in a given solution to that in the reference state. For the solvent the reference state is invariably taken to be the pure liquid, and, if this is at a pressure of 1 atmosphere and at a definite temperature, it is also the *standard state*. Since the mole fraction as well as the activity is unity, $\gamma_x = 1$.

Several choices are available in defining the standard state of the solute. If the solute is a liquid which is miscible with the solvent (as, for example, in a benzene-toluene mixture) then the standard state is again the pure liquid. Several different standard states have been used for solutions of solutes of limited solubility. In developing a relationship between drug activity and thermodynamic activity, Ferguson[2] used the pure substance as the standard state. The activity of the drug in solution was then taken to be the ratio of its concentration to its saturation solubility. The use of a pure substance as the standard state is of course of limited value since a different state is used for each compound. A more feasible approach is to use the infinitely dilute solution of the compound as the reference state. However, since the activity equals the concentration in such solutions it is not equal to unity as it should be for a standard state. This difficulty is overcome by defining the standard state as a hypothetical solution of unit concentration possessing, at the same time, the properties of an infinitely dilute solution. Some workers[3] have chosen to define the standard state in terms of an alkane solvent rather than water; one advantage of this solvent being the absence of specific solute–solvent interactions in the reference state which would be highly sensitive to molecular structure.

3.2.2 Activity of ionised drugs

A large proportion of the drugs that are administered in aqueous solution are salts which, on dissociation, behave as electrolytes. Simple salts such as ephedrine hydrochloride ($C_6H_5CH(OH)CH(NHCH_3)CH_3.HCl$) are 1:1 (or uni-univalent) electrolytes; that is, on dissociation each mole yields one cation, $C_6H_5CH(OH)CH(\overset{+}{N}H_2CH_3)CH_3$, and one anion, Cl^-. Other salts are more complex in their ionisation behaviour; for example, ephedrine sulphate is a 1:2 electrolyte, each mole giving two moles of the cation and one mole of $SO_4{}^{2-}$ ions. In general, we will consider a strong electrolyte which dissociates according to

$$C_{\nu_+}A_{\nu_-} \longrightarrow \nu_+ C^{z+} + \nu_- A^{z-} \qquad (3.2)$$

where ν_+ is the number of cations, C^{z+}, of valence $z+$, and ν_- is the number of anions, A^{z-}, of valence $z-$. The anion and cation may each have a different ionic activity in solution. Since it is not possible to determine individual ionic activities experimentally, a combined activity term, the mean ion activity, $a_\pm^\nu$, is used. The activity, a, of the electrolyte is then

$$\boxed{a = a_+^{\nu_+}\, a_-^{\nu_-} = a_\pm^\nu} \qquad (3.3)$$

where

$$\nu = \nu_+ + \nu_-$$

For example, for morphine sulphate which is a 1:2 electrolyte,

$$a = a_{\text{morph+}}^2 \times a_{SO_4^{2-}} = a_\pm^3 \qquad (3.4)$$

Similarly, we may also define a mean ion activity coefficient, $\gamma_\pm$, in terms of the individual ionic activity coefficients γ_+ and γ_-

$$\gamma_\pm^v = \gamma_+^{v+} \, \gamma_-^{v-} \tag{3.5}$$

or

$$\boxed{\gamma_\pm = (\gamma_+^{v+} \, \gamma_-^{v-})^{1/v}} \tag{3.6}$$

For a 1:1 electrolyte equation 3.6 reduces to

$$\gamma_\pm = (\gamma_+ \, \gamma_-)^{1/2} \tag{3.7}$$

Finally, we define a mean ionic molality, $m_\pm$, as

$$m_\pm^v = m_+^{v+} \, m_-^{v-} \tag{3.8}$$

or

$$\boxed{m_\pm = (m_+^{v+} \, m_-^{v-})^{1/v}} \tag{3.9}$$

For a 1:1 electrolyte, equation 3.9 reduces to

$$m_\pm = (m_+ \, m_-)^{1/2} = m \tag{3.10}$$

that is, mean ionic molality may be equated with the molality of the solution.

The activity of each ion is the product of its activity coefficient and its concentration

$$a_+ = \gamma_+ \, m_+ \quad \text{and} \quad a_- = \gamma_- \, m_-$$

so that

$$\gamma_+ = a_+/m_+ \quad \text{and} \quad \gamma_- = a_-/m_-$$

Expressed as the mean ionic parameters we have

$$\gamma_\pm = a_\pm/m_\pm \tag{3.11}$$

Substituting for $m_\pm$ from equation 3.9 gives

$$\boxed{\gamma_\pm = a_\pm/(m_+^{v+} \, m_-^{v-})^{1/v}} \tag{3.12}$$

This equation applies in any solution, whether the ions are added together, as a single salt, or separately as a mixture of salts. For a solution of a single salt of molality, m

$$m_+ = v_+ m \quad \text{and} \quad m_- = v_- m$$

Equation 3.12 reduces to

$$\gamma_\pm = \frac{a_\pm}{m(v_+^{v+} \, v_-^{v-})^{1/v}} \tag{3.13}$$

For example, for morphine sulphate, $v_+ = 2$, $v_- = 1$, and thus

$$\gamma_\pm = \frac{a_\pm}{(2^2 \, 1)^{1/3} \, m} = \frac{a_\pm}{4^{1/3} \, m}$$

Values of the mean ion activity coefficient may be determined experimentally using several methods including electromotive force measurement, solubility determinations and colligative properties. It is possible, however, to calculate $\gamma_\pm$ in very dilute solution using a theoretical method based on the Debye–Hückel theory. In this theory each ion is considered to be surrounded by an 'atmosphere' in which there is a slight excess of ions of opposite charge. The electrostatic energy due to this effect was related to the chemical potential of the ion to give a limiting expression for dilute solutions

$$\boxed{- \log \gamma_\pm = z_+ z_- A \sqrt{I}} \tag{3.14}$$

where z_+ and z_- are the valencies of each ion, A is a constant whose value is determined by the dielectric constant of the solvent and the temperature (A = 0.509 in water at 298 K), and I is the total ionic strength defined by

$$I = \tfrac{1}{2}\sum(mz^2) = \tfrac{1}{2}(m_1 z_1^2 + m_2 z_2^2 + \ldots) \tag{3.15}$$

where the summation is continued over all the different species in solution. It can be readily shown from equation 3.15 that, for a 1:1 electrolyte, the ionic strength is equal to its molality, for a 1:2 electrolyte $I = 3m$, and for a 2:2 electrolyte, $I = 4m$.

The Debye–Hückel expression as given by equation 3.14 is valid only in dilute solution ($I < 0.02$ mol kg^{-1}). At higher concentrations a modified expression has been proposed

$$\log \gamma_\pm = \frac{-A z_+ z_- \sqrt{I}}{1 + a_i \beta \sqrt{I}} \tag{3.16}$$

where a_i is the mean distance of approach of the ions or the mean effective ionic diameter, and β is a constant whose value depends on the solvent and temperature. As an approximation, the product $a_i \beta$ may be taken to be unity, thus simplifying the equation. Equation 3.16 is valid for $I \not> 0.1$ mol kg^{-1}.

Example 3.1

Calculate: (a) the mean ionic activity coefficient and the mean ionic activity of 0.002 mol kg^{-1} aqueous solution of ephedrine sulphate; (b) the mean ionic activity coefficient of an aqueous solution containing 0.002 mol kg^{-1} ephedrine sulphate and 0.01 mol kg^{-1} sodium chloride. Both solutions are at 25 °C.

(a) Ephedrine sulphate is a 1:2 electrolyte and hence the ionic strength is given by equation 3.15 as

$$I = \tfrac{1}{2}[(0.002 \times 2 \times 1^2) + (0.002 \times 2^2)] = 0.006$$

From the Debye–Hückel equation (equation 3.14)

$$-\log \gamma_\pm = 0.509 \times 1 \times 2 \times \sqrt{0.006}$$

$$\log \gamma_\pm = -0.0789$$

$$\gamma_+ = 0.834$$

The mean ionic activity may be calculated from equation 3.13

$$a_\pm = 0.834 \times 0.002 \times (2^2 \times 1)^{1/3}$$

$$= 0.00265$$

(b) Ionic strength of 0.01 mol kg^{-1} $NaCl = \frac{1}{2}(0.01 \times 1^2) + (0.01 \times 1^2)$

$$= 0.01$$

Total ionic strength $= 0.006 + 0.01 = 0.016$

$$-\log \gamma_{\pm} = 0.509 \times 2 \times \sqrt{0.016}$$

$$\log \gamma_{\pm} = -0.1288$$

$$\gamma_{\pm} = 0.743 \quad \blacksquare$$

3.2.3 Solvent activity

Although the activity of a solution usually refers to the activity of the solute in the solution as in the preceding section, we may also make reference to the activity of the solvent. Experimentally, solvent activity a_1 may be determined as the ratio of the vapour pressure p_1 of the solvent in a solution to that of the pure solvent $p_1^{\ominus}$, that is

$$a_1 = p_1/p_1^{\ominus} = \gamma_1 x_1 \tag{3.17}$$

where γ_1 is the solvent activity coefficient and x_1 is the mole fraction of solvent.

The relationship between the activities of the components of the solution is expressed by the Gibbs–Duhem equation

$$x_1 \, d \ln a_1 + x_2 \, d \ln a_2 = 0 \tag{3.18}$$

which provides a way of determining the activity of the solute from measurements of vapour pressure.

Water activity and bacterial growth

When the aqueous solution in the environment of a micro-organism is concentrated by the addition of solutes such as sucrose, the consequences for microbial growth result mainly from the change in water activity a_w. Every micro-organism has a limiting a_w below which it will not grow. The minimum a_w levels for growth of human bacterial pathogens such as streptococci, *Klebsiella, Escherichia coli, Corynebacterium, Clostridium perfringens* and other clostridia, and *Pseudomonas* is 0.91^4. *Staphylococcus aureus* can proliferate at an a_w as low as 0.86^5. Figure 3.1 shows the influence of a_w, adjusted by the addition of sucrose, on the growth rate of this micro-organism at $35\,^{\circ}C$ and pH 7.0^6. The control medium, with a water activity value of $a_w = 0.993$, supported the rapid growth of the test organism. Reduction of a_w of the medium by sucrose addition progressively increased generation times and lag periods and lowered the peak cell counts. Complete growth inhibition was achieved at an a_w of 0.858 (195 g sucrose per 100 g water) with cell numbers declining slowly throughout the incubation period. The results reported in this study explain why the old remedy of treating infected wounds with sugar, honey or molasses is so successful. When the wound is filled with sugar the sugar dissolves in the tissue water, creating an environment of low a_w, which inhibits bacterial growth. However, the difference in water activity between the tissue and the concentrated sugar solution causes migration

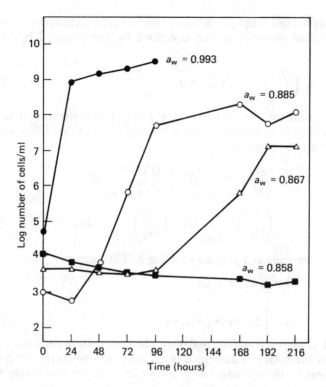

figure 3.1 Staphylococcal growth at 35 °C in medium alone (a_w = 0.993) and in media with a_w values lowered by additional sucrose. From reference 6 with permission

of water out of the tissue, hence diluting the sugar and raising a_w. Further sugar must then be added to the wound to maintain growth inhibition.

3.2.4 Chemical potential

Properties such as volume, enthalpy, free energy and entropy which depend on the quantity of substance are called *extensive* properties. In contrast, properties such as temperature, density and refractive index which are independent of the amount of material are referred to as *intensive* properties. The quantity denoting the rate of increase in the magnitude of an extensive property with increase in the number of moles of a substance added to the system at constant temperature and pressure is termed a partial molar quantity. Such quantities are distinguished by a bar above the symbol for the particular property. For example

$$\left(\frac{\partial V}{\partial n_2}\right)_{T,P,n_1} = \bar{V}_2 \qquad (3.19)$$

In practical terms the partial molar volume, $\bar{V}$, represents the change in the total volume of a large amount of solution when one additional mole of solute is added — it is the effective volume of one mole of solute in solution.

Of particular interest is the partial molar free energy, $\overline{G}$, which is also referred to as the chemical potential, μ, and is defined for component 2 in a binary system by

$$\left(\frac{\partial G}{\partial n_2}\right)_{T,P,n_1} = \overline{G}_2 = \mu_2 \tag{3.20}$$

Partial molar quantities are of importance when open systems, that is those involving transference of matter as well as energy, are considered. For an open system involving two components

$$dG = \left(\frac{\partial G}{\partial T}\right)_{P,n_1,n_2} dT + \left(\frac{\partial G}{\partial P}\right)_{T,n_1,n_2} dP$$

$$+ \left(\frac{\partial G}{\partial n_1}\right)_{T,P,n_2} dn_1 + \left(\frac{\partial G}{\partial n_2}\right)_{T,P,n_1} dn_2 \tag{3.21}$$

At constant temperature and pressure equation 3.21 reduces to

$$dG = \mu_1\,dn_1 + \mu_2\,dn_2 \tag{3.22}$$

$$G = \int dG = \mu_1 n_1 + \mu_2 n_2 \tag{3.23}$$

The chemical potential therefore represents the contribution per mole of each component to the total free energy. It is the effective free energy per mole of each component in the mixture and is always less than the free energy of the pure substance.

Chemical potential in two-phase systems

Consider a system of two phases, a and b, in equilibrium at constant temperature and pressure. If a small quantity of substance is transferred from phase a to phase b, then, because the over-all free energy change is zero, we have

$$dG_a + dG_b = 0 \tag{3.24}$$

where dG_a and dG_b are the free energy changes accompanying the transference of material for each phase.

From equation 3.22

$$dG_a = \mu_a dn_a \quad\text{and}\quad dG_b = \mu_b dn_b$$

and thus

$$\mu_a\,dn_a + \mu_b\,dn_b = 0 \tag{3.25}$$

A decrease of dn moles of component in phase a leads to an increase of exactly dn moles of this component in phase b, that is

$$dn_a = -dn_b \tag{3.26}$$

Substitution of equation 3.26 into equation 3.25 leads to the result

$$\boxed{\mu_a = \mu_b} \tag{3.27}$$

In general, the chemical potential of a component is identical in all the phases of a system at equilibrium at a fixed temperature and pressure.

Because of the need for equality of chemical potential at equilibrium, a substance in a system which is not at equilibrium will have a tendency to diffuse spontaneously from a phase in which it has a high chemical potential to another in which it has a low chemical potential. In this respect the chemical *potential* resembles electrical *potential* and hence its name is an apt description of its nature.

Chemical potential of a component in solution

(A) Non-electrolytes

In dilute solutions of non-volatile solutes, Raoult's law can usually be assumed to be obeyed and the chemical potential of the solute is given by

$$\mu_2 = \mu_2^{\ominus} + RT \, \ell n \, x_2 \tag{3.28}$$

It is usually more convenient to express solute concentration as molality, m, rather than mole fraction using

$$x_2 = m \, M_1 / 1000 \tag{3.29}$$

where M_1 = molecular weight of the solvent.

Thus

$$\mu_2 = \mu^{\ominus} + RT \, \ell n \, m \tag{3.30}$$

where $\mu^{\ominus} = \mu_2^{\ominus} + RT \, \ell n \, M_1 - RT \, \ell n \, 1000$.

At higher concentrations, the solution generally exhibits significant deviations from Raoult's law and mole fraction must be replaced by activity

$$\boxed{\mu_2 = \mu_2^{\ominus} + RT \, \ell n \, a_2} \tag{3.31}$$

or

$$\mu_2 = \mu_2^{\ominus} + RT \, \ell n \, \gamma_2 + RT \, \ell n \, x_2 \tag{3.32}$$

(B) Electrolytes

The chemical potential of a strong electrolyte, which may be assumed to be completely dissociated in solution, is equal to the sum of the chemical potentials of each of the component ions.

Thus

$$\mu_+ = \mu_+^{\ominus} + RT \, \ell n \, a_+ \tag{3.33}$$

and

$$\mu_- = \mu_-^{\ominus} + RT \, \ell n \, a_- \tag{3.34}$$

and therefore

$$\boxed{\mu_2 = \mu_2^{\ominus} + RT \, \ell n \, a} \tag{3.35}$$

where $\mu_2^{\ominus}$ is the sum of the chemical potentials of the ions, each in their respective standard states.

That is

$$\mu_2^{\ominus} = \nu_+ \mu_+^{\ominus} + \nu_- \mu_-^{\ominus}$$

where ν_+ and ν_- are the number of cations and anions, respectively, and a is the activity of the electrolyte as given in section 3.2.2.

For example, for a 1:1 electrolyte, from equation 3.3

$$a = a_\pm^2$$

Therefore $\quad \mu_2 = \mu_2^\ominus + 2RT \ln a_\pm$

From equation 3.11

$$a_\pm = m\,\gamma_\pm$$

Therefore $\quad \boxed{\mu_2 = \mu_2^\ominus + 2RT \ln m\gamma_\pm}$

3.3 Osmotic properties of drug solutions

A non-volatile solute added to a solvent affects the magnitude of the vapour pressure above the solvent, the freezing point and the boiling point to an extent that is proportional to the relative number of solute molecules present, rather than to the weight concentration of the solute. Properties that are dependent on the number of molecules in solution in this way are referred to as *colligative* properties and the most important of such properties from a pharmaceutical viewpoint is the osmotic pressure.

3.3.1 Osmotic pressure

Whenever a solution is separated from a solvent by a membrane, permeable only to solvent molecules (referred to as a semipermeable membrane) there is a passage of solvent across the membrane into the solution. This is the phenomenon of osmosis. If the solution is totally confined by a semipermeable membrane and immersed in the solvent, then a pressure differential develops across the membrane which is referred to as the osmotic pressure. Solvent passes through the membrane because of the inequality of the chemical potentials on either side of the membrane. Since the chemical potential of a solvent molecule in solution is less than that in pure solvent, solvent will spontaneously enter the solution until this inequality is removed. An equation which relates the osmotic pressure of the solution to the solution concentration can be derived in the following way.

At equilibrium, the chemical potential of the pure solvent $\mu_1^\ominus$ at 1 atmosphere may be equated with that of the solvent in the solution, μ_1; that is

$$\mu_1^\ominus = \mu_1 \tag{3.36}$$

There are two factors that affect the chemical potential of the solvent in solution – one is the presence of the solute molecules which lower the chemical potential, the other is the pressure, which increases the chemical potential. These two effects counterbalance, and equation 3.36 is obeyed when the applied pressure is equal to the osmotic pressure. The over-all effect of these two factors is given by

$$d\mu_1 = \left(\frac{\partial \mu_1}{\partial x_2}\right)_{T,P} dx_2 + \left(\frac{\partial \mu_1}{\partial P}\right)_{T,x_2} dP \tag{3.37}$$

But

$$\left(\frac{\partial G}{\partial P}\right)_{T, x_2} = V$$

Consequently

$$\left(\frac{\partial \mu_1}{\partial P}\right)_{T, x_2} = \bar{V}_1 \tag{3.38}$$

is the partial molal volume of the solvent.

An expression for $(\partial \mu_1 / \partial x_2)_{T, P}$ may be derived for ideal solutions by differentiation of the chemical potential equation

$$\mu_1 = \mu_1^{\ominus} + RT \ln x_1 = \mu_1^{\ominus} + RT \ln (1 - x_2)$$

Thus

$$\left(\frac{\partial \mu_1}{\partial x_2}\right)_{T, P} = \frac{-RT}{1 - x_2} \tag{3.39}$$

Substituting equations 3.38 and 3.39 into equation 3.37 yields

$$d\mu_1 = \bar{V}_1 \, dP - \frac{RT \, dx_2}{(1 - x_2)} \tag{3.40}$$

For equation 3.36 to be obeyed, $d\mu_1$ must be zero.

Therefore $\qquad \bar{V}_1 \, dP = RT \, dx_2/(1 - x_2) = -RT \, dx_1/x_1 \tag{3.41}$

Integrating over the pressure range 1 to $(1 + \pi)$, where π is the osmotic pressure, and from $x_1 = 1$ ($\ln x_1 = 0$) to $\ln x_1$, assuming that $\bar{V}_1$ is independent of pressure and concentration

$$\int_1^{1+\pi} \bar{V}_1 \, dP = -RT \int_0^{\ln x_1} d \ln x_1 \tag{3.42}$$

$$\bar{V}_1 \, \pi = -RT \ln x_1 = -RT \ln (1 - x_2) \tag{3.43}$$

Assuming that all terms but the first of the expansion of $\ln (1 - x_2)$ are negligible (that is, the solution is very dilute)

$$\bar{V}_1 \, \pi = RT \, x_2 \tag{3.44}$$

Again, in dilute solution, we may further approximate x_2 to n_2/n_1 and $\bar{V}_1$ to V/n_1 where V is the volume of the solution, giving

$$\boxed{\pi V = n_2 \, RT} \tag{3.45}$$

Equation 3.45, which bears a resemblance in form to the ideal gas equation, was derived empirically by van't Hoff and bears his name.

On application of the van't Hoff equation to the drug molecules in solution, consideration must be made of any ionisation of the molecules since osmotic pressure, being a colligative property, will be dependent on the total number of particles in solution (including the free counterions). To allow for what was at the time considered to be anomalous behaviour of electrolyte solutions, van't Hoff introduced a correction factor, i. The value of this factor approaches a number equal to that of the number of ions, v, into which each molecule dissociates as the solution is progressively diluted. The ratio i/v is termed the practical osmotic coefficient, ϕ.

For non-ideal solutions, the mole fraction x_1 in equation 3.43 must be replaced by the activity of the solvent, a_1.

Thus
$$\pi \, \overline{V}_1 = -RT \, \ln a_1 \tag{3.46}$$

Activity and osmotic pressure are related by the expression

$$\ln a_1 \equiv \frac{-v \, m \, M_1}{1000} \, \phi \tag{3.47}$$

where M_1 is the molecular weight of the solvent and m is the molality of the solution. The relationship between the osmotic pressure and the osmotic coefficient is thus

$$\pi = \left(\frac{RT}{\overline{V}_1}\right) \frac{v \, m \, M_1}{1000} \, \phi \tag{3.48}$$

3.3.2   Osmolality and osmolarity

The experimentally derived osmotic pressure is frequently expressed as the osmolality, ξ_m, which is the mass of solute which when dissolved in 1 kg of water will exert an osmotic pressure, π', equal to that exerted by a gram molecular weight of an ideal unionised substance dissolved in 1 kg of water.

According to the definition, $\xi_m = \pi/\pi'$. The value of π' may be obtained from equation 3.48 by noting that for an ideal unionised substance $v = \phi = 1$, and since m is also unity, equation 3.48 becomes

$$\pi' = \left(\frac{RT}{\overline{V}_1}\right) \frac{M_1}{1000}$$

Thus
$$\xi_m = v \, m\phi \tag{3.49}$$

Example 3.2

A 0.90% w/w solution of sodium chloride (mol. wt. = 58.5) has an osmotic coefficient of 0.928. Calculate the osmolality of the solution.

Osmolality is given by equation 3.49 as

$$\xi_m = v \, m \, \phi$$

so
$$\xi_m = 2 \times \frac{9.0}{58.5} \times 0.928 = 286 \text{ mosmol kg}^{-1} \quad \blacksquare$$

Pharmaceutical labelling regulations sometimes require a statement of the osmolarity; for example, the USP XIX requires that sodium chloride injection should be labelled in this way. Osmolarity is defined as the mass of solute which, when dissolved in 1 litre of solution, will exert an osmotic pressure equal to that exerted by a gram molecular weight of an ideal unionised substance dissolved in

1 litre of solution. The relationship between osmolality and osmolarity has been discussed by Streng *et al.*[7]. Table 3.1 lists the osmolalities of commonly used intravenous fluids.

table 3.1 Tonicities (osmolalities) of intravenous fluids*

Solution	Tonicity (mosmol kg^{-1})
Vamin 9	700
Vamin 9 Glucose	1350
Vamin 14	1145
Vamin 14 Electrolyte-free	810
Vamin 18	1130
Intralipid 10%	300
Intralipid 20%	350
Aminoplex 5	2415
Aminoplex 12	830
Aminoplex 14	875
Aminoplex 24	1570
Laevuflex 20	1330
Aminofusin L600	1300
Aminofusin L1000	2700
Aminofusin L Forte	1050

*From *ABPI Data Sheet Compendium*, Datapharm Publications, 1985

3.3.3 Clinical relevance of osmotic effects

Osmotic effects are particularly important from a physiological viewpoint since biological membranes, notably the red blood cell membrane, behave in a manner similar to that of semipermeable membranes. Consequently when red blood cells are immersed in a solution of greater osmotic pressure than that of their contents, they shrink as water passes out of the blood cells in an attempt to reduce the chemical potential gradient across the blood cell membrane. Conversely, on placing the cells in an aqueous environment of lower osmotic pressure, the cells swell as water enters and eventually lysis may occur. It is an important consideration, therefore, to ensure that the osmotic pressure of solutions for injection is approximately the same as that of blood serum. Such solutions are said to be *isotonic* with blood. Solutions with a higher osmotic pressure are *hypertonic* and those with a lower osmotic pressure are termed *hypotonic* solutions. Similarly, in order to avoid discomfort on administration of solutions to the delicate membranes of the body, such as the eyes, it is important to ensure that these solutions are made isotonic with the relevant tissues.

The osmotic pressures of many of the solutions of table 3.1 are in excess of that of plasma (291 mosmol ℓ^{-1}). It is generally recommended that any fluid with an osmotic pressure above 550 mosmol ℓ^{-1} should not be infused rapidly as this would increase the incidence of venous damage. The rapid infusion of marginally hypertonic solutions (in the range 300-500 mosmol ℓ^{-1}) would appear to be clinically practicable; the higher the osmotic pressure of the solution within this range, the slower should be its rate of infusion to avoid damage. Patients with centrally inserted lines are not normally affected by limits on tonicity as infusion is normally slow and dilution is rapid.

Certain oral medications commonly used in the intensive care of premature infants have very high osmolalities. The high tonicity of enteral feedings has been implicated as a cause of necrotising enterocolitis (NEC). A higher frequency of gastro-intestinal illness including NEC has been reported[8] amongst premature infants fed undiluted calcium lactate than amongst those fed no supplemental calcium or calcium lactate diluted with water or formula. White and Harkavy[9] have discussed a similar case of the development of NEC following medication with calcium glubionate elixir. These authors have measured osmolalities of several oral medications by freezing point depression and compared these with the osmolalities of analogous intravenous (i.v.) preparations (see table 3.2). Except in the case of digoxin, the osmolalities of the i.v. preparations were very much lower than those of the corresponding oral preparations despite the fact that the i.v. preparations contained at least as much drug per millilitre as did the oral forms. This striking difference may be attributed to the additives, such as ethyl alcohol, sorbitol and propylene glycol, which make a large contribution to the osmolalities of the oral preparations. The vehicle for the i.v. digoxin consists of 40 per cent propylene glycol and 10 per cent ethyl alcohol with calculated osmolalities of 5260 and 2174 mosmol kg^{-1} respectively, thus explaining the

table 3.2 **Measured and calculated osmolalities of drugs***

Drug (route)	Concentration of drug	Mean measured osmolality of full-strength drug (mosmol kg^{-1})	Calculated available milliosmoles in 1 kg of drug preparation†
Theophylline elixir (oral)	80 mg/15 mℓ	>3000	4980
Aminophylline (i.v.)	25 mg/mℓ	116	200
Phenobarbitone elixir (oral)	400 mg/dℓ	>3000	11 630
Phenobarbitone (i.v.)	65 mg/mℓ, diluted 400 mg/dℓ	114	134
Calcium glubionate (oral)	115 mg/5 mℓ	>3000	2270
Calcium gluceptate (i.v.)	90 mg/5 mℓ	507	950
Digoxin elixir (oral)	50 mg/ℓ	>3000	4420
Digoxin (i.v.)	100 mg/ℓ	>3000	9620
Dexamethasone elixir (oral)	0.5 mg/5 mℓ	>3000	3980
Dexamethasone sodium phosphate (i.v.)	4 mg/mℓ	284	312

*From reference 9 with permission
†This would be the osmolality of the drug if the activity coefficient were equal to 1 in the full-strength preparation. The osmolalities of serial dilutions of the drug were plotted against the concentrations of the solution, and a least-squares regression line was drawn. The value for the osmolality of the full-strength solution was then estimated from the line. This is the calculated available milliosmoles

unusually high osmolality of this i.v. preparation. These authors have recommended that extreme caution should be exercised in the administration of these oral preparations and perhaps any medication in a syrup or elixir form when the infant is at risk from necrotising enterocolitis. In some cases the osmolality of the elixir is so high that even mixing with infant formula does not reduce the osmolality to a tolerable level. For example, when clinically appropriate doses of dexamethasone and phenobarbitone elixirs were mixed in volumes of formula appropriate for a single feeding for a 1500 g infant, the osmolalities of the mixes increased by at least 300 per cent compared to formula alone (see table 3.3).

table 3.3 Osmolalities of drug–formula mixtures*

Drug (dose)	Volume of drug (mℓ) + volume of formula (mℓ)	Mean measured osmolality (mosmol kg^{-1})
Formula	–	292
Theophylline elixir, 1 mg kg^{-1}	0.3 + 15	392
	0.3 + 30	339
Calcium glubionate syrup, 0.5 mmol kg^{-1}	0.5 + 15	378
	0.5 + 30	330
Digoxin elixir, 5 μg kg^{-1}	0.15 + 15	347
	0.15 + 30	322
Phenobarbitone elixir, 2.66 mg kg^{-1}	1 + 15	934
	1 + 30	624
Dexamethasone elixir, 0.25 mg kg^{-1}	3.8 + 15	1149
	3.8 + 30	791

*From reference 9 with permission

Lerman and co-workers[10] have related the aqueous solubilities of several volatile anaesthetics to the osmolarity of the solution. The inverse relationship between solubility (expressed as the liquid/gas partition coefficient) of those anaesthetics and the osmolarity is shown in table 3.4. These findings have practical applications for the clinician. Although changes in serum osmolarity within the physiological range (209–305 mosmol ℓ^{-1}) have only a small effect on the liquid/gas partition coefficient, changes in the serum osmolarity and the concentration of serum constituents at the extremes of the physiological range may significantly decrease the liquid/gas partition coefficient. For example, the blood/gas partition coefficient of isoflurane decreases significantly after an infusion of mannitol[11]. This may be attributed to both a transient increase in the osmolarity of the blood and a more prolonged decrease in the concentration of serum constituents caused by the influx of water due to the osmotic gradient.

3.3.4 Preparation of isotonic solutions

Since osmotic pressure is not a readily measurable quantity, it is usual to make use of the inter-relation of colligative properties and to calculate the osmotic pressure from a more easily measured property such as the freezing point depres-

table 3.4 **Liquid/gas partition coefficients of anaesthetics in four aqueous solutions at 37 °C***

Solution	Osmolarity (mosmol ℓ^{-1})	Isoflurane	Enflurane	Halothane	Methoxy- flurane
Distilled H_2O	0	0.626 ± 0.05	0.754 ± 0.06	0.859 ± 0.02	4.33 ± 0.5
Normal saline	308	0.590 ± 0.01	0.713 ± 0.01	0.825 ± 0.02	4.22 ± 0.30
Isotonic heparin (1000 U mℓ^{-1})	308	0.593 ± 0.01	0.715 ± 0.01	–	4.08 ± 0.22
Mannitol (20 per cent)	1098	0.476 ± 0.023	0.575 ± 0.024	0.747 ± 0.03	3.38 ± 0.14

*From reference 10 with permission

sion. In so doing, however, it is important to realise that the red blood cell membrane is not a perfect semipermeable membrane and allows through small molecules such as urea and ammonium chloride. Therefore, although the quantity of each substance required for an isotonic solution may be calculated from freezing point depression values, these solutions, when administered, may cause cell lysis.

It has been shown that a solution which is isotonic with blood has a freezing point depression, ΔT_f, of 0.52 °C. One has therefore to adjust the freezing point of the drug solution to this value to give an isotonic solution. *The Pharmaceutical Codex* lists freezing point depressions of a series of compounds and it is a simple matter to calculate the concentration required for isotonicity from these values. For example, a 1 per cent NaCl solution has a freezing point depression of 0.576 °C. The percentage concentration of NaCl required to make isotonic saline solution is therefore $(0.52/0.576) \times 1.0 = 0.90\%$ w/v.

With a solution of a drug, it is not of course possible to alter the drug concentration in this manner, and an adjusting substance must be added to achieve isotonicity. The quantity of adjusting substance can be calculated as follows:

If the drug concentration is x g per 100 mℓ solution then

$$\Delta T_f \text{ for drug solution}$$

$$= x \times (\Delta T_f \text{ of 1 per cent drug solution}) = a$$

Similarly, if w is the weight in grams of adjusting substance to be added to 100 mℓ of drug solution to achieve isotonicity, then

$$\Delta T_f \text{ for adjusting solution}$$

$$= w \times (\Delta T_f \text{ of 1 per cent adjusting substance})$$

$$= w \times b$$

For an isotonic solution

$$a + (w \times b) = 0.52$$

Therefore $$w = \frac{0.52 - a}{b}$$ (3.50)

Extensive lists of freezing point depressions are given in standard reference texts[1,12].

Example 3.3

Calculate the amount of sodium chloride which should be added to 50 mℓ of a 0.5% w/v solution of lignocaine hydrochloride to make the solution isotonic with blood serum.

From reference lists, the values of b for sodium chloride and lignocaine hydrochloride are 0.576 °C and 0.130 °C, respectively.
In equation 3.50 we use

$$a = 0.5 \times 0.130 = 0.065$$

Therefore
$$w = \frac{0.52 - 0.065}{0.576}$$

$$= 0.790 \text{ g}$$

Therefore the weight of sodium chloride to be added to 50 mℓ of solution

$$= \frac{0.790}{2} = 0.395 \text{ g} \quad \blacksquare$$

3.4 Ionisation of drugs in solution

Many drugs are either weak organic acids (for example, acetylsalicylic acid), or weak organic bases (for example, cocaine), or their salts (for example, ephedrine hydrochloride). The degree to which these drugs are ionised in solution is highly dependent on the pH. The exceptions to this general statement are the non-electrolytes, for example, the steroids, and the quaternary ammonium compounds which are completely ionised at all pH values and in this respect behave as strong electrolytes. The extent of ionisation of a drug has an important effect on its absorption, distribution and elimination and there are many examples of where pH is altered to change these properties. The pH of urine may be adjusted (for example by administration of ammonium chloride or sodium bicarbonate) in cases of overdosing with amphetamines, barbiturates, narcotics and salicylates, to ensure that these drugs are completely ionised and hence readily excreted. Conversely the pH of the urine may be altered to prevent ionisation of a drug in cases where reabsorption is required for therapeutic reasons. Sulphonamide crystalluria may also be avoided by making the urine alkaline. An understanding of the relationship between pH and drug ionisation is of use in the prediction of the causes of precipitation in admixtures, in the calculation of the solubility of drugs and in the attainment of optimum bioavailability by maintaining a certain ratio of ionised to unionised drug. Table 3.5 shows the nominal pH of some body fluids and sites which are useful in the prediction of the percentage ionisation of drugs *in vivo*.

table 3.5 **Nominal pH values of some body fluids and sites***

Site	Nominal pH
Aqueous humour	7.21
Blood, arterial	7.40
Blood, venous	7.39
Blood, maternal umbilical	7.25
Cerebrospinal fluid	7.35
Duodenum	5.5
Faeces†	7.15
Ileum, distal	8.0
Intestine, microsurface	5.3
Lacrimal fluid (tears)	7.4
Milk, breast	7.0
Muscle, skeletal‡	6.0
Nasal secretions	6.0
Prostatic fluid	6.45
Saliva	6.4
Semen	7.2
Stomach	1.5
Sweat	5.4
Urine, female	5.8
Urine, male	5.7
Vaginal secretions, pre-menopause	4.5
Vaginal secretions, post-menopause	7.0

*From D. W. Newton and R. B. Kluza. *Drug Intell. clin. pharm.*, 12, 547 (1978) with permission
†Value for normal soft, formed stools; hard stools tend to be more alkaline, whereas watery, unformed stools are acidic
‡Studies conducted intracellularly in the rat

3.4.1 Dissociation of weakly acidic and basic drugs and their salts

According to the Lowry-Brönsted theory of acids and bases, an acid is a substance which will donate a proton and a base is a substance which will accept a proton. Thus the dissociation of a weak acid such as acetylsalicylic acid could be represented by

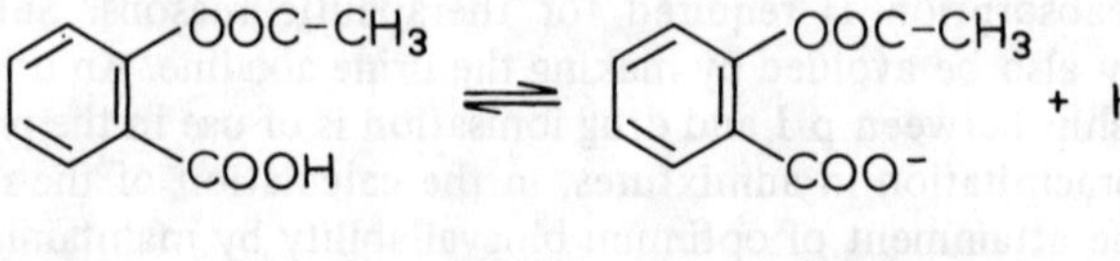

$$\text{acetylsalicylic acid} \rightleftharpoons \text{acetylsalicylate} + H^+ \quad (3.51)$$

In the equilibrium represented above, acetylsalicylic acid acts as an acid because it donates a proton and the acetylsalicylate ion acts as a base because it accepts a proton to yield an acid. An acid and base represented by such an equilibrium is said to be a conjugate acid-base pair. Equation 3.51 is not a realistic expression, however, since protons are too reactive to exist independently and are rapidly taken up by the solvent. The proton accepting entity, by the Lowry-Brönsted definition, is a base, and the product formed when the proton has been accepted by the solvent is an acid. Thus a second acid-base equilibrium occurs when the solvent accepts the proton, and this may be represented by

$$H_2O + H^+ \rightleftharpoons H_3O^+ \tag{3.52}$$

The over-all equation on summing equations 3.51 and 3.52 is

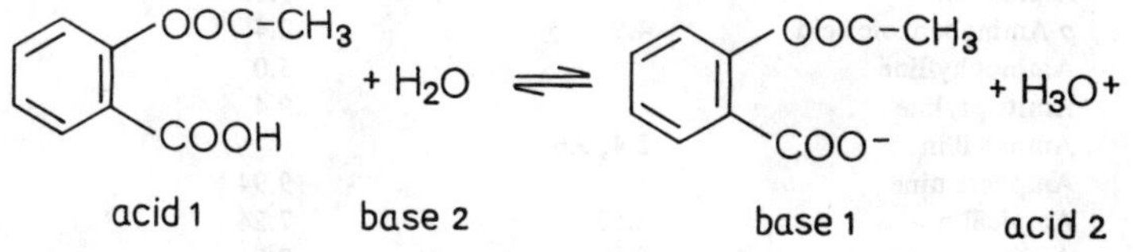

or, in general

$$HA + H_2O \rightleftharpoons A^- + H_3O^+ \tag{3.53}$$

By a similar reasoning, the dissociation of a weak base (for example, benzocaine) may be represented by the equilibrium

$$\underset{\text{base 1}}{NH_2C_6H_5COOC_2H_5} + \underset{\text{acid 2}}{H_2O} \rightleftharpoons \underset{\text{acid 1}}{NH_3^+C_6H_5COOC_2H_5} + \underset{\text{base 2}}{OH^-} \tag{3.54}$$

or, in general

$$B + H_2O \rightleftharpoons BH^+ + OH^- \tag{3.55}$$

Comparison of equation 3.53 with equation 3.55 shows that H_2O can act as either an acid or a base. Such solvents are called amphiprotic solvents.

Salts of weak acids or bases are essentially completely ionised in solution. For example, ephedrine hydrochloride (salt of the weak base, ephedrine, and the strong acid HCl) exists in aqueous solution in the form of the conjugate acid of the weak base, $C_6H_5CH(OH)CH(CH_3)\overset{+}{N}H_2CH_3$, together with its Cl^- counter-ions. In a similar manner, when sodium salicylate (salt of the weak acid, salicylic acid, and the strong base NaOH) is dissolved in water, it ionises almost entirely into the conjugate base of salicylic acid, $HOC_6H_5COO^-$, and Na^+ ions.

The conjugate acids and bases formed in this way are, of course, subject to acid-base equilibria described by equations 3.53 and 3.55.

3.4.2 The effect of pH on the ionisation of weakly acidic or basic drugs and their salts

If the ionisation of a weak acid is represented by equation 3.53 we may express an equilibrium constant as follows

$$K_a = \frac{a_{H_3O^+} \times a_{A^-}}{a_{HA}} \tag{3.56}$$

table 3.6   pK$_a$ values of some medicinal compounds*

Compound	pK$_a$ Acid	Base
Acebutolol		9.20
Acetylsalicylic acid	3.49	
Acyclovir		2.27, 9.25
Adrenaline	9.9	8.5
Adriamycin		8.2
Alphaprodine		8.7
Alprenolol		9.63
p-Aminobenzoic acid	4.9	2.4
Aminophylline		5.0
Amitriptyline		9.4
Amoxicillin	2.4, 9.6	
Amphetamine		9.94
Ampicillin	2.53	7.24
Apomorphine	8.92	7.0
Ascorbic acid	4.17, 11.57	
Atropine		9.25
Barbitone	7.91	
Benzocaine		2.78
Bromazepam	11.0	2.9
Bupivacaine		8.1
Caffeine	14.0	0.6
Chlorcyclizine		8.15
Chlordiazepoxide		4.76
Chloroquine		8.10, 9.94
Chlorpheniramine		8.99
Chlorpromazine		9.3
Chlorprothixene		8.84
Cimetidine		6.80
Clindamycin		7.45
Cocaine		8.5
Codeine		8.2
Cyclobarbitone	7.50	
Daunomycin		8.2
Desipramine		10.2
Dextromethorphan		8.3
Diazepam		3.4
Dibucaine		8.3
L-Dopa	2.31, 9.74, 13.40	
Doxepin		8.0
Doxycycline	7.7	3.4, 9.3
Ephedrine		9.63
Erythromycin		8.8
Ethopropazine		9.6
Fenoprofen	4.5	
Flucloxacillin	2.7	
Fluopromazine		9.2
Fluorouracil	8.0, 13.0	
Fluphenazine		3.9, 8.1
Flurazepam	8.16	1.90
Guanethidine		11.9
Guanoxan		12.3
Hexobarbitone	8.34	
Hydralazine		0.5, 7.1
Ibuprofen	5.2	
Imipramine		9.5

Compound	pK_a	
	Acid	Base
Indomethacin	4.5	
Isoniazid		2.0, 3.85
Labetalol		9.45
Lincomycin		7.5
Maprotiline		10.2
Meclofenamic acid	4.0	
Methadone		8.25
Methotrexate	3.76, 4.83	5.60
Minocycline	7.8	2.8, 5.0, 9.5
Minoxidil		4.6
Morphine	9.85	7.87
Nafcillin	2.7	
Nalorphine		7.8
Naloxone		7.94
Naltrexone	9.51	8.13
Naproxen	4.15	
Nitrazepam	10.8	3.2
Nortriptyline		9.73
Novobiocin	4.3, 9.1	
Oxycodone		8.9
Oxytetracycline	7.3	3.3, 9.1
Penicillin G	2.76	
Pentazocine		8.76
Pentobarbitone	8.11	
Phenazocine		8.50
Phenobarbitone	7.41	
Physostigmine		1.96, 8.08
Pilocarpine		1.63, 7.05
Piperazine		5.55, 9.82
Practolol		9.5
Procaine		8.8
Promazine		9.4
Promethazine		9.1
Propranolol		9.45
Quinidine		4.21, 8.34
Quinine		4.2, 8.8
Ranitidine		2.7, 8.2
Sotalol	8.3	9.8
Strychnine		2.3, 8.0
Sulphadiazine	6.48	2.0
Sulphaguanidine	12.05	2.75
Sulphamerazine	7.06	2.26
Sulphathiazole	7.12	2.36
Tamoxifen		8.85
Tetracaine		8.39
Tetracycline	7.7	3.3, 9.5
Theophylline	8.6	3.5
Thiopentone	7.45	
Timolol		9.21
Triflupromazine		9.2
Verapamil		8.75
Warfarin	5.05	

*For a more complete list see: D. W. Newton and R. B. Kluza. *Drug Intell. clin. Pharm.*, 12, 547 (1978) and G. C. Raymond and J. L. Born. *ibid.*, 20, 683 (1986)

Assuming the activity coefficients approach unity in dilute solution, the activities may be replaced by concentrations

$$K_a = \frac{[H_3O^+]\,[A^-]}{[HA]} \tag{3.57}$$

K_a is variously referred to as the ionisation, dissociation, or acidity constant for the weak acid. The negative logarithm of K_a is referred to as pK_a, just as the negative logarithm of the hydrogen ion concentration is called the pH.

Thus

$$pK_a = -\log K_a \tag{3.58}$$

Similarly, the dissociation constant or basicity constant for a weak base as derived from equation 3.55 is

$$K_b = \frac{a_{OH^-} \times a_{BH^+}}{a_B} \simeq \frac{[OH^-]\,[BH^+]}{[B]} \tag{3.59}$$

and

$$pK_b = -\log K_b \tag{3.60}$$

The pK_a and pK_b values provide a convenient means of comparing the strength of weak acids and bases. The lower the pK_a, the stronger the acid, the lower the pK_b, the stronger is the base. The pK_a values of a series of drugs are given in table 3.6. pK_a and pK_b values of conjugate acid–base pairs are linked by the expression

$$pK_a + pK_b = pK_w \tag{3.61}$$

where pK_w is the negative logarithm of the dissociation constant for water, K_w, which is defined by the following sequence of equations:

$$H_2O + H_2O \rightleftharpoons H_3O^+ + OH^- \tag{3.62}$$

where one molecule of water is behaving as the weak acid or base and the other is behaving as the solvent

$$K = \frac{a_{H_3O^+} \times a_{OH^-}}{a_{H_2O}^2} \simeq \frac{[H_3O^+]\,[OH^-]}{[H_2O]^2} \tag{3.63}$$

The concentration of molecular water is considered to be virtually constant for dilute aqueous solutions.

Therefore

$$K_w = [H_3O^+]\,[OH^-] \tag{3.64}$$

where the dissociation constant for water now incorporates the term for molecular water and has the values given in table 3.7.

table 3.7 **Ionic product for water**

Temperature (°C)	$K_w \times 10^{14}$	pK_w
0	0.1139	14.94
10	0.2920	14.53
20	0.6809	14.17
25	1.008	14.00
30	1.469	13.83
40	2.919	13.54
50	5.474	13.26
60	9.614	13.02
70	15.1	12.82
80	23.4	12.63
90	35.5	12.45
100	51.3	12.29

When the pH of an aqueous solution of the weakly acidic or basic drug approaches the pK_a or pK_b there is a very pronounced change in the ionisation of that drug. An expression that enables predictions of the pH dependence of the degree of ionisation to be made can be derived as follows.

Taking logarithms of the expression for the dissociation constant of a weak acid (equation 3.57)

$$-\log K_a = -\log [H_3O^+] - \log \frac{[A^-]}{[HA]}$$

Therefore

$$\boxed{pH = pK_a + \log \frac{[A^-]}{[HA]}} \qquad (3.65)$$

Equation 3.65 may be rearranged to facilitate the direct determination of the molar percentage ionisation as follows

$$[HA] = [A^-] \text{ antilog } (pK_a - pH)$$

Therefore

$$\text{percentage ionisation} = \frac{[A^-]}{[HA] + [A^-]} \times 100$$

$$\boxed{\text{percentage ionisation} = \frac{100}{1 + \text{antilog } (pK_a - pH)}} \qquad (3.66)$$

An analogous series of equations for the percentage ionisation of a weak base may be derived as follows.

Taking logarithms of equation 3.59 and rearranging gives

$$-\log K_b = -\log [OH^-] - \log \frac{[BH^+]}{[B]}$$

Therefore

$$pH = pK_w - pK_b - \log \frac{[BH^+]}{[B]}$$ (3.67)

Rearranging to facilitate calculation of the percentage ionisation leads to

$$\text{percentage ionised} = \frac{100}{1 + \text{antilog}\,(pH - pK_w + pK_b)}$$ (3.68)

The influence of pH on the percentage ionisation may be determined for drugs of known pK_a using table 3.8.

Example 3.4

Calculate the percentage of cocaine existing as the free base in a solution of cocaine hydrochloride at pH 4.5, and at pH 8.0. The pK_b of cocaine is 5.6.

From equation 3.66

$$\text{Percentage ionisation at pH 4.5} = \frac{100}{1 + \text{antilog}\,(4.5 - 14.0 + 5.6)}$$

$$= \frac{100}{1.000\,126}$$

$$= 99.99 \text{ per cent}$$

Thus the percentage existing as cocaine base = 0.01 per cent.

$$\text{Percentage ionisation at pH 8.0} = \frac{100}{1 + \text{antilog}\,(8.0 - 14.0 + 5.6)}$$

$$= \frac{100}{1.398}$$

$$= 71.53 \text{ per cent}$$

Thus the percentage existing as cocaine base = 28.47 per cent. ∎

3.4.3 Ionisation of amphoteric electrolytes

Amphoteric electrolytes (ampholytes) are electrolytes which can function as either acids or bases. A group of such compounds of particular interest are the amino acids and proteins. Over the pH range 3-9, glycine exists in solution predominantly in the form $^+NH_3\,CH_2\,COO^-$. Such a structure, having both positive and negative charges on the same molecule, is referred to as a zwitterion and can react both as an acid

$$^+NH_3\,CH_2\,COO^- + H_2O \rightleftharpoons NH_2\,CH_2\,COO^- + H_3O^+$$

table 3.8 **Percentage ionisation of anionic and cationic compounds as a function of pH**

	At pH above pK_a			At pH below pK_a	
pH–pK_a	if anionic	if cationic	pK_a–pH	if anionic	if cationic
6.0	99.99990	0.0000999	0.1	44.27	55.73
5.0	99.99900	0.0009999	0.2	38.68	61.32
4.0	99.9900	0.0099990	0.3	33.39	66.61
			0.4	28.47	71.53
			0.5	24.03	75.97
3.5	99.968	0.0316			
3.4	99.960	0.0398			
3.3	99.950	0.0501	0.6	20.07	79.93
3.2	99.937	0.0630	0.7	16.63	83.37
3.1	99.921	0.0794	0.8	13.70	86.30
			0.9	11.19	88.81
			1.0	9.09	90.91
3.0	99.90	0.0999			
2.9	99.87	0.1257			
2.8	99.84	0.1582	1.1	7.36	92.64
2.7	99.80	0.1991	1.2	5.93	94.07
2.6	99.75	0.2505	1.3	4.77	95.23
			1.4	3.83	96.17
			1.5	3.07	96.93
2.5	99.68	0.3152			
2.4	99.60	0.3966			
2.3	99.50	0.4987	1.6	2.450	97.55
2.2	99.37	0.6270	1.7	1.956	98.04
2.1	99.21	0.7879	1.8	1.560	98.44
			1.9	1.243	98.76
2.0	99.01	0.990	2.0	0.990	99.01
1.9	98.76	1.243	2.1	0.7879	99.21
1.8	98.44	1.560	2.2	0.6270	99.37
1.7	98.04	1.956	2.3	0.4987	99.50
1.6	97.55	2.450	2.4	0.3966	99.60
			2.5	0.3152	99.68
1.5	96.93	3.07			
1.4	96.17	3.83			
1.3	95.23	4.77	2.6	0.2505	99.75
1.2	94.07	5.93	2.7	0.1991	99.80
1.1	92.64	7.36	2.8	0.1582	99.84
			2.9	0.1257	99.87
1.0	90.91	9.09	3.0	0.0999	99.90
0.9	88.81	11.19			
0.8	86.30	13.70	3.1	0.0794	99.921
0.7	83.37	16.63	3.2	0.0630	99.937
0.6	79.93	20.07	3.3	0.0501	99.950
			3.4	0.0398	99.960
0.5	75.97	24.03	3.5	0.0316	99.968
0.4	71.53	28.47			
0.3	66.61	33.39	4.0	0.0099990	99.9900
0.2	61.32	38.68	5.0	0.0009999	99.99900
0.1	55.73	44.27	6.0	0.0000999	99.99990
0	50.00	50.00			

or as a base

$$^+NH_3\,CH_2\,COO^- + H_2O \rightleftharpoons\ ^+NH_3\,CH_2\,COOH + OH^-$$

At a certain pH, known as the isoelectric point, the two ionisations proceed to the same extent so that the solution contains equivalent amounts of $NH_2\,CH_2\,COO^-$ and $^+NH_3\,CH_2\,COOH$. Many of the physical properties of the amino acids, such as viscosity and solubility, show either a maximum or a minimum at this pH.

3.4.4 Ionisation of polyprotic drugs and microdissociation constants

In the examples we have considered so far the acidic drugs have donated a single proton. There are several acids, for example citric, phosphoric and tartaric acid, that are capable of donating more than one proton and these compounds are referred to as polyprotic or polybasic acids. Similarly, a polyprotic base is one capable of accepting two or more protons. Examples include physostigmine, pilocarpine, morphine, mepyramine, strychnine and quinine. Each stage of the dissociation may be represented by an equilibrium expression and hence each stage has a distinct pK_a or pK_b value. The dissociation of phosphoric acid, for example, occurs in three stages; thus

$$H_3PO_4 + H_2O \rightleftharpoons H_2PO_4^- + H_3O^+ \quad K_1 = 7.5 \times 10^{-3}$$
$$H_2PO_4^- + H_2O \rightleftharpoons HPO_4^{2-} + H_3O^+ \quad K_2 = 6.2 \times 10^{-8}$$
$$HPO_4^{2-} + H_2O \rightleftharpoons PO_4^{3-} + H_3O^+ \quad K_3 = 2.1 \times 10^{-13}$$

The experimentally determined dissociation constants for the various stages of dissociation are referred to as macroscopic values. It is not, however, always easy to assign macroscopic dissociation constants to the ionisation of specific groups of the molecule, particularly when the pK_a values are close together. The diprotic drug, morphine, has macroscopic pK_a values of 8.3 and 9.5, arising from ionisation of amino and phenolic groups. Experience suggests that the first pK_a value is probably associated with the ionisation of the amino group and the second with that of the phenolic group. However, it is not possible to assign unequivocally the values of these groups and for a more complete picture of the dissociation it is necessary to take into account all possible ways in which the molecule may be ionised and all the possible species present in solution. We may represent the most highly protonated form of morphine ^+HMOH as +0, where the + refers to the protonated amino group and the 0 refers to the uncharged phenolic group. Dissociation of the amino proton only produces an uncharged form MOH represented by 00, whilst dissociation of the phenolic proton gives a zwitterion $^+HMO^-$ represented by +−. The completely dissociated form MO^- is 0−. The entire dissociation scheme is given by

The constants k_1, k_2, k_{12}, and k_{21} are termed microdissociation constants and are defined by

$$k_1 = \frac{[+-]\,[H_3O^+]}{[+0]} \qquad k_2 = \frac{[00]\,[H_3O^+]}{[+0]}$$

$$k_{12} = \frac{[0-]\,[H_3O^+]}{[+-]} \qquad k_{21} = \frac{[0-]\,[H_3O^+]}{[00]}$$

The micro- and macrodissociation constants are related by the following expressions

$$K_1 = k_1 + k_2 \tag{3.69}$$

$$1/K_2 = 1/k_{12} + 1/k_{21} \tag{3.70}$$

and

$$K_1 K_2 = k_1 k_{12} = k_2 k_{21} \tag{3.71}$$

Various methods have been proposed whereby the microdissociation constants for the morphine system may be evaluated[13,14]. Other drugs for which microdissociation constants have been derived include the tetracyclines[15], doxorubicin[16], cephalosporin[17,18], and dopamine[19].

3.4.5 Preparation of buffer solutions

A mixture of a weak acid and its salt (that is, a conjugate base), or a weak base and its conjugate acid, has the ability to reduce the large changes in pH which would otherwise result from the addition of small amounts of acid or alkali to the solution. The reason for the buffering action of a weak acid HA and its ionised salt (for example, NaA) is that the A^- ions from the salt combine with the added H^+ ions, removing them from solution as undissociated weak acid.

$$A^- + H_3O^+ = H_2O + HA \tag{3.72}$$

Added OH^- ions are removed by combination with the weak acid to form undissociated water molecules.

$$HA + OH^- = H_2O + A^- \tag{3.73}$$

The buffering action of a mixture of a weak base and its salt arises from a removal of H^+ ions by the base B to form the salt and removal of OH^- ions by the salt to form undissociated water.

$$B + H_3O^+ = H_2O + BH^+ \tag{3.74}$$

$$BH^+ + OH^- = H_2O + B \tag{3.75}$$

The concentrations of buffer components required to maintain a solution at the required pH may be calculated using equation 3.65. Since the acid is weak and therefore only very slightly ionised, the term [HA] in this equation may be equated with the total acid concentration. Similarly, the free A^- ions in solution may be considered to originate entirely from the salt and the term $[A^-]$ may be replaced by the salt concentration.

Equation 3.65 now becomes

$$pH = pK_a + \log \frac{[\text{salt}]}{[\text{acid}]} \qquad (3.76)$$

By a similar reasoning equation 3.67 may be modified to facilitate the calculation of the pH of a solution of a weak base and its salt giving

$$pH = pK_w - pK_b + \log \frac{[\text{base}]}{[\text{salt}]} \qquad (3.77)$$

Equations 3.76 and 3.77 are often referred to as the Henderson–Hasselbalch equations.

Example 3.5

Calculate the amount of sodium acetate to be added to 100 mℓ of a 0.1 mol ℓ$^{-1}$ acetic acid solution to prepare a buffer of pH 5.20.

The pK_a of acetic acid is 4.76 and substitution in equation 3.76 gives

$$5.20 = 4.76 + \log \frac{[\text{salt}]}{[\text{acid}]}$$

The molar ratio of [salt]/[acid] is 2.754. Since 100 mℓ of 0.1 mol ℓ$^{-1}$ acetic acid contains 0.01 moles we would require 0.02754 moles of sodium acetate (2.258 g) ignoring dilution effects. ∎

Equations 3.76 and 3.77 are also useful in calculating the change in pH which results from the addition of a specific amount of acid or alkali to a given buffer solution.

Example 3.6

Calculate the change in pH following the addition of 10 mℓ of 0.1 mol ℓ$^{-1}$ NaOH to the buffer solution described in example 3.5.

The added 10 mℓ of 0.1 mol ℓ$^{-1}$ NaOH (equivalent to 0.001 moles) combines with 0.001 moles of acetic acid to produce 0.001 moles of sodium acetate. Re-applying equation 3.76 using the revised salt and acid concentrations gives

$$pH = 4.76 + \log \frac{(0.02754 + 0.001)}{(0.01 - 0.001)} = 5.26$$

The pH of the buffer has been increased by only 0.06 units following the addition of the alkali. ∎

If, instead of a single weak monobasic acid, a suitable mixture of polybasic and monobasic acids is used, it is possible to produce a buffer which is effective over a wide pH range. Such solutions are referred to as universal buffers. A typical example is a mixture of citric acid (pK_{a1} = 3.06, pK_{a2} = 4.78 and pK_{a3} = 5.40), Na_2HPO_4 (pK_a of conjugate acid, $H_2PO_4^-$ = 7.2), diethylbarbituric acid (pK_{a1} =

7.43) and boric acid (pK_{a1} = 9.24). This buffer is effective over the range pH 2.4-12).

The effectiveness of a buffer in reducing changes in pH is expressed as the buffer capacity, β. The buffer capacity is defined by the ratio

$$\beta = dc/d(\text{pH}) \tag{3.78}$$

where dc is the number of moles of alkali (or acid) needed to change the pH of 1 litre of solution by an amount $d(\text{pH})$. If the addition of 1 mole of alkali to 1 litre of buffer solution produces a pH change of 1 unit the buffer capacity is unity.

Equation 3.65 may be rewritten in the form

$$\text{pH} = pK_a + \frac{1}{2.303} \, \ell n \left(\frac{c}{c_0 - c} \right) \tag{3.79}$$

where c_0 is the total initial buffer concentration and c is the amount of alkali added.

Rearrangement and subsequent differentiation yields

$$c = c_0 / \left\{ 1 + \exp \left[-2.303 \, (\text{pH} - pK_a) \right] \right\} \tag{3.80}$$

Therefore $$\beta = \frac{dc}{d(\text{pH})} = \frac{2.303 \, c_0 \exp \left[2.303 \, (\text{pH} - pK_a) \right]}{\left\{ 1 + \exp \left[2.303 \, (\text{pH} - pK_a) \right] \right\}^2}$$

$$\boxed{\beta = \frac{2.303 \, c_0 K_a \, [H_3 O^+]}{([H_3 O^+] + K_a)^2}} \tag{3.81}$$

Example 3.7

Calculate the buffer capacity of the acetic acid–acetate buffer of example 3.5 at pH 4.0.

The total amount of buffer components in 100 mℓ of solution = 0.01 + 0.02754 = 0.03754 moles.

Therefore c_0 = 0.3754 mol ℓ^{-1}

pK_a of acetic acid = 4.76

Therefore K_a = 1.75 × 10^{-5}

pH of the solution = 4.0

Therefore $[H_3 O^+]$ = 10^{-4}

Substituting in equation 3.81

$$\beta = \frac{2.303 \times 0.3754 \times 1.75 \times 10^{-5} \times 10^{-4}}{(10^{-4} + 1.75 \times 10^{-5})^2} = 0.1096$$

The buffer capacity of the acetic acid–acetate buffer is 0.1096 mol ℓ^{-1} per pH unit. ■

Figure 3.2 shows the variation of buffer capacity with pH for the acetic acid–acetate buffer used in the numerical examples above (c_0 = 0.3754 mol ℓ^{-1}) as calculated from equation 3.81. It should be noted that β is at a maximum when

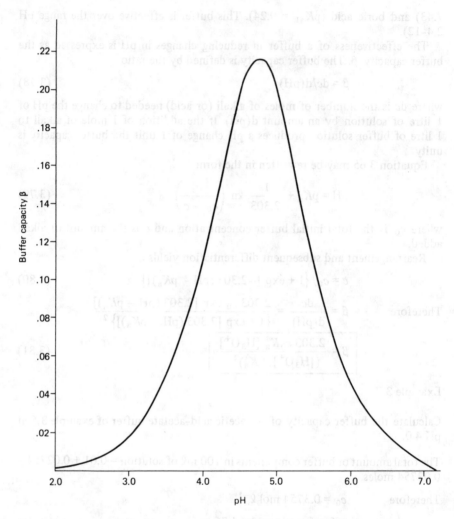

figure 3.2 Buffer capacity of acetic acid–acetate buffer (initial concentration = 0.3754 mol ℓ^{-1}) as a function of pH

pH = pK_a (that is, at pH 4.76). When selecting a weak acid for the preparation of a buffer solution one should therefore be chosen with a pK_a as close as possible to that of the pH required. Substituting pH = pK_a into equation 3.81 gives the useful result that the maximum buffer capacity β_{max} is

$$\beta_{max} = 0.576 \, c_0$$

where c_0 is the total buffer concentration.

Buffer solutions are widely used in pharmacy to adjust the pH of aqueous solutions to that required for maximum stability or that needed for optimum physiological effect. Solutions for application to delicate tissues, particularly the eye, should also be formulated at a pH not too far removed from that of the appropriate tissue fluid otherwise irritation may be caused on administration. The pH of tears lies between 7 and 8, with an average value of 7.4. Fortunately, the buffer capacity of tears is high and, provided that the solutions to be administered have a low buffer capacity, a reasonably wide range of pH may be tolerated, although there is a difference in the irritability of the various ionic species that are commonly used as buffer components. The pH of blood is maintained at about 7.4 by primary buffer components in the plasma (carbonic acid–carbonate and the acid–sodium salts of phosphoric acid) and secondary buffer components (oxyhaemoglobin–haemoglobin and acid–potassium salts of phosphoric acid) in the erythrocytes. Values of 0.025 and 0.039 mol ℓ^{-1} per pH unit have been quoted for the buffer capacity of whole blood. Parenteral solutions are not normally buffered, or alternatively are buffered at a very low capacity, since the buffers of the blood are usually capable of bringing them within a tolerable pH range.

3.5 Diffusion of drugs in solution

Diffusion is the process by which a concentration difference is reduced by a spontaneous flow of matter. Consider the simplest case of a solution containing a single solute. The solute will spontaneously diffuse from a region of high chemical potential to one of low chemical potential, that is from a region of higher concentration to one of lower concentration, whilst the solvent molecules move in the reverse direction. Although the driving force for diffusion is the gradient of chemical potential, it is more usual to think of the diffusion process in terms of the concentration gradient. Imagine the solution to be divided into volume elements. Although no individual solute particle in a particular volume element shows a preference for motion in any particular direction, a definite fraction of the molecules in this element may be considered to be moving in say the x direction. In an adjacent volume element the same fraction may be moving in the reverse direction. If the concentration in the first volume element is greater than that in the second, the over-all effect is that more particles are leaving the first element for the second and hence there is a net flow of solute in the x direction, the direction of decreasing concentration. The expression which relates the flow of material to the concentration gradient (dc/dx) is referred to as Fick's first law equation

$$J = -D\,(dc/dx)$$ (3.82)

where J is the flux of a component across a plane of unit area and D is the diffusion coefficient (or diffusivity). The negative sign indicates that the flux is in the direction of decreasing concentration. J is in mol m^{-2} s^{-1}, c is in mol m^{-3} and x is in m; therefore, the units of D are m^2 s^{-1}.

Fick's first law equation describes the diffusion process under conditions of steady state; that is, the concentration gradient, dc/dx, does not change with time.

In many of the experimental methods used to study diffusion, however, the variation of c with both time and distance is of interest. In such cases equation

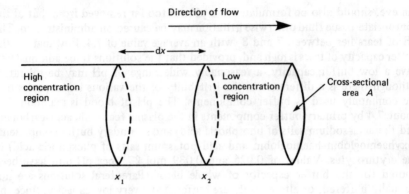

figure 3.3 Hypothetical volume element

3.82 may be converted into a second-order partial differential equation as follows.

In the hypothetical model of the solution depicted in figure 3.3 we consider a volume element lying between two planes perpendicular to the x axis, situated at x_1 and x_2. The rate of entry of solute, dm/dt, into the volume element at x_1 is

$$\left(\frac{\partial m}{\partial t}\right)_{x_1} = -DA\left(\frac{\partial c}{\partial x}\right)_{x_1} \tag{3.83}$$

and the mass per unit time leaving the element at x_2 is

$$\left(\frac{\partial m}{\partial t}\right)_{x_2} = -DA\left(\frac{\partial c}{\partial x}\right)_{x_2} \tag{3.84}$$

The rate of change of mass in the volume element is the mass per unit time entering at x_1 less the mass per unit time leaving at x_2. The mass in the volume element is the product of the volume of the element and the concentration c within the region (that is, $A\,dx\,c$). Hence, equations 3.83 and 3.84 may be combined to give

$$\frac{\partial (A\,dx\,c)}{\partial t} = A\,dx\,\frac{\partial c}{\partial t} = -DA\left(\frac{\partial c}{\partial x}\right)_{x_1} + DA\left(\frac{\partial c}{\partial x}\right)_{x_2} \tag{3.85}$$

The concentration gradient at x_2 may be expressed as the gradient at x_1 plus the rate at which the gradient is changing with x, multiplied by the distance dx.

That is

$$\left(\frac{\partial c}{\partial x}\right)_{x_2} = \left(\frac{\partial c}{\partial x}\right)_{x_1} + \frac{\partial(\partial c/\partial x)}{\partial x}\,dx$$

$$= \left(\frac{\partial c}{\partial x}\right)_{x_1} + \left(\frac{\partial^2 c}{\partial x^2}\right)dx \tag{3.86}$$

Substituting equation 3.86 into equation 3.85 gives

$$A \, dx \, \frac{\partial c}{\partial t} = -DA \left(\frac{\partial c}{\partial x} \right)_{x_1} + DA \left(\frac{\partial^2 c}{\partial x^2} \right) \, dx + DA \left(\frac{\partial c}{\partial x} \right)_{x_1}$$

(3.87)

which reduces to

$$\boxed{\frac{\partial c}{\partial t} = D \frac{\partial^2 c}{\partial x^2}}$$

(3.88)

Equation 3.88 is Fick's second law for unidimensional flow. A more general form of the equation for the other diffusional vectors is

$$\frac{\partial c}{\partial t} = D \left(\frac{\partial^2 c}{\partial x^2} + \frac{\partial^2 c}{\partial y^2} + \frac{\partial^2 c}{\partial z^2} \right)$$

(3.89)

which in vector notation is

$$\boxed{\frac{\partial c}{\partial t} = D \nabla^2 c}$$

(3.90)

Fick's second law states that the rate of change of concentration in a volume element within the diffusional field is proportional to the rate of change in the concentration gradient at that point in the field, the proportionality 'constant' being the diffusion coefficient or diffusivity, D.

The form of Fick's law which is applicable to a particular problem in diffusion depends on the boundary conditions imposed by the diffusion problem (see review by Flynn et al.[20]).

Table 3.9 shows typical diffusion coefficients of drugs in aqueous solution. The relationship between the radius, a, of the drug molecule and its diffusion

table 3.9 **Diffusion coefficients of typical drugs in aqueous media**

Drug substance	$D \, (10^{10} \, m^2 \, s^{-1})$
Amphetamine	7.2
Ephedrine	5.2
Isoniazid	7.6
Paracetamol	7.4
Sulphathiazole	7.3
Chloramphenicol	5.8
Benzoic acid	12.0
Salicylic acid	11.3
Potassium chloride	18.4
Acetylcholine	8.0
Thioridazine	8.0
Prostaglandin $F_{2\alpha}$	5.6

coefficient (assuming spherical particles or molecules) is given by the Stokes–Einstein equation as

$$D = \frac{RT}{6\pi\eta\, a\, N_A} \qquad (3.91)$$

Equation 3.91 also shows the influence of the viscosity of the medium, η, on the diffusion coefficient.

The diffusional properties of a drug have relevance in pharmaceutical systems in a consideration of such processes as the dissolution of the drug and its membrane transport.

References

1. *The Pharmaceutical Codex*, 11th Edition, The Pharmaceutical Press, London, 1979
2. J. Ferguson. *Proc. R. Soc. B*, 127, 387 (1939)
3. J. H. Rytting, S. S. Davis and T. Higuchi. *J. pharm. Sci.*, 61, 816 (1972)
4. J. H. B. Christian. In *Water Activity: Influences on Food Quality* (ed. L. B. Rockland and G. F. Stewart), Academic Press, New York, 1981, p. 825
5. W. J. Scott. *Aust. J. biol. Sci.*, 6, 549 (1953)
6. J. Chirife, G. Scarmato and C. Herszage. *Lancet*, i, 560 (1982)
7. W. H. Streng, H. E. Huber and J. T. Carstensen. *J. pharm. Sci.*, 67, 384 (1978)
8. D. M. Willis, J. Chabot and I. C. Radde. *Pediatrics*, 60, 535 (1977)
9. K. C. White and K. L. Harkavy. *Am. J. Dis. Child.*, 136, 931 (1982)
10. J. Lerman, M. M. Willis, G. A. Gregory and E. I. Eger. *Anesthesiology*, 59, 554 (1983)
11. L. H. Laasberg and J. Hedley-Whyte. *Physiologist*, 12, 279 (1969)
12. *The Merck Index*, 10th Edition, Merck, 1983
13. G. Schill and K. Gustavii. *Acta pharm. suec.*, 1, 24 (1964)
14. P. J. Niebergall, R. L. Schnaare and E. T. Sugita. *J. pharm. Sci.*, 61, 232 (1972)
15. L. J. Leeson, J. E. Krueger and R. A. Nash. *Tetrahedron Lett.*, 18, 1155 (1963)
16. R. J. Sturgeon and S. G. Schulman. *J. pharm. Sci.*, 66, 958 (1977)
17. W. H. Streng, H. E. Huber, J. L. DeYoung and M. A. Zoglio. *ibid.*, 65, 1034 (1976)
18. W. H. Streng, H. E. Huber, J. L. DeYoung and M. A. Zoglio. *ibid.*, 66, 1357 (1977)
19. T. Ishimitsu, S. Hirose and H. Sakurai. *Chem. pharm. Bull.*, 26, 74 (1978)
20. G. L. Flynn, S. H. Yalkowsky and T. J. Roseman. *J. pharm. Sci.*, 63, 479 (1974)

4 Chemical Stability of Drugs

Most drugs are subject to some form of chemical decomposition, particularly when formulated as liquid dosage forms. Some of the consequences of degradation are that the aged medicinal preparation no longer has the desired pharmacological potency. It may also exhibit physical manifestations of decomposition such as the discolouration that often follows photodegradation; or, more seriously (but fortunately, more rarely), it may contain harmful decomposition products.

In this chapter some of the causes of chemical degradation are examined and methods of minimising decomposition are discussed. The kinetics of drug decomposition, both in the liquid and solid states, and methods of prediction of drug stability are also presented.

Although the control of the chemical degradation of the active principles of medicinal preparations and the prediction of the shelf-life are of prime importance in formulation, it is, of course, also necessary to ensure the physical stability of the dosage form itself. This form of stability will be examined in chapter 7, which deals with emulsions and suspensions. Other aspects of stability, such as enzyme-catalysed decomposition and growth of micro-organisms, although important, are outside the scope of this book.

4.1 The chemical decomposition of drugs

The two most common causes of the decomposition of drugs are hydrolysis and oxidation. Drugs may also lose their activity due to isomerisation or photochemical decomposition. All of these decomposition mechanisms will be considered here and some methods of preventing degradation will be discussed.

4.1.1 Hydrolysis

The main classes of drugs that are susceptible to hydrolytic cleavage are the esters, amides, and lactams. Hydrolysis is frequently catalysed by hydrogen ions (specific acid catalysis) or hydroxyl ions (specific base catalysis) and also by other acidic or basic species that are commonly encountered as components of buffers. This latter type of catalysis is referred to as general acid–base catalysis. Both types of catalysis will be dealt with in greater depth in section 4.4.

The most common type of ester hydrolysis is a bimolecular reaction involving acyl-oxygen cleavage. The hydrolysis of procaine, which was studied by Higuchi et al.[1], may be represented as

$$NH_2\text{—}\langle\bigcirc\rangle\text{—}\underset{\underset{O}{\|}}{C}\text{—O—CH}_2\text{—CH}_2\text{—N(C}_2\text{H}_5)_2 \xrightarrow[\text{OH}^-]{\text{H}^+} NH_2\text{—}\langle\bigcirc\rangle\text{—}\underset{\underset{O}{\|}}{\overset{\overset{\text{OH}}{|}}{C}} + \text{HO—CH}_2\text{—CH}_2\text{—N(C}_2\text{H}_5)_2$$

Further examples of drugs with ester linkages include cocaine, physostigmine, tetracaine and methyldopate.

Hydrolysis of amides involves cleavage of the amide linkage. The hydrolysis of the local anaesthetic, dibucaine, for example, proceeds as follows[2]

Other drugs with amide linkages include ergometrine, benzylpenicillin sodium, and chloramphenicol.

Examples of lactam ring hydrolysis are shown in figures 4.9 and 4.10 for nitrazepam and chlordiazepoxide, respectively. Other drugs, apart from the benzodiazepines, that are susceptible to this form of hydrolysis include the penicillins and cephalosporins.

Several methods are available to stabilise a solution of a drug which is susceptible to acid–base-catalysed hydrolysis. The usual method is to determine the pH of maximum stability from kinetic experiments at a range of pH values and to formulate the product at this pH (section 4.4.1). Alteration of the dielectric constant by the addition of non-aqueous solvents such as alcohol, glycerin or propylene glycol may in many cases reduce hydrolysis (section 4.4.1). Since only that portion of the drug which is in solution will be hydrolysed, it is possible to suppress degradation by making the drug less soluble. Swintosky et al.[3], for example, reduced the solubility of penicillin in procaine penicillin suspensions using additives such as citrates, dextrose, sorbitol and gluconate and, in so doing, significantly increased the stability. Higuchi and Lachman[4] suggested adding a compound that would form a complex with the drug, as a means of increasing stability. The addition of caffeine to aqueous solutions of benzocaine, procaine and amethocaine was shown to decrease the base-catalysed hydrolysis of these local anaesthetics in this way. Solubilisation of a drug by surfactants in many cases protects against hydrolysis, as discussed in section 4.4.

The control of drug stability by modifying chemical structure using appropriate substituents has been suggested for drugs for which such a modification does not reduce therapeutic efficiency. The Hammett linear free energy relationship for the effect of substituents on the rates of aromatic side-chain reactions, such as the hydrolysis of esters, is given by[5]

$$\log k = \log k_0 + \sigma\rho \qquad (4.1)$$

where k and k_0 are the rate constants for the reaction of the substituted and unsubstituted compounds, respectively, σ is the Hammett substituent constant which is determined by the nature of the substituents and is independent of the reaction, and ρ is the reaction constant which is dependent on the reaction, the conditions of reaction and the nature of the side-chains undergoing reaction. Thus, a plot of $\log k$ against the Hammett constant (values are readily available in the literature[6]) is linear if this relationship is obeyed, with a slope of ρ. Carstensen et al.[7] have used this concept in the production of the best substituents for allylbarbituric acids to obtain optimum stability.

4.1.2 Oxidation

Whereas the hydrolytic degradation of drugs has been thoroughly studied, their oxidative degradation has received comparatively little attention. Indeed, in cases where simultaneous hydrolytic and oxidative degradation can occur, the oxidative process has usually been eliminated by storage under anaerobic conditions without an investigation of the oxidative mechanism. Oxidative degradation is a major cause of drug instability and drugs that are affected include phenolic compounds such as morphine and phenylephrine, catecholamines such as dopamine and adrenaline, steroids, antibiotics, vitamins, oils and fats.

Oxidation involves the removal of an electropositive atom, radical or electron, or the addition of an electronegative atom or radical. Many pharmaceutical oxidations are chain reactions which proceed quite slowly under the influence of molecular oxygen. Such a reaction process is referred to as autoxidation. In the oxidation of fats and oils[8] the initiation can be via free radicals formed from organic compounds by the action of light, heat or trace metals. The propagation stage of the reaction involves the combination of molecular oxygen with the free radical to form a peroxy radical ROO•, which then removes H from a molecule of the organic compound to form a hydroperoxide, ROOH, and in so doing creates a new free radical.

Initiation $\quad$ R• + R′—CH$_2$—CH=CH—R″ $\longrightarrow$ R′—ĊH—CH=CH—R″ + RH

Propagation

$$\text{R′—ĊH—CH=CH—R″ + O}_2 \longrightarrow \text{R′—CH(O—O•)—CH=CH—R″}$$

$$\text{R′—CH(O—O•)—CH=CH—R″ + R′—CH}_2\text{—CH=CH—R″}$$

$$\longrightarrow \text{R′—CH(O—OH)—CH=CH—R″ + R′—ĊH—CH=CH—R″}$$

The reaction proceeds until the free radicals are destroyed by inhibitors or by side-reactions which eventually break the chain. The rancid odour which is a characteristic of oxidised fats and oils is due to aldehydes, ketones and short-chain fatty acids which are the breakdown products of the hydroperoxides.

A further type of oxidation involves the reversible loss of electrons without the addition of oxygen. A typical example is the oxidative degradation of morphine (figure 4.1), which was studied in detail by Yeh and Lach[9]. The undissociated and the protonated forms of morphine, M, are both oxidised by atmospheric oxygen to give a free radical peroxide and a semiquinone, SQ. The latter is further transformed to a free radical quinone, Q. The free radical quinone is then thought to undergo coupling with undissociated or protonated morphine to give

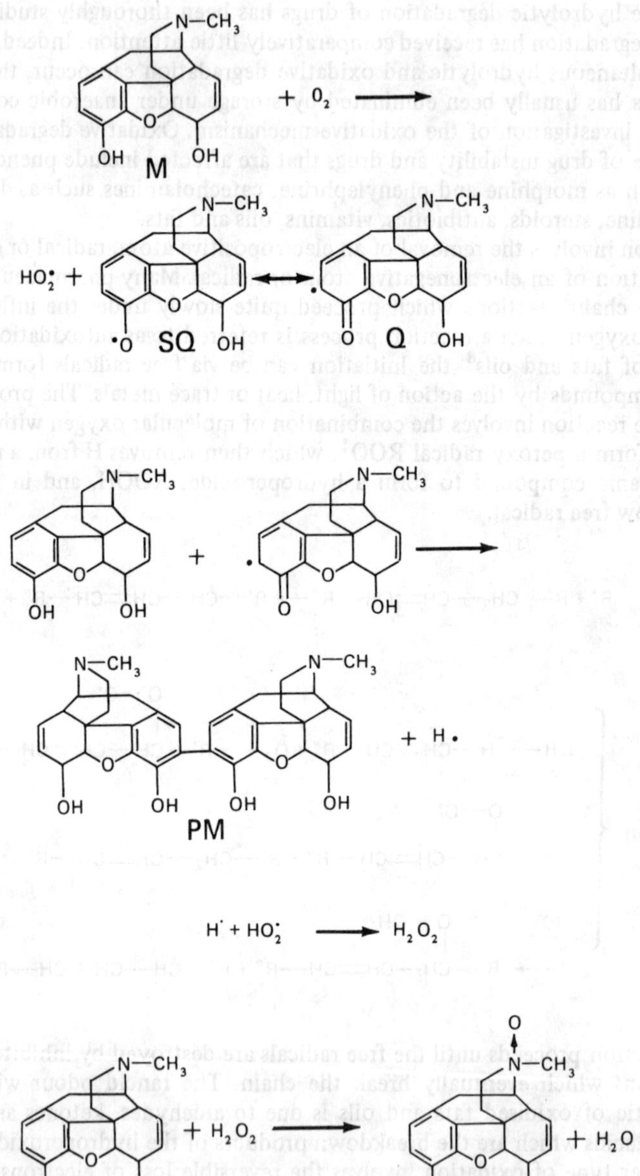

figure 4.1 Oxidative degradation of morphine. M, morphine; SQ, semiquinone; Q, quinone, PM, pseudomorphine; MNO, morphine *N*-oxide. From reference 9

the dimer pseudomorphine, PM, a process which results in the elimination of a hydrogen free radical. The hydrogen free radical reacts with peroxide free radical to form hydrogen peroxide. The hydrogen peroxide so formed can react with morphine to form morphine N-oxide, MNO, or may decompose to produce a free radical oxygen which can also react with morphine base to give the N-oxide.

Stabilisation of drugs against oxidation involves observing a number of precautions during manufacture and storage. The oxygen in pharmaceutical containers should be replaced with nitrogen or carbon dioxide; contact of the drug with heavy metal ions which catalyse oxidation should be avoided and storage should be at reduced temperatures.

It is very difficult to remove all of the oxygen from a container and even traces of oxygen are sufficient to initiate the oxidation chain. The propagation of the chain reaction may be prevented or delayed by adding low concentrations of compounds which act as inhibitors. Such compounds are called antioxidants and interrupt the propagation by interaction with the free radical. The antioxidant free radical so formed is not sufficiently reactive to maintain the chain reaction and is eventually annihilated. Commonly used antioxidants include the tocopherols, butylated hydroxyanisole, butylated hydroxytoluene, gallic acid and the gallates, for example propyl gallate. Reducing agents such as sodium metabisulphite may also be added to formulations to prevent oxidation. These compounds are more readily oxidised than the drug and so protect it from oxidation. Oxidation is catalysed by unprotonated amines such as aminophylline, and hence admixture of susceptible drugs with such compounds should be avoided.

4.1.3 Isomerisation

Isomerisation is the process of conversion of a drug into its optical or geometric isomers. Since the various isomers of a drug are frequently of different activity, such a conversion may be regarded as a form of degradation, often resulting in a serious loss of therapeutic activity. For example, the appreciable loss of activity of solutions of adrenaline at low pH has been attributed[10] to racemisation – the conversion of the therapeutically active form, in this case the laevo-rotary form, into its less-active isomer.

The tetracyclines undergo epimerisation at carbon atom 4, in acidic conditions, to form an equilibrium mixture of tetracycline and the epimer, which is referred to as 4-epi-tetracycline[11].

Partial structure
of 4–epi–tetracycline

Partial structure of
natural tetracycline

The epi-tetracyclines have been isolated and found to possess a much reduced therapeutic activity than the natural isomers. Remmers *et al.*[12] have shown that the epimerisation follows the kinetics of a first-order reversible reaction (see equation 4.21). The degradation rate is pH-dependent (maximum epimerisation occurring at pH 3.2) and also catalysed by phosphate and citrate ions.

Pilocarpine has been reported[13] to undergo simultaneous base-catalysed hydrolysis and epimerisation in aqueous solution. The postulated reactions (figure 4.2) were the specific base-catalysed hydrolysis of pilocarpine to pilocarpate (P → PA⁻), the base-catalysed epimerisation of pilocarpine (P → P⁻ and P⁻ → IP) and the hydrolysis of isopilocarpine (IP → IPA⁻). Both the hydrolysis and the epimerisation reactions followed pseudo first-order rate equations and the individual rate constants were expressed by equations of similar form to equations 4.30 and 4.31.

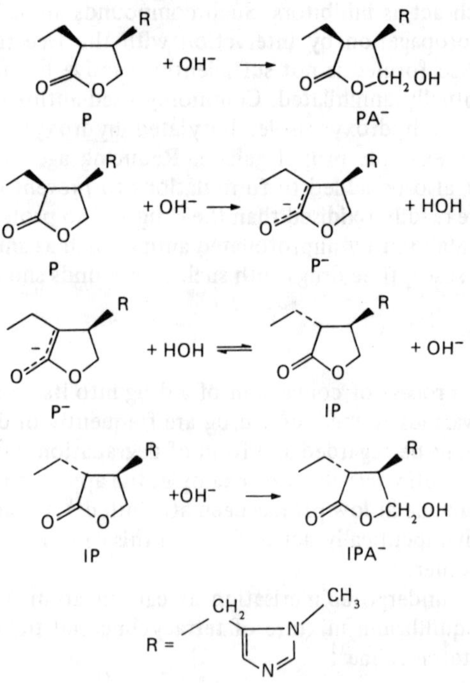

figure 4.2 Hydroxide ion catalysed hydrolysis and epimerisation of pilocarpine (P). PA⁻, pilocarpate; P⁻, delocalised carbanion = IP⁻; IP, isopilocarpine; and IPA⁻, isopilocarpate. From reference 13

Cis-trans isomerisation may be a cause of loss of potency of a drug if the two geometric isomers have different therapeutic activities. Lehman *et al.*[14] report such isomerisation in aqueous preparations of vitamin A. Isomerisation of this vitamin to form *cis* isomers at the 2 and 6 positions leads to a decreased activity compared with the all-*trans* molecule.

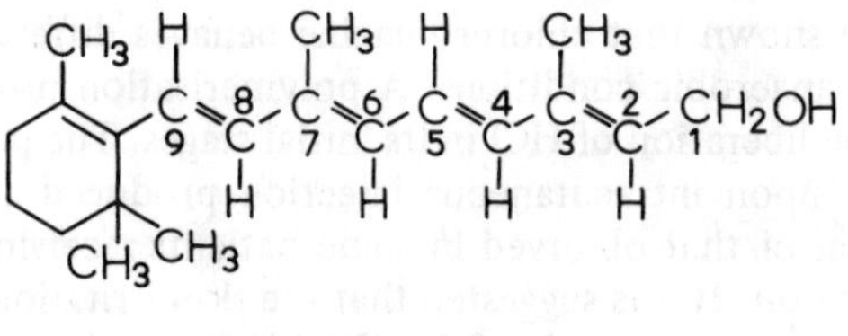

all-trans vitamin A

4.1.4 Photochemical decomposition

Many pharmaceutical compounds, including the phenothiazine tranquillisers, hydrocortisone, prednisolone, riboflavine, ascorbic acid and folic acid, are light-sensitive. The mechanisms of photodegradation, however, are of such complexity as to have been fully elucidated in only a few cases. The phenothiazine, chlorpromazine (CLP), for example, is rapidly decomposed under the action of ultraviolet light, the decomposition being accompanied by discolouration of the solutions. The first step of the photodegradation is the loss of an electron to yield the semiquinone free radical R. Further stages in the degradation yield the phenazathonium ion P which is thought to react with water to yield chlorpromazine sulphoxide[15] (CPO) (figure 4.3). The chlorpromazine sulphoxide is itself photolabile and further decomposition occurs. Other products of the photo-oxidation include chlorpromazine *N*-oxide and hydroxychlorpromazine. Huang

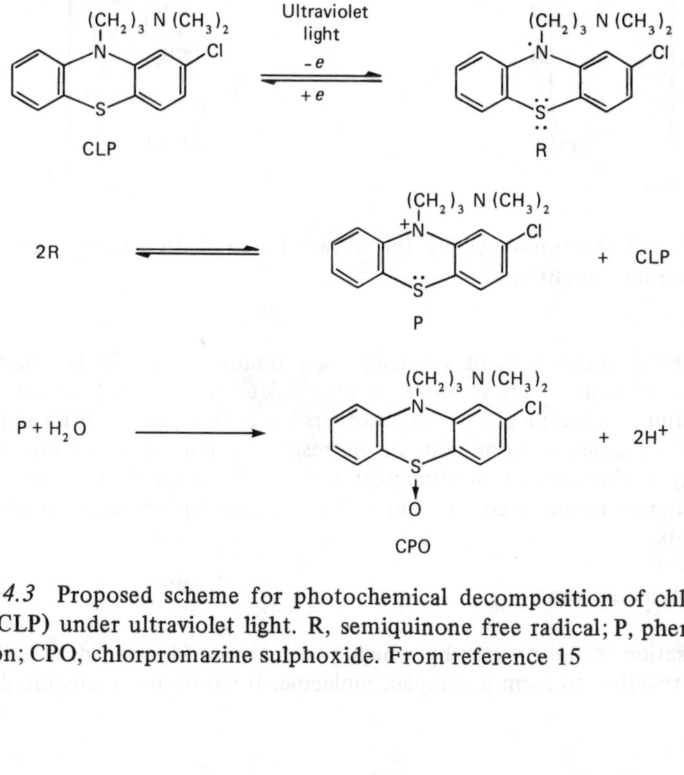

figure 4.3 Proposed scheme for photochemical decomposition of chlorpromazine (CLP) under ultraviolet light. R, semiquinone free radical; P, phenazathonium ion; CPO, chlorpromazine sulphoxide. From reference 15

and Sands[16] have shown that chlorpromazine behaves differently to ultraviolet irradiation under anaerobic conditions. A polymerisation process was proposed which involved the liberation of HCl in its initial stages. The polymer (figure 4.4) was isolated and upon intracutaneous injection produced a bluish-purple discolouration typical of that observed in some patients receiving prolonged chlorpromazine medication. It was suggested that the skin irritation that accompanies the discolouration may be a result of the HCl liberation during photodecomposition.

Pharmaceutical products can be adequately protected from photo-induced decomposition by the use of coloured glass containers and storage in the dark.

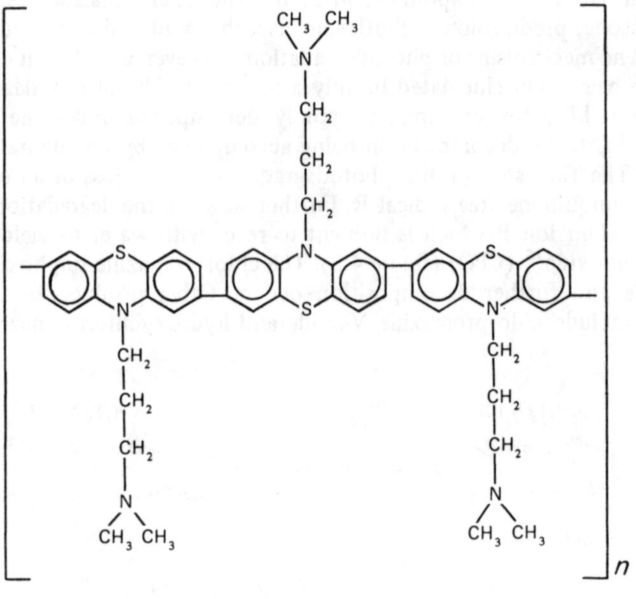

figure 4.4 Polymer produced by the ultraviolet irradiation of chlorpromazine under anaerobic conditions

Amber glass excludes light of wavelength < 470 mm and so affords considerable protection of compounds sensitive to ultraviolet light. Coating tablets with a polymer film containing ultraviolet absorbers has been suggested as an additional method for protection from light. In this respect, a film coating of vinyl acetate containing oxybenzone as an ultraviolet absorber has been shown[17] to be effective in minimising the discolouration and photolytic degradation of sulphasomidine tablets.

4.1.5 Polymerisation

Polymerisation is the process by which two or more identical drug molecules combine together to form a complex molecule. It has been demonstrated that a polymerisation process occurs during the storage of concentrated aqueous solutions of amino-penicillins, for example ampicillin sodium. The reactive β-lactam bond of the ampicillin molecule is opened by reaction with the side-chain of a second ampicillin molecule[18] and a dimer is formed (figure 4.5). The process can continue to form higher polymers. Such polymeric substances have been shown to be highly antigenic in animals and they are considered to play a part in eliciting penicilloyl-specific allergic reactions to ampicillin in man[19].

The dimerising tendency of the amino-penicillins was shown to increase with the increase in the basicity of the side-chain amino group, the order, in terms of increasing rates, being cyclacillin $\ll$ ampicillin $<$ epicillin $<$ amoxycillin[20]. It was suggested that prevention of dimerisation and hence reduction in allergic response to the amino-penicillins might be achieved by rational design of amino-penicillins based on a consideration of these findings.

The hydrate of formaldehyde, $HOCH_2 OH$, may under certain conditions polymerise in aqueous solution to form paraformaldehyde, $HOCH_2 (OCH_2)_n OCH_2 OH$, which appears as a white deposit in the solution. The polymerisation may be prevented by adding to the solution 10–15 per cent of methanol.

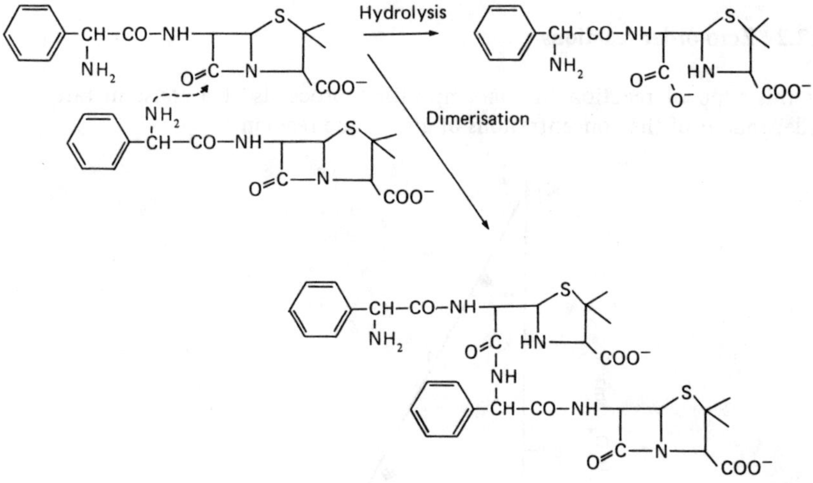

figure 4.5 Concomitant self-aminolysis (dimerisation) and hydrolysis of ampicillin (α-aminobenzylpenicillin). From reference 18

4.2 Kinetics of chemical decomposition in solution

4.2.1 Order of reaction

Decomposition reactions can be conveniently classified in terms of their order of reaction. If the concentration in mol dm^{-3} of a reactant A is denoted by [A], then the rate of reaction may be expressed in terms of the rate of change of the concentration of this reactant with time, that is $-d[A]/dt$. The differential equation expressing the rate as a function of the concentration of each of the species which affect the rate is called the general rate equation. For example, kinetic studies may show the general rate equation for the reaction between two compounds A and B to be

$$-d[A]/dt = k[A]^2 [B]$$

The reaction is then said to be second-order with respect to A and first-order with respect to B; that is, the order with respect to each component is the exponent of the concentration term for that reactant. The over-all order of reaction is the sum of the exponents of the concentration terms which affect the reaction rate. The reaction specified above is thus a third-order reaction. The proportionality constant, k, is termed the rate constant.

If the initial concentration of reactant is a and the amount decomposed in time t is x, then the amount remaining at time t is $(a - x)$. Thus

$$-d[A]/dt = -d(a - x)/dt = dx/dt$$

since the initial concentration, a, is constant. It is often convenient to express the general rate equation in terms of dx/dt rather than $-d[A]/dt$, as in the following illustrations of the various orders of reaction.

4.2.2   Zero-order reactions

In this type of reaction the decomposition proceeds at a constant rate and is independent of the concentrations of any of the reactants.

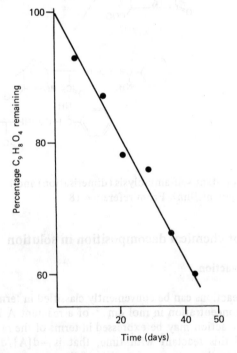

figure 4.6 Hydrolysis of a suspension of acetylsalicylic acid at 34 °C. From K. C. James. *J. Pharm. Pharmacol.*, 10, 363 (1958) with permission

That is $$\frac{dx}{dt} = k \tag{4.2}$$

Integrating, noting that $x = 0$ at $t = 0$ gives

$$\int_0^x dx = k \int_0^t dt \tag{4.3}$$

i.e. $$\boxed{x = kt} \tag{4.4}$$

A plot of the amount remaining (as ordinate) against time (as abscissa) is linear with a slope of k (concentration time^{-1}).

Many decomposition reactions in the solid phase (section 4.3), or in suspensions apparently follow zero-order kinetics. Figure 4.6 shows the hydrolysis of acetylsalicylic acid suspensions.

4.2.3 First-order reactions

The rate of a first-order reaction is determined by one concentration term. The general rate equation is

$$-d[A]/dt = k[A] \tag{4.5}$$

Since $[A] = (a - x)$, equation 4.5 may be written as

$$dx/dt = k(a - x) \tag{4.6}$$

Since $x = 0$ at the start of the measurements (that is, when $t = 0$)

$$\int_0^x \frac{dx}{(a-x)} = k \int_0^t dt \tag{4.7}$$

$$k = \frac{2.303}{t} \log \frac{a}{a-x} \tag{4.8}$$

Or, rearranging into a linear form

$$\boxed{t = \frac{2.303}{k} \log a - \frac{2.303}{k} \log (a - x)} \tag{4.9}$$

According to equation 4.9, a plot of the logarithm of the amount of drug remaining (as ordinate) as a function of time (as abscissa) is linear if the decomposition follows first-order kinetics. The first-order rate constant may be obtained from the slope of the plot (slope $= -k/2.303$). k has the dimensions of time^{-1}.

The time taken for half of the reactant to decompose is referred to as the half-life of the reaction, $t_{0.5}$. An expression for $t_{0.5}$ for a first-order reaction may be derived from equation 4.8 noting that when $t = t_{0.5}$, $x = a/2$.

Thus $$t_{0.5} = \frac{2.303}{k} \log \frac{a}{a/2}$$

$$\boxed{t_{0.5} = \frac{2.303}{k} \log 2 = 0.693/k} \tag{4.10}$$

The half-life is therefore independent of the initial concentration of reactants.

Although a reaction may involve more than one reacting species, its rate may still follow first-order kinetics. The most common example of this occurs when one of the reactants is in such a large excess that any change in its concentration is negligible compared with changes in the concentration of the other reactants. This type of reaction is termed a pseudo first-order reaction. Such reactions are often met in stability studies of drugs that hydrolyse in solution, the water being in such excess that changes in its concentration are negligible and hence the rate of reaction is dependent solely on the drug concentration.

Example 4.1

The following data were obtained for the hydrolysis of homatropine in 0.226 mol ℓ^{-1} HCl at 90 °C (M. H. Krasowska *et al. Dansk Tidsskr. Farm.*, 42, 170 (1968)).

Percentage homatropine remaining	93.4	85.2	75.9	63.1	52.5	41.8
Time (h)	1.38	3.0	6.0	8.6	12	17

Show that the hydrolysis follows first-order kinetics and calculate (a) the rate constant, and (b) the half-life.

(a) The reaction will be first-order if a plot of the logarithm of the amount of homatropine remaining against time is linear.

Log percentage remaining	1.97	1.93	1.88	1.80	1.72	1.62
Time	1.38	3.0	6.0	8.6	12	17

Figure 4.7 shows a linear plot with a slope = $-\dfrac{1.96 - 1.55}{20 - 2} = -2.278 \times 10^{-2}$

$$\text{Slope} = -\frac{k}{2.303}$$

Therefore $k = 5.25 \times 10^{-2} \text{ h}^{-1}$

The first-order rate constant (k) is $5.25 \times 10^{-2} \text{ h}^{-1}$.

(b) From equation 4.10

$$t_{0.5} = \frac{0.693}{k} = 13.2 \text{ h}$$

The half-life of the reaction is 13.2 h. ∎

Photolytic decomposition of drugs in solid dosage forms is frequently first-order. For example Eble and Garrett[21] have reported a pseudo first-order photolysis of fumagillin (an anti-amoebic substance, not in current use), and DeMerre and Wilson[22] found the photodecomposition of cyanocobalamin to be first-order.

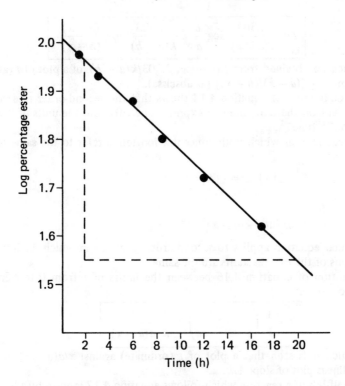

figure 4.7 First-order plot for hydrolysis of homatropine in hydrochloric acid (0.226 mol ℓ^{-1}) at 90 °C. Data from M. H. Krasowska, S. Schytt Larsen and K. Ilver. *Dansk Tidsskr. Farm.*, 42, 170 (1968) with permission

4.2.4 Second-order reactions

The rate of a second-order reaction is determined by the concentrations of two reacting species. The general rate equation is

$$-d[A]/dt = k[A][B] \qquad (4.11)$$

If the initial concentrations of reactants A and B are a and b, respectively, equation 4.11 may be written

$$dx/dt = k(a - x)(b - x) \qquad (4.12)$$

where x is the amount of A and B decomposed in time t. Integration of equation 4.12 by the method of partial fractions yields

$$k = \frac{2.303}{t(a - b)} \log \frac{b(a - x)}{a(b - x)} \qquad (4.13)$$

Rearranging into a linear form suitable for plotting gives

$$t = \frac{2.303}{k(a-b)} \log \frac{b}{a} + \frac{2.303}{k(a-b)} \log \frac{(a-x)}{(b-x)}$$

(4.14)

k can then be obtained from the slope, $2.303/k(a-b)$, of a plot of t (as ordinate) against $\log[(a-x)/(b-x)]$ (as abscissa).

An examination of equation 4.14 shows that the second-order rate constant is dependent on the units used to express concentration; the units of k are concentration^{-1} time^{-1}.

For reactions in which both concentration terms refer to the same reactant we may write

$$-d[A]/dt = k[A]^2$$

(4.15)

and

$$dx/dt = k(a-x)^2$$

(4.16)

A similar equation applies to second-order reactions in which the initial concentrations of the two reactants are the same.

Integration of equation 4.16 between the limits of t from 0 to t and of x from 0 to x yields

$$t = \frac{1}{k}\left(\frac{1}{a-x} - \frac{1}{a}\right) = \frac{1}{k}\frac{x}{a(a-x)}$$

(4.17)

from which it is seen that a plot of t (ordinate) against $x/a(a-x)$ (abscissa) yields a linear plot of slope $1/k$.

The half-life of a reaction which follows equation 4.17 is given by

$$t_{0.5} = 1/ka$$

(4.18)

Unlike $t_{0.5}$ for the first-order reactions, the half-life of the second-order reaction is dependent on the initial concentration of reactants. It is not possible to derive a simple expression for the half-life of a second-order reaction with unequal initial concentrations.

Second-order reactions are encountered in the study of the acid- or base-catalysed hydrolysis. The hydrolysis of chlorbutol, for example, was shown to be first-order with respect to both chlorbutol and hydroxyl ions, the over-all rate being second-order[23].

4.2.5 Third-order reactions

Third-order reactions are only rarely encountered in drug stability studies involving, as they do, the simultaneous collision of three reactant molecules. Bundgaard[18] has shown that the over-all rate of ampicillin breakdown by simultaneous hydrolysis and polymerisation may be represented by an equation of the form

$$-\frac{d[A]_T}{dt} = k_1[A]_T + k_2[A]_T^2 + k_3[A]_T^3$$

(4.19)

where k_1, k_2 and k_3 are the pH-dependent apparent rate constants for hydrolysis, uncatalysed polymerisation and the general acid–base-catalysed polymerisation of ampicillin, respectively. As seen from equation 4.19 the decomposition rate shows both second-order and third-order dependency on the total ampicillin concentration $[A]_T$.

4.2.6 Determination of the order of reaction

The most obvious method is to determine the amount of drug decomposed after various intervals and to substitute the data into the integrated equations for zero-, first- and second-order reactions. The equation giving the most consistent value of k for a series of time intervals is that corresponding most closely to the order of the reaction. Alternatively, the data may be displayed graphically according to the linear equations for the various orders of reaction until a straight line plot is obtained. Thus, for example, if the data yield a linear graph when plotted as t vs log $(a - x)$ the reaction is then taken to be first-order.

Fitting data to the standard rate equations may, however, produce misleading results if a fractional order of reaction applies. An alternative method of determining the order of reaction, which avoids this problem, is based on equation 4.20:

$$\log t_{0.5} = \log \left[\frac{2^{n-1} - 1}{k(n - 1)}\right] + (1 - n) \log C_0 \qquad (4.20)$$

The half-life of the reaction is determined for a series of initial drug concentrations, C_0, and the order, n, is calculated from the slope of plots of log $t_{0.5}$ as a function of log C_0.

Example 4.2

The kinetics of decomposition of a drug in aqueous solution were studied using a series of solutions of different initial drug concentration, C_0. For each solution the time taken for half the drug to decompose (that is, $t_{0.5}$) was determined with the following results.

C_0 (mol ℓ^{-1})	4.625	1.698	0.724	0.288
$t_{0.5}$ (min)	87.17	240.1	563.0	1414.4

Determine the order of reaction and calculate the rate constant.

Application of equation 4.20 requires values for log C_0 and log $t_{0.5}$; thus

log C_0	0.665	0.230	−0.140	−0.540
log $t_{0.5}$	1.94	2.38	2.75	3.15

A plot of log $t_{0.5}$ against log C_0 is linear (see figure 4.8).

$$\text{Slope} = -\frac{3.01 - 2.00}{0.6 - (-0.4)} = -1.01$$

$$\text{Hence } 1 - n = -1.01$$
$$n = 2.01$$

That is, the reaction is second-order.

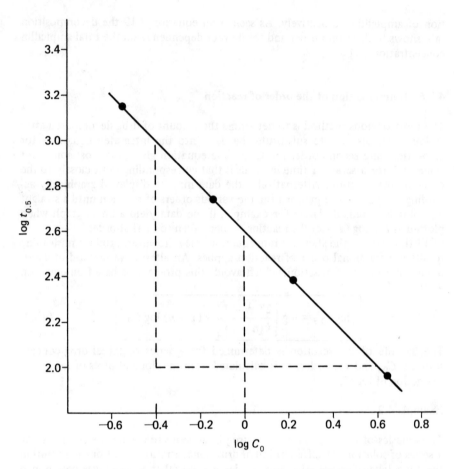

figure 4.8 Plot of log of half-life ($t_{0.5}$) against log of initial drug concentration (C_0)

The intercept of the graph (that is, the value of log $t_{0.5}$ at log $C_0 = 0$) is 2.60.

Thus, from equation 4.20

$$\log\left(\frac{2-1}{k}\right) = 2.60$$

and $k = 2.51 \times 10^{-3} \; \ell \quad mol^{-1} \; min^{-1}$ ∎

4.2.7 Complex reactions

There are many examples of drugs in which decomposition occurs simultaneously by two or more pathways, or involves a sequence of decomposition steps or a reversible reaction. Indeed, the degradative pathways of some drugs include examples of each of these types of complex reactions. Modification of the rate equations is necessary whenever such reactions are encountered.

Reversible reactions

Treatment of the kinetics of a reversible reaction involves two rate constants; one, k_f, to describe the rate of the forward reaction and the other, k_r, to describe the rate of the reverse reaction. For the simplest example in which both of these reactions are first-order, that is

$$A \underset{k_r}{\overset{k_f}{\rightleftharpoons}} B \qquad (4.21)$$

the rate of decomposition of reactant is

$$-d[A]/dt = k_f[A] - k_r[B] \qquad (4.22)$$

The integrated form of the rate equation is

$$t = \frac{2.303}{(k_f + k_r)} \log \frac{A_0 - A_{eq}}{A - A_{eq}} \qquad (4.23)$$

where A_0, A and A_{eq} represent the initial concentration, the concentration at time t and the equilibrium concentration of reactant A, respectively. Equation 4.23 indicates that a plot of t (as ordinate) against $\log [(A_0 - A_{eq})/(A - A_{eq})]$ should be linear with a slope of $2.303/(k_f + k_r)$. k_f and k_r may be calculated separately if the equilibrium constant K is also determined, since

$$K = B_{eq}/A_{eq} = (1 - A_{eq})/A_{eq} = k_f/k_r \qquad (4.24)$$

where B_{eq} is the equilibrium concentration of product B.

The epimerisation of tetracycline (see section 4.1.3) is an example of a first-order reversible decomposition reaction.

Parallel reactions

The decomposition of many drugs involves two or more pathways, the preferred route of reaction being dependent on reaction conditions. Nitrazepam (I) decomposes in two pseudo first-order parallel reactions[24] as illustrated in figure 4.9. The main decomposition product is 2-amino-5-nitrobenzophenone (II) in aqueous solution and 3-amino-6-nitro-4-phenyl-2(1*H*)-quinolone (III) in the solid state. Decomposition of nitrazepam tablets in the presence of moisture will occur by both routes, the ratio of II to III being dependent on the amount of water present.

In other cases, decomposition may occur simultaneously by two different decomposition processes as in the simultaneous hydrolysis and epimerisation of pilocarpine (see section 4.1.3). The over-all rate equation for a parallel reaction is the sum of the constants for each pathway. For example, for a decomposition of a drug X involving two pathways, each of which is first-order,

$$X \overset{k_1}{\underset{k_2}{\diagup \diagdown}} \begin{matrix} A \\ B \end{matrix} \qquad (4.25)$$

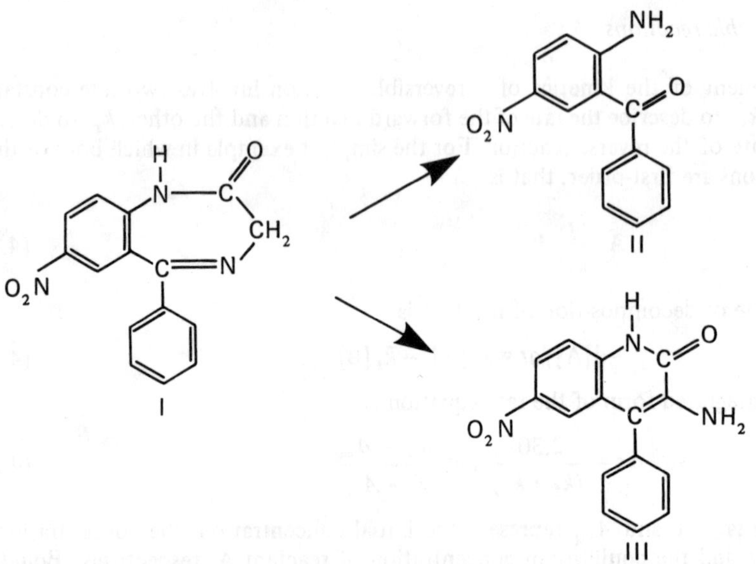

figure 4.9 Simplified decomposition scheme for nitrazepam (I). II, 2-amino-5-nitrobenzophenone; III, 3-amino-6-nitro-4-phenyl-2(1*H*)-quinolone. From reference 24

the rate equation is

$$-d[X]/dt = (k_1 + k_2)[X] = k_{exp}[X] \qquad (4.26)$$

where k_1 and k_2 are the rate constants for the formation of A and B, respectively, and k_{exp} is the experimentally determined rate constant. Values of the rate constants k_1 and k_2 may be evaluated separately by determining the ratio R of the concentration of products formed by each reaction.

$$R = [A]/[B] = k_1/k_2 \qquad (4.27)$$

Since $\qquad k_{exp} = k_1 + k_2 \qquad (4.28)$

$$k_{exp} = k_1 + k_1/R \qquad (4.29)$$

Solving for k_1 gives

$$k_1 = k_{exp}\left(\frac{R}{R+1}\right) \qquad (4.30)$$

Similarly $\qquad k_2 = k_{exp}/(R+1) \qquad (4.31)$

Example 4.3

Pilocarpine has been shown to undergo simultaneous hydrolysis and epimerisation in aqueous solution[13]. The experimentally determined rate constant, k_{exp}, at 25 °C is 6.96×10^2 ℓ mol^{-1} h^{-1}. Analysis has shown that the percentage of the epimerised form of pilocarpine (isopilocarpine) at 25 °C is 20.62 per cent. Calculate the rate constants for hydrolysis, k_H, and epimerisation, k_E.

The ratio R of pilocarpine to isopilocarpine is

$$R = \frac{79.38}{20.62} = 3.85$$

From equation 4.30

$$k_H = 6.96 \times 10^2 \left(\frac{3.85}{4.85}\right)$$

$$k_H = 5.48 \times 10^2 \ \ell \ mol^{-1} \ h^{-1}$$

From equation 4.31

$$k_E = \frac{6.96 \times 10^2}{4.85}$$

$$k_E = 1.44 \times 10^2 \ \ell \ mol^{-1} \ h^{-1}$$

Thus the rate constants for hydrolysis and epimerisation are 5.48×10^2 and $1.44 \times 10^2 \ \ell \ mol^{-1} \ h^{-1}$, respectively. ∎

Consecutive reactions

The simplest example of a consecutive reaction is that described by a sequence

$$A \xrightarrow{\ k_1\ } B \xrightarrow{\ k_2\ } C \tag{4.32}$$

where each step is a non-reversible first-order reaction. The hydrolysis of chlordiazepoxide[25] follows a first-order decomposition scheme similar to that described in equation 4.32. The neutral or cationic chlordiazepoxide (I or IH^+) is transformed to the lactam (II) and, finally, in acidic solutions, to the yellow benzophenone (III) (figure 4.10).

The rate of decomposition of A is

$$-d[A]/dt = k_1 [A] \tag{4.33}$$

The rate of change of concentration of B is

$$d[B]/dt = k_1 [A] - k_2 [B] \tag{4.34}$$

and that of C is

$$d[C]/dt = k_2 [B] \tag{4.35}$$

Integration of the rate equation 4.33 yields

$$[A] = [A_0] \ e^{-k_1 t} \tag{4.36}$$

Substitution of equation 4.36 into equation 4.34 gives

$$d[B]/dt = k_1 [A_0] \ e^{-k_1 t} - k_2 [B] \tag{4.37}$$

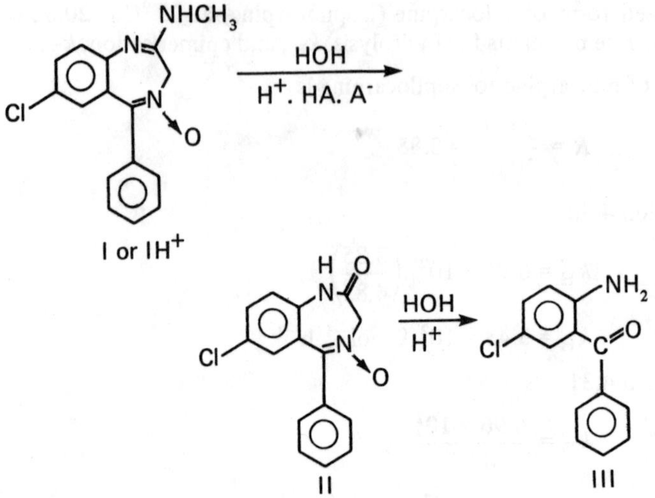

figure 4.10   Decomposition scheme for chlordiazepoxide (I or IH$^+$). II, lactam; III, benzophenone. From reference 25

which upon integration gives

$$[B] = \frac{k_1 [A_0]}{(k_2 - k_1)} [e^{-k_1 t} - e^{-k_2 t}] \tag{4.38}$$

Since, at any time

$$[A_0] = [A] + [B] + [C] \tag{4.39}$$

then

$$[C] = [A_0] - [A] - [B]$$

$$= [A_0] \times \left[1 + \frac{1}{k_1 - k_2} (k_2 e^{-k_1 t} - k_1 e^{-k_2 t}) \right] \tag{4.40}$$

Equations 4.36, 4.38 and 4.40 may be used to estimate the rate constants k_1 and k_2 and also the concentration of the breakdown product C.

4.3   Kinetics of chemical decomposition in solid dosage forms

In spite of the importance of solid dosage forms there have been relatively few attempts to evaluate the detailed kinetics of decomposition. A lot of earlier work was carried out with the sole objective of predicting stability and data were treated using the rate equations derived for reactions in solution. Many systems were reported to decompose by first-order kinetics; for example, chlor-tetracycline capsules[26], vitamin A tablets[27], tablets of vitamins B, C and A[28], aspirin in tablets made in a microcrystalline cellulose base[29]. In some systems it was found difficult to distinguish between first- and zero-order as, for example, in the decomposition of solid dosage forms containing thiamine, riboflavine and niacinamide[30-32], thiamine mononitrate[33] and ascorbic acid[34].

More recently the mechanisms that were developed to describe the kinetics of decomposition of pure solids have been applied to pharmaceutical systems and some rationalisation of decomposition behaviour has been possible. A comprehensive account of this topic has been presented by Carstensen[35].

It is convenient to divide single component systems into two categories. Consider first those solids, such as *p*-aminosalicylic acid, that decompose to a solid product and a gas.

$$NH_2 C_6 H_3 (OH)COOH \rightarrow NH_2 C_6 H_4 OH + CO_2$$

The decomposition curves which result from such a reaction show either (a) an initial rapid decomposition followed by a more gradual decomposition rate, or (b) an initial lag period, giving a sigmoidal appearance. The shape produced by (a) can usually be accounted for by *topochemical* (or contracting geometry) reactions and that produced in (b) by *nucleation* theories.

The model used in the treatment of topochemical decomposition is that of a cylinder or sphere (figure 4.11) in which it is assumed that the radius of the intact chemical substance decreases linearly with time. The various theoretical treatments of this type of decomposition have been reviewed by Jacobs and Tompkins[36]. For the contracting cylinder model, the mole fraction x decomposed at time t is given by

$$(1 - x)^{1/2} = 1 - (k/r_0)t \qquad (4.41)$$

For the contracting sphere model

$$(1 - x)^{1/3} = 1 - (k/r_0)t \qquad (4.42)$$

There are a few pharmaceutical examples of compounds that decompose by topochemical reaction. Thus, the decomposition of aspirin at elevated temperatures has been shown[37] to conform to equation 4.41 as shown in figure 4.12.

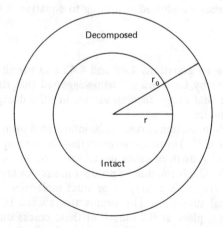

figure 4.11 Model of sphere or cylinder used in theoretical treatment of topochemical reactions. From reference 35

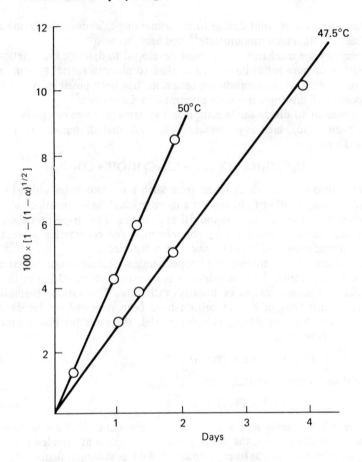

figure 4.12 Decomposition of aspirin at elevated temperatures in tablets containing sodium bicarbonate plotted according to equation 4.41. From reference 37 with permission

A similarity between equations 4.41 and 4.42 and the first-order rate equations was pointed out by Carstensen[35] who suggested that this similarity might account for the fact that many decompositions in solid dosage forms appear to follow first-order kinetics.

The sigmoidal decomposition curves can be interpreted using a model proposed by Prout and Tompkins[38]. The model assumes that the decomposition is governed by the formation and growth of active nuclei which occur on the surface as well as inside the crystals. The formation of product molecules sets up further strains in the crystal since the surface array of product molecules has a different unit cell from the original substance. The strains are relieved by the formation of cracks. Reaction takes place at the mouth of these cracks due to lattice imperfections and spreads down into the crevices. Decomposition on these surfaces produces further cracking and so the chain reaction spreads.

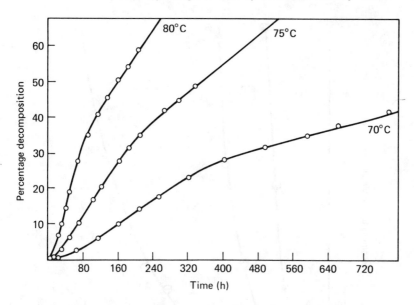

figure 4.13 Degradation of powdered *p*-aminosalicylic acid in a dry atmosphere at elevated temperatures. From reference 39 with permission

The equation proposed to describe decomposition by this process is of the form

$$\ln\left[x/(1-x)\right] = (k/r_0)t + C \qquad (4.43)$$

where C is a lag-time term.

The decomposition curves of *p*-aminosalicylic acid are sigmoidal (see figure 4.13) and linear plots are produced when the data are plotted according to equation 4.43. Stability measurements made inadvertently in the lag periods of this type of decomposition would suggest zero-order kinetics.

A second category of solids comprises those that decompose to give a liquid and a gas. Aminobenzoic acid, for example, decomposes into aniline and carbon dioxide. Decomposition causes a layer of liquid to form around the solid which dissolves the solid. The decomposition curves shown an initial lag period (figure 4.14) which corresponds to the establishment of the liquid layer. Beyond this region, the plot is linear, representing first-order decomposition of the solid in solution in its liquid decomposition products. There are thus two rate constants, that for the initial decomposition of the solid itself, k_s, and that for the decomposition of the solid in solution, k_ϱ.

4.4 Factors affecting the rate of chemical decomposition

As we have seen from section 4.1 the hydrolysis of drugs in solution may be catalysed by H^+ or OH^- ions and consequently the stability of solutions of these drugs will be appreciably affected by pH. Controlling the pH, however,

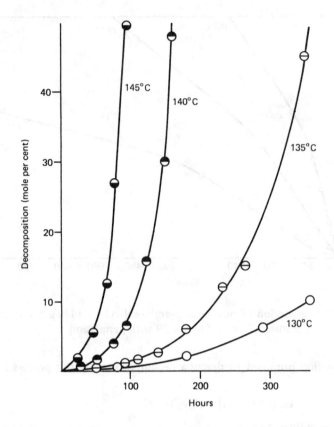

figure 4.14   Decomposition curves of *p*-aminobenzoic acid. From J. T. Carstensen and M. N. Musa *J. pharm. Sci.*, 61, 1113 (1972)

involves the addition of buffer components which may themselves catalyse hydrolysis. There are many other examples of cases where the addition of compounds that are normally considered to be inert leads to increased decomposition. For example, sodium chloride, added to adjust the tonicity of a parenteral solution, may lead to increased hydrolysis due to an increase in ionic strength. Organic solvents added to the solution to increase the drug solubility also alter the dielectric constant and this too is a factor controlling drug stability. Surfactants which are commonly present in a variety of formulations may also have a modifying effect on the decomposition rate.

In solid forms one of the most important factors affecting stability is the presence of moisture since this may have a significant effect not only on the decomposition rate but also on the kinetics of decomposition. Stability may also be appreciably reduced if reactions occur between the ingredients of the solid dosage form, for example, between the tablet excipients and active principle. Similarly, interaction between an ointment or cream base and the active principle may seriously affect the stability of semisolid preparations.

One environmental factor that significantly affects the stability of both solid and liquid dosage forms is temperature. An examination of the effect of temperature on stability is routinely carried out during storage testing (see section 4.5) and is of particular importance for products destined for use in tropical countries or which are subjected to heat sterilisation.

The various factors outlined above will now be examined in detail for liquid, semisolid and solid dosage forms.

4.4.1 Liquid dosage forms

Temperature

The effect of temperature on the rate of decomposition may be described by an empirical equation proposed by Arrhenius

$$k = Ae^{-E_a/RT} \tag{4.44}$$

or

$$\log k = \log A - \frac{E_a}{2.303RT} \tag{4.45}$$

The pre-exponential factor A is often referred to as the frequency factor. E_a is the activation energy and is the energy which must be exceeded if the collision of two reactant molecules is to lead to reaction. According to equation 4.45 a plot of $\log k$ as a function of reciprocal temperature should be linear with a slope of $-E_a/2.303R$ from which E_a may be calculated.

Example 4.4

Calculate the activation energy, E_a, and the frequency factor, A, for the hydrolysis of atropine in 0.099 mol ℓ^{-1} $HClO_4$ using the following data (A. A. Kondritzer and P. Zvirblis. *J. Am. pharm. Ass.*, 46, 531 (1957)).

Temperature (°C)	70	80	90
Rate constant (k, min^{-1} × 10^5)	1.73	3.38	6.45

Converting the data into a form suitable for plotting according to the Arrhenius equation

$1/T$ (K^{-1} × 10^3)	2.915	2.833	2.755
$\log k$	−4.76	−4.47	−4.19

A plot of $\log k$ versus $1/T$ (figure 4.15) has a slope of -3.60×10^3 K.

From equation 4.45

$$\text{Slope} = \frac{-E_a}{2.303R} = -3.60 \times 10^3$$

Therefore

$$-E_a = 2.303 \times 8.314 \times -3.60 \times 10^3$$

$$E_a = 68.9 \text{ kJ mol}^{-1}$$

From figure 4.15

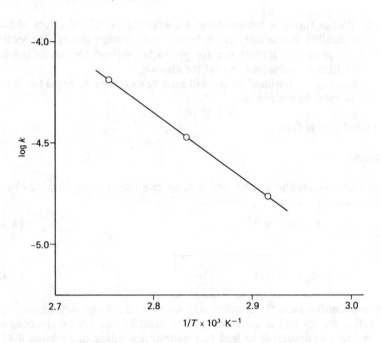

figure 4.15 Arrhenius plot for the acid-catalysed hydrolysis of atropine in 0.099 mol ℓ^{-1} HClO$_4$. Data from A. A. Kondritzer and P. Zvirblis. *J. Am. pharm. Ass.*, 46, 531 (1957)

when $1/T = 2.83 \times 10^{-3}$ K^{-1}, log $k = -4.45$

Substituting into equation 4.45

$$-4.45 = \log A - \frac{68\,900 \times 2.83 \times 10^{-3}}{2.303 \times 8.314}$$

$$\log A = 10.18 - 4.45 = 5.73$$

$$A = 5.37 \times 10^5 \text{ min}^{-1}. \quad \blacksquare$$

Differentiation of equation 4.45 yields

$$\frac{d \ln k}{dT} = \frac{E_a}{RT^2} \qquad (4.46)$$

Integration of equation 4.46 over the temperature range T_1 to T_2, assuming E_a remains constant, gives

$$\boxed{\log \frac{k_2}{k_1} = \frac{E_a}{2.303R} \left(\frac{T_2 - T_1}{T_1 T_2} \right)} \qquad (4.47)$$

Equation 4.47 is a useful form of the Arrhenius equation for calculating the rate constant k_2 at a temperature T_2 from a knowledge of its value at a second temperature, T_1.

Temperature increase often causes an appreciable increase of the decomposition rate. For many reactions the increase in k is of the order of two to three times for a 10 K rise in temperature.

pH

A general expression which accounts for both specific acid and base catalysis and also catalysis by unionised water molecules (solvent catalysis) is

$$\frac{d[S]}{dt} = (k_0 + k_{H^+}[H^+] + k_{OH^-}[OH^-])\,[S] \qquad (4.48)$$

where S represents the drug undergoing hydrolysis and k_0, k_{H^+} and k_{OH^-} are the rate constants for solvent catalysis, acid catalysis and base catalysis, respectively. The simplest form of pH dependence of log k for a drug undergoing both acid and base catalysis is typified by the hydrolysis of procaine and tetracaine (figure 4.16). In solutions of low pH the $k_{H^+}[H^+]$ term will be greater than either k_0 or $k_{OH^-}[OH^-]$ and will dictate the hydrolysis rate. Thus, as pH is increased in acid

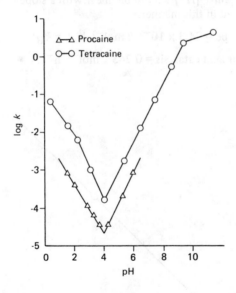

figure 4.16 Over-all velocity constant for hydrolysis of procaine (lower curve) and of tetracaine (upper curve) at 50 °C as a function of pH. From J. Anschel, J. A. Mollica and K. S. Lin. *Bull. parenteral Drug Ass.*, 26, 271 (1972), with permission

solutions, the hydrolysis rate gradually decreases. In solutions of high pH, the $k_{OH^-}[OH^-]$ term will influence the hydrolysis rate and thus the rate will increase with increasing $[OH^-]$. The pH at which minimum hydrolysis occurs depends on the relative magnitudes of k_{H^+} and k_{OH^-}.

Example 4.5

The following rate constants were obtained (Krasowska *et al., Dansk Tidsskr. Farm.*, 42, 170 (1968)) for the acid-catalysed hydrolysis of homatropine at 90 °C.

HCl concentration (mol ℓ^{-1})	0.226	0.414	0.624	0.832
k_{obs} (h^{-1} × 10^2)	5.3	10.5	15.0	19.5

Calculate the rate constant for acid catalysis, k_{H^+}, assuming solvent and hydroxyl ion catalysis to be negligible.

For catalysis by H$^+$ ions only, equation 4.48 becomes

$$\frac{d[S]}{dt} = (k_{H^+}[H^+])[S] = k_{obs}[S]$$

That is, the observed first-order rate constant, k_{obs}, is

$$k_{obs} = k_{H^+}[H^+]$$

A plot of k_{obs} against [H$^+$] should be linear with a slope of k_{H^+}. Figure 4.17 shows the data plotted in this manner.

$$\text{Slope} = 24.3 \times 10^{-2} \, \ell \, \text{mol}^{-1} \, \text{h}^{-1} = k_{H^+}$$

The rate constant for acid catalysis = 0.243 ℓ mol^{-1} h^{-1}. ∎

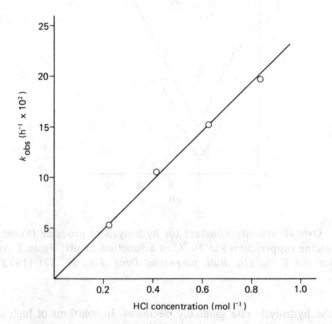

figure 4.17 Plot of observed rate constant, k_{obs}, against HCl concentration for the acid-catalysed hydrolysis of homatropine at 90 °C. Data from Krasowska *et al. Dansk Tidsskr. Farm.*, 42, 170 (1968)

The pH–rate profile of many drugs, however, is more complex than the simple example illustrated in figure 4.16. As an example of the types of complexities that may arise we will consider the specific acid–base-catalysed hydrolysis of mecillinam which has been studied in detail by Larsen and Bundgaard[40]. Mecillinam, which is an antimicrobially active amidinopenicillamic acid, has the structure shown below. It may exist either as a cation (MH_2^+), a zwitterion ($MH^\pm$), or an anion (M^-), the apparent pK_a of MH_2^+ and $MH^\pm$ being 2.64 and

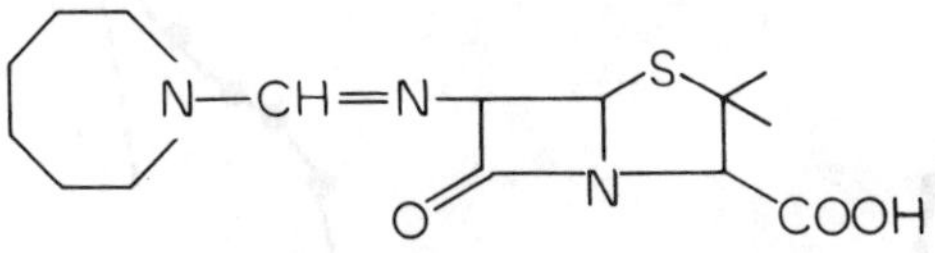

Mecillinam

8.79, respectively (35 °C, ionic strength = 0.5). Each of these species may undergo specific acid or base hydrolysis. It has been suggested that the following reactions contribute to the over-all velocity of mecillinam degradation:

(1) $MH_2^+ + H^+ \xrightarrow{k_H}$ products

(2) $MH^\pm + H^+ \xrightarrow{k'_H}$ products

(3) $MH^\pm + H_2O \xrightarrow{k_0}$ products

(4) $MH^\pm + OH^- \xrightarrow{k'_{OH}}$ products

(5) $M^- + OH^- \xrightarrow{k_{OH}}$ products

The pH–rate profile is illustrated in figure 4.18. At pH < 1.5 mecillinam exists predominantly in the cationic form and the rate of degradation is dictated exclusively by reaction 1. At pH ~ pK_{a1}, the zwitterionic form $MH^\pm$ appears in solution and reaction 2 occurs simultaneously with reaction 1. At pH 4–6 the over-all degradation consists chiefly of the water-catalysed reaction 3 and is consequently independent of pH. Between pH 6.5 and the pK_a for ionisation of the amidino side-chain group (pK_{a2} = 8.79) the pH–rate profile is dictated by reaction 4. In the pH range 8–11 the rate of reaction is influenced by the ionic state of the amidino side-chain. Above pH 12, reaction 5 predominates.

pH–rate profiles are routinely determined in preformulation studies on drugs susceptible to hydrolysis. Although such studies indicate the pH of maximum stability it may not always be possible to formulate at this pH due to solubility problems or reduced therapeutic activity at this pH. This is particularly true of the weakly basic drugs, physostigmine, pilocarpine and atropine, which are widely used in eye-drop preparations. Although these drugs are most stable in acid solution, they are most active when present in solution as the free base rather than as the salt form, as in acid solutions. In such cases a compromise must be sought between the opposing effects.

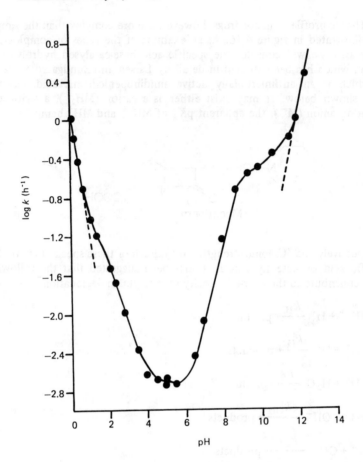

figure 4.18 Log k–pH profile for the degradation of mecillinam in aqueous solution at 35 °C (ionic strength = 0.5), where k is the apparent first-order rate constant for degradation in buffer-free solutions or in buffers showing no effect on rate of degradation. From reference 40, with permission

In general acid–base catalysis the acidic and basic species that are responsible for catalysis may be provided by the buffer components. Adjustment of the pH of the solution may therefore significantly affect the stability because of changes in the concentration of buffer components in solution. The relationship between the ability of a buffer component to catalyse hydrolysis, denoted by the catalytic coefficient, k, and its dissociation constant, K, is expressed by the Brönsted catalysis law as

$$k_A = aK_A^\alpha \text{ for a weak acid} \tag{4.49}$$

and $\qquad\qquad k_B = bK_B^\beta \text{ for a weak base} \tag{4.50}$

where a, b, α and β are constants characteristic of the solvent and temperature. α and β are positive and vary between 0 and 1.

Lindsay and Helm[41] demonstrated the catalytic effect of monohydrogen and dihydrogen citrate ions on the decomposition of potassium penicillin G in buffered aqueous solutions. Below pH 6.5, the over-all rate constant for the catalysed hydrolysis was expressed in terms of the specific acid-catalysed hydrolysis term $k_{H^+}[H^+]$ and the general acid-catalysed terms $k_{C_6H_6O_7{}^{2-}}[C_6H_6O_7{}^{2-}]$ and $k_{C_6H_7O_7{}^-}[C_6H_7O_7{}^-]$. At pH 6.5, specific acid-catalysed, specific base-catalysed hydrolysis and monohydrogen citrate ion-catalysed hydrolysis all determine the hydrolysis and the over-all rate constant thus contains the terms $k_{H^+}[H^+]$ $k_{OH^-}[OH^-]$ and $k_{C_6H_6O_7{}^{2-}}[C_6H_6O_7{}^{2-}]$.

Although a buffer system may provide excellent control over pH it may also, unfortunately, contribute to the degradation of the drug. Buffer systems should therefore be selected on the basis of adequate control of pH and also on the absence of catalytic effects.

The effect of pH on drug decomposition is not restricted entirely to acid–base catalysis of drugs susceptible to hydrolysis. The oxidative degradation of some drugs has been reported to be pH-dependent and this is often a consequence of the effect of pH on the oxidation or reduction potential of the system. Morphine, the decomposition scheme of which was given in section 4.1, deteriorates rapidly in alkaline or neutral solution following pseudo first-order kinetics[9]. Acidic solutions are relatively stable. Other drugs susceptible to oxidation, the stability of which have been reported to be pH-dependent, include adrenaline, antibiotics and vitamins.

Ionic strength

The relationship between the rate constant and ionic strength I is given by the Brönsted–Bjerrum equation, which for aqueous solutions at 25 °C is

$$\log k = \log k_0 + 1.018 \, z_A z_B \sqrt{I} \tag{4.51}$$

The effect of ionic strength on reaction rate may readily be ascertained by varying the ionic strength by the addition of an inert electrolyte such as sodium chloride. From equation 4.51 a plot of $\log k$ versus $\sqrt{I}$ should be linear with a slope of $1.018 \, z_A z_B$. Extrapolation to zero ionic strength yields k_0.

Equation 4.51 is valid for ionic solutions up to $I = 0.01$. For solutions of higher concentration (for $I \leqslant 0.1$) a modified form of the Brönsted–Bjerrum equation has been proposed

$$\log k = \log k_0 + 1.018 \, z_A z_B \, \frac{\sqrt{I}}{1 + \beta \sqrt{I}} \tag{4.52}$$

β depends on the ionic diameter of the reacting species and is often approximated to unity.

Figure 4.19 shows the effect of ionic strength on the degradation of penicillin in phosphate buffer, plotted according to equations 4.51 and 4.52. The slope, calculated using the modified Brönsted–Bjerrum equation, is seen to give a much better approximation to the theoretical value of 2.04.

For reactions between ions of like charge, for example that between hydronium ions and protonated substrates, the slope is positive. For reaction between ions of opposite charge, such as that between hydroxyl ions and protonated sub-

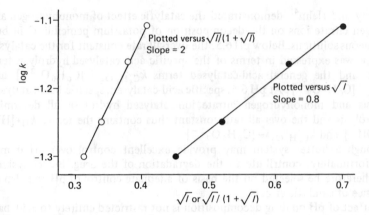

figure 4.19   Kinetic salt effect on degradation of penicillin in phosphate buffer at pH 6.8 and 60 °C, plotted according to equation 4.51 (o) and equation 4.52 (●). Plots in borate buffer at pH 8.75 parallel the lines shown in the figure. From reference 42 with permission

strates, the slope will be negative. If one of the reactants is uncharged as, for example, with neutral esters, equation 4.51 predicts an absence of ionic strength effects since the product $z_A z_B = 0$. Although this is essentially true in dilute solution, an effect due to ionic strength is, in fact, observed at higher concentrations. This effect is attributable to an influence of ionic strength on the activity coefficient of the neutral molecule. Under these circumstances the effect of the ionic strength on the rate of reaction is best described by

$$\log k = \log k_0 + bI \tag{4.53}$$

where b is an empirical constant.

Kinetic salt effects on drug degradation have been reviewed by Carstensen[42]. Salt effects arise in practice from addition of electrolyte to adjust the tonicity, to prevent oxidation, or to buffer the solution against pH change.

The ionic effects outlined above are referred to as primary salt effects. Salts may also affect reaction rate through their influence on the catalytic action of weak acids or bases present in solution as buffer components. The addition of electrolyte affects the dissociation constant of the weak acid or base and hence, from the Brönsted catalysis law (equations 4.49 and 4.50), alters its catalytic action. This effect is referred to as a secondary salt effect.

Solvent

Many drugs that are susceptible to hydrolysis may be stabilised by formulating them, where possible, in non-aqueous solution. Commonly used solvents are ethanol, glycerol, or propylene glycol. Injections containing such solvent mixtures include diazepam injection BNF* and phenobarbitone injection BP. The

*British National Formulary

reason why a change to non-aqueous solvent is often successful in reducing hydrolysis is explicable in terms of the dielectric constant.

The effect of the dielectric constant of the solvent, ϵ, on stability, where decomposition involves the reaction of an ion and a charged drug (for example, specific acid or base catalysis of a charged drug) is given by an equation of the form

$$\log k = \log k_{\epsilon=\infty} - Kz_A z_B/\epsilon \qquad (4.54)$$

where z_A and z_B are the charges on the ion and the drug and K is a constant for a given system at a fixed temperature. $k_{\epsilon=\infty}$ is the rate constant in a solvent of infinite dielectric constant and is obtained by extrapolation of a plot of $\log k$ versus $1/\epsilon$ to $1/\epsilon = 0$. Whether or not a decrease in dielectric constant will stabilise such a reaction is thus dependent on the charges of the two ions. If both the drug and the attacking ion are similarly charged, as in the hydroxyl ion catalysed hydrolysis of an anionic drug or the hydrogen ion catalysed hydrolysis of a cationic drug, the slope of the $\log k$ versus $1/\epsilon$ plot is negative and so formulation in solvents of low dielectric constant will decrease the decomposition rate. Conversely, if the charges on the attacking ion and the drug molecule are of opposite sign, as in the general or specific base-catalysed hydrolysis of protonated drugs, the slope of the plot is positive and hence stabilisation will not be achieved by formulation in a solvent of low dielectric constant.

Figure 4.20 shows an increase in the rate of the acid-catalysed hydrolysis of chloramphenicol in aqueous solutions containing increasing proportions of propylene glycol.

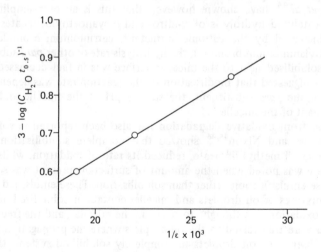

figure 4.20 Rate of hydrolysis of chloramphenicol at 91.3 °C as a function of the dielectric constant, ϵ, of the propylene glycol–water solution. The rate is expressed as the log of the reciprocal of the product of the half-life, $t_{0.5}$, and the water concentration C_{H_2O}. The integer 3 is used merely to avoid negative logarithms. From A. D. Marcus and A. J. Tarazka. *J. Am. pharm. Ass.*, 48, 77 (1959) with permission

Effect of surfactants

As might be expected, the presence of surfactants in micellar form has a modifying effect on the rate of hydrolysis of drugs. The magnitude of the effect depends on the difference in the rate constant when the drug is in aqueous solution and when it is solubilised within the micelle, and also on the extent of solubilisation.

Thus $$k_{obs} = k_m f_m + k_w f_w \qquad (4.55)$$

where k_{obs}, k_m and k_w are the observed, micellar and aqueous rate constants, respectively, and f_m and f_w are the fractions of drug associated with the micelles and aqueous phase, respectively. The value of k_m is dependent on the location of the drug within the micelle. As discussed in chapter 6 (section 6.6), a solubilisate may be incorporated into the micelle in a variety of locations. Non-polar compounds are thought to be solubilised within the lipophilic core and, as such, are likely to be more effectively removed from the attacking species than those compounds that are located close to the micellar surface. Sheth and Parrott[43] noted a greater stabilisation of benzocaine compared with homatropine when solubilised by non-ionic surfactants, and attributed this effect to a deeper penetration of the less-polar benzocaine into the micellar palisade layer, making it less susceptible to attack by hydroxyl ions. Where the drug is located near to the micellar surface, and therefore still susceptible to attack, the ionic nature of the surfactant is an important influence on decomposition rate. For base-catalysed hydrolysis, solubilisation into anionic micelles affords an effective stabilisation due to repulsion of OH^- by the micelles. Conversely, solubilisation into cationic micelles might be expected to cause an enhanced base-catalysed hydrolysis. Winterborn *et al.*[44] have shown, however, that this is an over-simplification. Whilst base-catalysed hydrolysis of *p*-nitro- and *p*-cyanoethylbenzoates was, as predicted, increased by the cationic surfactant, cetrimonium bromide (hexadecyltrimethylammonium bromide), the hydrolysis rate of other *para*-substituted esters also solubilised near to the micellar surface was in fact decreased. These authors have suggested that modification of the reaction rate was dependent on the nature of the *para* substituent, the surface pH of the micelle and the dielectric constant of the micelle.

Protection from oxidative degradation has also been reported in solubilised systems. Carless and Nixon[45,46] showed that complete solubilisation of the water-insoluble oil methyl linoleate, reduced its rate of oxidation, whilst a high oxidation rate was noted when the amount of surfactant present was sufficient only to cause emulsification rather than solubilisation. These emulsified systems contained mixtures of oil droplets and micelles containing solubilised oil. Initiation of the oxidation was thought to occur in the micelles, and the free radicals so produced were transferred to the oil droplets where the propagation rate was higher. An absence of oil droplets in completely solubilised systems therefore resulted in a reduced propagation rate and a greatly stabilised system.

Many drugs associate to form micelles in aqueous solution (see chapter 6) and several studies have been reported of the effect of this self-association on stability. Ong and Kostenbauder[47] reported that in micellar solutions of penicillin G (500 000 units $m\ell^{-1}$) the apparent rate of the hydrogen ion catalysed degradation was increased twofold, but that of water and hydroxyl ion catalysed hydrolysis was decreased two- to threefold. Consequently, the pH profile was shifted to higher pH values and the pH of minimum degradation was found to be 7.0 compared to 6.5 for monomeric solutions (8000 units $m\ell^{-1}$). When compared at the respective pH–rate profile minima, micellar penicillin G was reported to be 2.5 times as stable as the monomeric solutions under conditions of constant pH and ionic strength.

4.4.2 Semisolid dosage forms

The chemical stability of active ingredients incorporated into ointments or creams is frequently dependent on the nature of the ointment or cream base used in the formulation. Allen and Das Gupta[48,49] evaluated the stability of hydrocortisone in a series of commercially available bases and reported maximum decomposition in polyethylene glycol base. The reported shelf-life was only six months in this base, which makes manufacture on a commercial basis an unreasonable proposition considering the length of time involved in distribution of the drug from wholesaler to patient.

Not only should possible stability problems be borne in mind in the choice of ointment base at the formulation stage, but also similar care should be exercised if the ointment is diluted at a later stage. Such dilution is, unfortunately, common practice in cases where the practitioner wishes to reduce the potency of highly active topical preparations, particularly steroids. The pharmaceutical and biopharmaceutical dangers of this procedure have been stressed by Busse[50] (section 9.5.4). Of particular interest here are the problems of drug stability, which can occur through the use of unsuitable diluents. The example has been cited by this author of the dilution of betamethasone valerate cream with a cream base having a neutral to alkaline pH. Under such conditions, conversion of the 17-ester to the less-active 21-ester can occur. Similarly, diluents containing oxidising agents could cause chemical degradation of fluocinolone acetate to less-active compounds.

Incorporation of drugs into gel structures frequently leads to a change in their stability. Ullman *et al.*[51] reported an increased degradation of penicillin G sodium in hydrogels of various natural and semisynthetic polymers. Testa and Etter[52], in a study of the reversible hydrolysis of pilocarpine at pH 6 in Carbopol hydrogels, showed that the percentage of undecomposed pilocarpine at equilibrium was a simple function of the apparent viscosity of the medium. The rate constant for degradation was not, however, significantly affected by changes in viscosity. Poust and Colaizzi[53] noted little influence of viscosity on the rate of oxidation of ascorbic acid in solutions of gels of Polysorbate 80.

4.4.3 Solid dosage forms

Temperature

Although the kinetics of decomposition of the solid state are different to those in solution, the temperature dependence of the rate constant can nevertheless generally be described by the Arrhenius equation (equation 4.44). Exceptions to this generalisation are those solids in which decomposition exhibits an approach to equilibrium. Examples of such systems include vitamin A in gelatin beadlets and vitamin E in lactose base tablets[54]. The effect of temperature should in these cases be described by the van't Hoff equation rather than by the Arrhenius equation. Equilibrium concentrations of products and reactants are determined at a series of temperatures and the logarithm of K, the equilibrium constant is plotted as a function of $1/T$ according to the van't Hoff equation (see figure 4.21).

$$\ln K = -\frac{\Delta H}{RT} + \text{constant}$$

Equilibrium kinetics in solid dosage forms can also arise as a result of the presence of impurities on the surface of solid excipients[55].

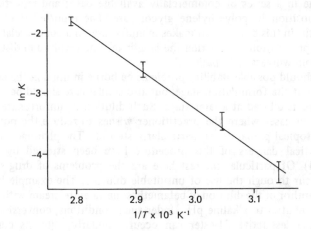

figure 4.21 Van't Hoff plot for vitamin E succinate decomposition in lactose base tablets. From reference 54 with permission

Moisture

Moisture can have a significant effect on the kinetics of decomposition. Kornblum and Sciarrone[39] showed that added moisture decreased the lag time and increased the zero-order rate constant for the decomposition of aminosalicylic acid. The logarithm of the rate constant for the decomposition of a 1 per cent dilution of nitrazepam in microcrystalline cellulose was reported to vary directly with relative humidity (figure 4.22). When the moisture content is very high, decomposition of solid dosage forms may often be treated by solution kinetics of a saturated solution[56]. In such cases, decomposition generally follows zero-order kinetics.

Access of moisture during storage can be minimised by the correct selection of packaging. Tingstad and Dudzinski[57] quote an example of tablets containing a water-labile component. At 50 °C this component was considerably more stable in a water-permeable blister package than in a sealed glass bottle, but at room temperature and 70 per cent humidity the situation was reversed. The reason for this behaviour was attributed to the loss of considerable amounts of water through the film at 50 °C, so improving stability, the reverse diffusion at room temperature decreasing stability.

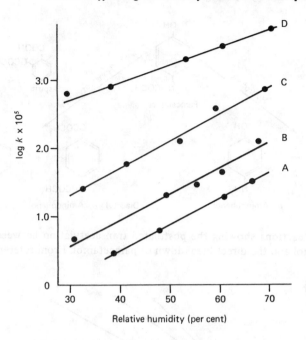

figure 4.22 Logarithm of the nitrazepam decomposition constant, k, as a function of relative humidity at various temperatures: A, 42.5 °C; B, 55.0 °C; C, 73.8 °C; and D, 83.4 °C. From D. Genton and U. W. Kesselring. *J. pharm. Sci.*, 66, 676 (1977) with permission

Chemical interactions

Chemical interaction between components in solid dosage forms may lead to increased decomposition. Replacement of the phenacetin in compound codeine and APC tablets by paracetamol in NHS formulations in Australia in the 1960s, because of its undesirable side-effects, led to an unexpected decreased stability of the tablets. The cause was later attributed[58] to a transacetylation reaction between aspirin and paracetamol and also a possible direct hydrolysis of the paracetamol (figure 4.23). Figure 4.24 shows the increased generation of free salicylic acid at 37 °C in the tablets containing paracetamol. It is interesting to note from this figure, the effect on stability of tablet excipients. Addition of 1 per cent talc caused only a minimal increase in the decomposition, whilst 0.5 per cent magnesium stearate increased the breakdown rate dramatically.

Many cases of the effects of tablet excipients on drug decomposition have been reported. In a study of the effect of tablet lubricants on the hydrolysis rate of aspirin, Kornblum and Zoglio[55] reported an increased degradation effect in the order: hydrogenated vegetable oil < stearic acid < talc < aluminium stearate. These authors suggested that stearate salts should be avoided as tablet lubricants if the active component is subject to hydroxyl ion catalysed degradation. The degradative effect of the alkali stearates was inhibited in the presence of malic,

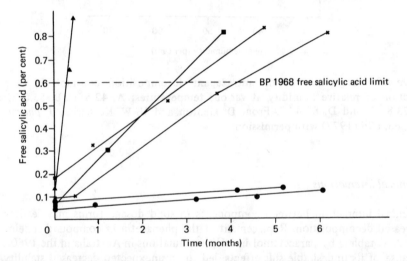

figure 4.23   Reactions showing the postulated transacetylation between aspirin and paracetamol and the direct breakdown of paracetamol. From reference 58

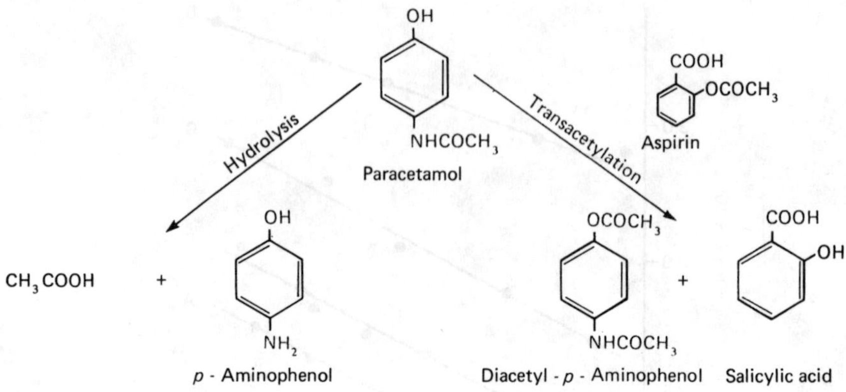

figure 4.24   Development of free salicylic acid in aspirin–paracetamol–codeine and aspirin–phenacetin–codeine tablets at 37°C. ●, Aspirin–phenacetin–codeine; x, aspirin–paracetamol–codeine; ■, aspirin–paracetamol–codeine–talc; ▲, aspirin–paracetamol–codeine–talc–magnesium stearate. From reference 58 with permission

hexamic or maleic acids[59]. In addition to the pH effects contributed by these organic acids, the mechanism operative in inhibiting the degradation was thought to involve a competition for the lubricant cation between aspirin and the additive acid. Maudling *et al.*[60] reported that the degree of hydrolysis of aspirin in tablet or suspension form caused by magnesium trisilicate was considerably greater than that caused by other excipients including aluminium hydroxide, calcium stearate and magnesium stearate. The effect was attributed to the high water content of this excipient.

The base used in the formulation of suppositories can often affect the rate of decomposition of the active ingredients. Jun *et al.*[61] studied the decomposition of aspirin in several polyoxyethylene glycols which are often incorporated into suppository bases. Degradation was shown to be due in part to transesterification giving the decomposition products, salicylic acid and acetylated polyethylene glycol. The rate of decomposition, which followed pseudo first-order kinetics, was considerably greater than when a fatty base such as cocoa butter was used[62].

Analysis of commercial batches of 100 mg indomethacin–polyethylene glycol suppositories[63] showed that approximately 2, 3.5 and 4.5 per cent of the original amount of indomethacin was esterified with polyoxyethylene glycol 300 (see figure 4.25) after storage times of 1, 2 and 3 years, respectively.

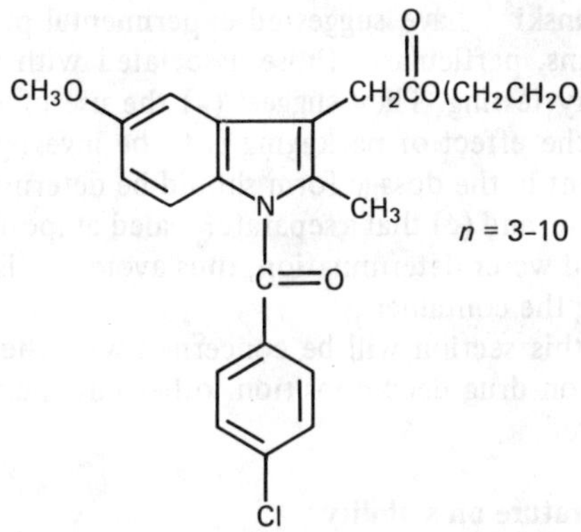

figure 4.25 Polyethylene glycol esters of indomethacin identified in stored suppositories (n = number of ethylene oxide units)

4.5 Stability testing of drugs

We have seen in section 4.1 that drug deterioration may be caused by a wide range of environmental factors such as temperature, light, and oxygen. In this section we examine the ways in which the effect of any particular factor on the drug breakdown under conditions of normal storage may be evaluated from a series of experiments in which this factor is greatly exaggerated. Such testing methods, referred to as accelerated storage tests, have for many years proved an effective replacement at the development stage, for the original time-consuming practice of storing the product at room temperature for periods corresponding to the normal time the products would be likely to remain in stock.

Some of the particular problems that are involved in stability testing of solid dosage forms have been discussed by Tingstad and Dudzinski[57]. The main problems arise because (a) the analytical results tend to have more scatter because tablets and capsules are distinct dosage units rather than the true aliquots encountered with stability studies on drugs in solution, and (b) these dosage forms are heterogeneous systems often involving a gas phase (air and water vapour), a liquid phase (adsorbed moisture) and the solid phase itself. The composition of all of these phases can vary during an experiment.

The first of these problems can be overcome by ensuring uniformity of the dosage form before commencing the stability studies. The problems arising from the heterogeneity are more difficult to overcome. The main complicating factor is associated with the presence of moisture. As we have seen in section 4.4.3, moisture can have a significant effect on the kinetics of decomposition and this may produce many experimental problems during stability testing. For example, with gelatin capsules the water in the capsule shell must equilibrate with that in the formulation and surrounding air and this may require an appreciable time. The prediction of stability is difficult in solid dosage forms in which there is chemical interaction between components, or chemical equilibrium phenomena. In fact, the data for stability studies involving the latter are often plotted using a van't Hoff plot rather than an Arrhenius plot (see first part of section 4.4.3).

Tingstad and Dudzinski[57] have suggested experimental procedures whereby some of these problems, particularly those associated with moisture, may be reduced during stability testing. They suggest (a) the use of tightly sealed containers except where the effect of packaging is to be investigated, (b) that the amount of water present in the dosage form should be determined, preferably at each storage temperature, and (c) that a separate, sealed ampoule should be taken for each assay point and water determination, thus avoiding disturbance of water equilibrium on opening the container.

Although most of this section will be concerned with the prediction of the effect of temperature on drug decomposition, other environmental factors will also be briefly considered.

4.5.1 Effect of temperature on stability

From the Arrhenius equation (equation 4.45) a plot of $\log k$ versus $1/T$ should be linear (see figure 4.15). The rate constant at any selected storage temperature may thus be extrapolated from measurements at a series of elevated temperatures where the reaction proceeds at an accelerated rate. The time taken in conducting the stability tests is considerably reduced compared to that for a simple experiment in which the product is maintained at the required storage temperature and sampled over a period corresponding to the normal storage time.

An alternative method of data treatment is to plot the logarithm of the half-life, $t_{0.5}$, as a function of reciprocal temperature since, from equation 4.10, $t_{0.5} = 0.693/k$.

Therefore
$$\log k = \log 0.693 - \log t_{0.5} \tag{4.56}$$

and substituting into equation 4.45 gives

$$\log t_{0.5} = \log 0.693 - \log A + \frac{E_a}{2.303RT} \tag{4.57}$$

Once the rate constant is known at the required storage temperature it is a simple matter to calculate a shelf-life for the product based on an acceptable degree of decomposition. The usual decomposition level is taken as 90 per cent of the initial concentration of the drug, although this may vary, depending, for example,

on whether the decomposition products produce discolouration or have undesirable side-effects.

Example 4.6

The initial concentration of active principle in an aqueous preparation was
5.0×10^{-3} g cm^{-3}. After 20 months the concentration was shown by analysis
to be 4.2×10^{-3} g cm^{-3}. The drug is known to be ineffective after it has decomposed to 70 per cent of its original concentration. Assuming that decomposition follows first-order kinetics, calculate the expiry date of the drug preparation.

Substituting into the first-order equation (equation 4.8)

$$k = \frac{2.303}{20} \log \frac{5 \times 10^{-3}}{4.2 \times 10^{-3}}$$

$$k = 8.719 \times 10^{-3} \text{ month}^{-1}$$

70 per cent of the initial concentration = 3.5×10^{-3} g cm^{-3}

$$t = \frac{2.303}{8.719 \times 10^{-3}} \log \frac{5 \times 10^{-3}}{3.5 \times 10^{-3}}$$

$$t = 40.9 \text{ months}$$

The expiry date is thus 40.9 months after initial preparation. ∎

It should be noted that, although decomposition in the solid state follows
different kinetic mechanisms to those in solution, it is still generally possible to
treat rate constants that have been determined using equations 4.41–4.43 by the
normal Arrhenius equation and hence to predict stability at room temperature in
the manner described above (for exceptions see section 4.4.3).

Although accelerated storage testing based on the use of the Arrhenius equation has resulted in a very significant saving of time, it still involves the time-
consuming step of the initial determination of the order of reaction for the
decomposition. Whilst most investigators have emphasised the need for a knowledge of the exact kinetic pathway of degradation, some have bypassed this initial
step by assuming a particular decomposition model. Londi and Scott[64] have
indicated that at less than 10 per cent degradation and within the limits of experimental error involved in stability studies, it is not possible to distinguish
between zero-, first- or simple second-order kinetics using curve-fitting techniques; consequently these authors have suggested that the assumption of first-
order kinetics for any decomposition reaction should involve minimum error.
Amirjahed[65], in fact, established a linear relationship between the logarithm of
$t_{0.9}$ (the time taken for the concentration of the reactant to decompose to 90
per cent of its original value) and the reciprocal temperature, which was independent of the order of reaction for the decomposition of a series of drugs. On the
basis of these findings it was suggested that the use of such linear plots to determine $t_{0.9}$ at the required temperature would provide a rapid, and yet sufficiently
accurate, means of studying decomposition rate during the development stage.

Even with the modifications suggested above, the method of stability testing based on the Arrhenius equation is still time-consuming, involving as it does the separate determination of rate constants at a series of elevated temperatures. Experimental techniques have been developed[66,67] which enable the decomposition rate to be determined from a single experiment. Such methods involve raising the temperature of the product in accordance with a predetermined temperature–time programme and are consequently referred to as non-isothermal stability studies.

Any suitable temperature–time programme may be used. In the method proposed by Rogers[66] the rise of temperature was programmed so that the reciprocal of the temperature varied logarithmically with time according to

$$1/T_0 - 1/T_t = 2.303 \, b \, \log (1 + t) \tag{4.58}$$

where T_0 and T_t are the temperatures at zero time and at time t, respectively, and b is any suitable proportionality constant. Applying the Arrhenius equation at both temperatures and subtracting gives

$$\log k_t = \log k_0 + \left(\frac{E_a}{2.303R}\right)\left(\frac{1}{T_0} - \frac{1}{T_t}\right) \tag{4.59}$$

Substituting equation 4.58 into equation 4.59 gives

$$\log k_t = \log k_0 + (E_a b/R) \log (1 + t) \tag{4.60}$$

Therefore
$$k_t = k_0 (1 + t)^{E_a b/R} \tag{4.61}$$

For first-order reactions $-dc/dt = kc$, where c is concentration. Substituting for k from equation 4.61 and integrating gives

$$-\int_{c_0}^{c_t} \frac{dc}{c} = k_0 \int_0^t (1 + t)^{E_a b/R} \, dt \tag{4.62}$$

where c_0 and c_t are the concentrations at zero time and at time t, respectively.

Therefore
$$\log f(c) = \log k_0 - \log [1 + (E_a b/R)] + [1 + (E_a b/R)]$$
$$\times \log (1 + t) + \log [1 - (k_0/k_t)^{1 + R/E_a b}] \tag{4.63}$$

where
$$f(c) = 2.303 \log (c_0/c_t) \tag{4.64}$$

A similar equation applies to second-order reactions with

$$f(c) = \frac{2.303}{a_0 - b_0} \log \frac{a_t}{b_t} + \frac{2.303}{a_0 - b_0} \log \frac{b_0}{a_0} \tag{4.65}$$

where a_0 and b_0 are the concentrations of the reactants at the beginning of the experiment, and a_t and b_t their concentrations at time t. The value of the final term of equation 4.63 rapidly tends to zero as k_t becomes greater than k_0. Thus a graph of $\log f(c)$ against $\log (1 + t)$ will be linear from that time after which k_0 is negligible in comparison with k_t. The slope of the line is $(1 + E_a b/R)$, enabling E_a to be determined. The rate constant k_0 may then be calculated from the intercept when $\log (1 + t) = 0$, which is equal to $\log k_0 - \log (1 + E_a b/R)$. The rate constant at any other temperature may be calculated from k_0 and E_a.

Example 4.7

In a study of the first-order decomposition of riboflavine in 0.05 mol ℓ^{-1} NaOH using accelerated storage techniques[65,66], the temperature was programmed to rise from 12.5 to 55 °C using a programme constant, b, of 2.171×10^{-4} K^{-1}. The initial concentration, c_0, of riboflavine was 10^{-4} mol ℓ^{-1}, and the concentration remaining at time t, c_t, was as follows:

t (h)	0.585	0.996	1.512	2.163	2.982	4.013	5.312	6.946
c_t (mol $\ell^{-1} \times 10^{-5}$)	9.881	9.763	9.532	9.109	8.371	6.902	4.931	2.435

Calculate the activation energy and the rate constant at 20 °C.

For first-order reactions the data are plotted according to equations 4.63 and 4.64.

t	$\log(1 + t)$	$\log[2.303 \log(c_0/c_t)]$
0.585	0.2	−1.94
0.996	0.3	−1.62
1.512	0.4	−1.32
2.163	0.5	−1.03
2.982	0.6	−0.75
4.013	0.7	−0.43
5.312	0.8	−0.15
6.946	0.9	+0.15

A plot of $\log[2.303 \log(c_0/c_t)]$ against $\log(1 + t)$ is linear (see figure 4.26) with a slope of 2.95.

From equation 4.63

$$\text{Slope} = 1 + E_a b/R$$

Therefore

$$E_a = \frac{1.95 \times 8.314}{2.171 \times 10^{-4}} = 74.68 \text{ kJ mol}^{-1}$$

Intercept at $\log(1 + t) = 0$ is −2.55.

From equation 4.63

$$\text{Intercept} = \log k_0 - \log\left[1 + \frac{E_a b}{R}\right]$$

where the final term on the right of equation 4.63 is neglected.

Therefore $\quad \log k_0 = -2.55 + \log 2.95 = -2.08$

and $\quad k_0 = 0.0083 \text{ h}^{-1}$

That is the rate constant at temperature T_0 (12.5 °C) is 0.0083 h^{-1}.

The rate constant at 20 °C may then be calculated from the Arrhenius equation in the form of equation 4.59:

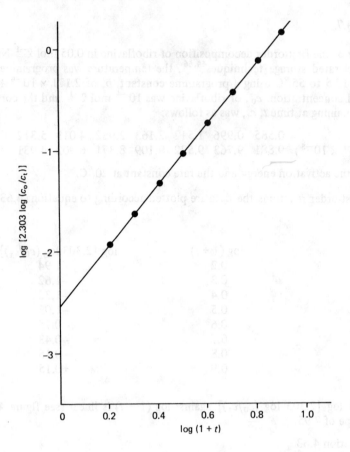

figure 4.26 Accelerated storage plot for the decomposition of riboflavine in 0.05 mol ℓ^{-1} NaOH using data from reference 66

$$\log k_t = -2.08 + \left(\frac{74680}{2.303 \times 8.314}\right)\left(\frac{1}{285.5} - \frac{1}{293}\right)$$

$$\log k_t = -2.08 + 0.3497 = -1.730$$

$$\text{and } k_t = 1.86 \times 10^{-2} \text{ h}^{-1}$$

The rate constant at $20\,^\circ\text{C}$ is thus 0.0186 h^{-1}. ■

The advantages of this method over the conventional method of stability testing are that (a) the data required to calculate the stability are obtained in a single one-day experiment rather than from a series of experiments which may last for several weeks, (b) no preliminary experiments are required to determine the optimum temperatures for the accelerated storage test, and (c) the linearity of the plot of $\log f(c)$ against $\log (1 + t)$ confirms that the correct order of reaction has been assumed.

Using a different theoretical approach, Eriksen and Stelmach[67] derived a general expression which, for cases where k_0 is very small (usually the case with effective drug formulations), reduces to

$$\log f(c) = b \ \frac{E_a t}{2.303R} + \log \frac{Rk_0}{bE_a} \qquad (4.66)$$

Both E_a and k_0 may be found directly from a plot of $\log f(c)$ as a function of time t. Equations 4.63 and 4.66 give the same result when the t_0 value is very small.

Several improvements on the original non-isothermal stability testing methods have been suggested. Rather than subjecting the drug formulation to a predetermined fixed time–temperature profile, Maudling and Zoglio[68] have suggested a method by which the temperature may be changed during the course of the experiment at a rate consistent with the analytical results from the experiment. The resultant time–temperature data are fitted to a polynomial expression of sufficient degree to describe the changes. This relationship and the experimental data are then combined and utilised to compute a series of degradation pathways corresponding to a series of values of activation energy. The curves are matched with the experimental analytical data to obtain the correct activation energy for the reaction. Using this activation energy and the analytical data, the reaction rate and stability may be calculated. Computational procedures whereby the activation energy and frequency factor of the Arrhenius equation may be determined from simple non-isothermal experiments with a fixed temperature–time profile have been described[69,70].

A recent improvement in the design of stability tests, which avoids the difficulties inherent in the non-linear curve-fitting procedures outlined above, has been suggested by Zoglio *et al.*[71]. The experimental procedure involves changing the temperature of the samples being studied until degradation is rapid enough to proceed at a convenient rate for isothermal studies to be carried out. The analytical information obtained during the non-isothermal and isothermal portions of the experiment is utilised in calculating the activation energy and determining the order of reaction and the reaction rate and predicting stability at any required temperature.

Investigations of the influence of temperature on the deterioration of drugs in *suspension* form are complicated by the change in drug solubility with temperature. The concentration of drug in solution usually remains constant because, as the reaction proceeds, more of the suspended drug dissolves and the solution remains saturated with respect to the undegraded reactant. The system thus appears to follow zero-order kinetics; that is, $k_0 = -dc/dt$, where k_0 is the apparent zero-order rate constant. If the actual decomposition of the drug is first-order, we should write

$$-dc/dt = ks \qquad (4.67)$$

Thus $$k_0 = ks \qquad (4.68)$$

where s is the solubility of the drug at the given temperature.

Increase of temperature not only causes the usual increase in k but also an increase in s. In order to calculate the first-order rate constant, k, it is therefore necessary to determine the solubility at various reaction temperatures and then to calculate k from the experimentally determined values of k_0 and s. The stability at room temperature may then be estimated from the enthalpy of activation and the solubility at room temperature[72].

An alternative treatment of data whereby the room temperature stability can be estimated without determining the temperature dependence of solubility of the drug was proposed by Tingstad et al.[73]. The proposed treatment is applicable when (a) the drug degradation takes place only in solution and obeys first-order kinetics, (b) the drug–solvent system obeys the classical relationships of temperature with respect to solubility and the rate of reaction, and (c) the dissolution–degradation kinetics are not limited by dissolution rate.

Taking logarithms of equation 4.68

$$\log k_0 = \log k + \log s \qquad (4.69)$$

The Arrhenius equation may be written in the form

$$\log k = -\frac{E_a}{2.303RT} + \text{constant} \qquad (4.70)$$

A similar relationship exists between solubility and temperature, thus

$$\log s = -\frac{\Delta H_f}{2.303RT} + \text{constant} \qquad (4.71)$$

Substituting equations 4.70 and 4.71 into equation 4.69 gives

$$\log k_0 = -\frac{E_a + \Delta H_f}{2.303RT} + \text{constant} \qquad (4.72)$$

where E_a and ΔH_f are the molar activation energy and the molar heat of fusion, respectively. Thus a plot of $\log k_0$ versus $1/T$ should be linear with a slope of $-[(E_a + \Delta H_f)/2.303R]$. By this method, stability at room temperature can be estimated from studies at elevated temperatures without determining solubilities or first-order rate constants.

4.5.2   Other factors affecting stability

Light

As Rogers[74] has pointed out, the prediction of decomposition rate from experiments in which the effect of light is exaggerated may be complicated by the fact that in many preparations many of the molecules may be masked by others, particularly those away from the surface in solid dosage forms. As a result the illumination may not reach all molecules and consequently there may not be a simple relationship between the flux of light and the degradation rate. Just as the data from accelerated temperature tests are extrapolated to a temperature representative of normal storage conditions, so accelerated tests of light sensitivity are extrapolated to conditions regarded as representative of the light flux which might be expected during storage in the selected containers.

Oxygen

The stability of an oxidisable drug in a liquid dosage form is generally a function of the efficiency of any antioxidant included in the formulation. Exaggeration of the effect of oxygen on stability may be achieved by an increase in the partial pressure of oxygen in the system. It is not often easy, however, to make decisions on what would be the normal access of oxygen during storage and a meaningful extrapolation of the acquired data may be difficult.

Moisture content

The stability of solid dosage forms is usually very susceptible to the moisture content of the atmosphere in the container in which they are stored (see section 4.4.3). Lachman[75] demonstrated a linear relationship between log k and the water vapour pressure for vitamin A palmitate beadlets in sugar-coated tablets. Similarly, a linear relationship between the logarithm of the rate constant for the decomposition of nitrazepam in the solid state and the relative humidity has been established (figure 4.22). The need for consideration of the effect of moisture on stability has been stressed by Carstensen, who stated that stability programmes should always include samples that have been artificially stressed by addition of moisture. One purpose of a stability programme should be to define the stability of the dosage form as a function of moisture content.

References

1. T. Higuchi, A. Havinga and L. W. Busse. *J. Am. pharm. Ass.*, 39, 405 (1950)
2. N. G. Londi and J. E. Christian. *ibid.*, 45, 300 (1956)
3. T. V. Swintosky, E. Rosen, M. J. Robinson, R. E. Chamberlain and J. R. Guarini. *ibid.*, 45, 34, 37 (1956)
4. T. Higuchi and C. Lachman. *ibid.*, 44, 521 (1955)
5. C. P. Hammett. *J. Am. chem. Soc.*, 59, 96 (1937)
6. P. R. Wells. *Chem. Rev.*, 63, 171 (1963)
7. J. T. Carstensen, E. G. Serenson and J. J. Vance. *J. pharm. Sci.*, 53, 1547 (1964)
8. E. H. Farmer. *Trans. Farad. Soc.*, 38, 340 (1942)
9. S. Y. Yeh and J. L. Lach. *J. pharm. Sci.*, 50, 35 (1961)
10. L. C. Schroeter and T. Higuchi. *J. Am. pharm. Ass.*, 47, 426 (1958)
11. J. R. D. McCormick *et al. J. Am. chem. Soc.*, 79, 2849 (1957)
12. E. G. Remmers, G. M. Sieger and A. D. Doerschuk. *J. pharm. Sci.*, 52, 752 (1963)
13. M. A. Nunes and E. Brochmann-Hanssen. *ibid.*, 63, 716 (1974)
14. R. W. Lehman, J. M. Dieterle, W. T. Fisher and S. R. Ames. *J. Am. pharm. Ass.*, 49, 363 (1960)
15. F. H. Merkle and C. A. Discher. *J. pharm. Sci.*, 53, 620 (1964)
16. C. L. Huang and F. L. Sands. *ibid.*, 56, 259 (1967)
17. Y. Matsuda, H. Inouye and R. Nakanishi. *ibid.*, 67, 196 (1978)
18. H. Bundgaard. *Acta pharm. suec.*, 13, 9 (1976)

19. J. M. Dewdney, H. Smith and A. W. Wheeler. *Immunology*, 21, 517, 527 (1971)
20. H. Bundgaard. *Acta pharm. suec.*, 14, 67 (1977)
21. T. Eble and E. R. Garrett. *J. Am. pharm. Ass.*, 43, 536 (1954)
22. L. J. DeMerre and C. Wilson. *ibid.*, 45, 129 (1956)
23. D. Nair and J. Lach. *ibid.*, 48, 390 (1959)
24. W. Mayer, S. Erbe and R. Voigt. *Pharmazie*, 27, 32 (1972)
25. H. V. Maudling, J. P. Nazareno, J. E. Pearson and A. F. Michaels. *J. pharm. Sci.*, 64, 278 (1975)
26. J. Haynes, J. T. Carstensen, J. Calahan and R. Card. *Stevens Symp. statist. Meth. chem. Ind.*, 3, 1 (1959)
27. J. T. Carstensen. *J. pharm. Sci.*, 53, 839 (1964)
28. R. Tardiff. *ibid.*, 54, 281 (1965)
29. G. Enezian. *Rev. prod. Probl. Pharm.*, 23, 1 (1968)
30. J. Campbell. *Can. med. Ass. J.*, 68, 103 (1953)
31. J. Campbell. *J. Am. pharm. Ass.*, 44, 598 (1955)
32. J. Campbell and H. McLeod. *ibid.*, 44, 263 (1955)
33. J. Bird and R. Shelton. *ibid.*, 39, 500 (1950)
34. S. Seth and H. Mital. *Indian J. Pharm.*, 27, 119 (1965)
35. J. T. Carstensen. *J. pharm. Sci.*, 63, 1 (1974)
36. P. W. M. Jacobs and F. C. Tompkins. In *Chemistry of the Solid State* (ed. W. E. Garner), Butterworths, London, 1955, p. 201
37. E. Nelson, D. Eppich and J. T. Carstensen. *J. pharm. Sci.*, 63, 755 (1974)
38. E. Prout and F. C. Tompkins. *Trans. Farad. Soc.*, 40, 448 (1944)
39. S. S. Kornblum and B. J. Sciarrone. *J. pharm. Sci.*, 53, 935 (1964)
40. C. Larsen and H. Bundgaard. *Arch. pharm. Chem.*, 5, 66 (1977)
41. R. E. Lindsay and S. C. Helm. *J. pharm. Sci.*, 61, 202 (1972)
42. J. T. Carstensen. *ibid.*, 59, 1141 (1970)
43. P. B. Sheth and E. L. Parrott. *ibid.*, 56, 983 (1967)
44. I. K. Winterborn, B. J. Meakin and D. J. G. Davies. *ibid.*, 63, 64 (1974)
45. J. E. Carless and J. R. Nixon. *J. Pharm. Pharmacol.*, 9, 963 (1957)
46. J. E. Carless and J. R. Nixon. *ibid.*, 12, 348 (1960)
47. J. T. H. Ong and H. B. Kostenbauder. *J. pharm. Sci.*, 64, 1378 (1975)
48. A. A. Allen and V. Das Gupta. *ibid.*, 63, 107 (1974)
49. V. Das Gupta. *ibid.*, 67, 299 (1978)
50. M. J. Busse. *Pharm. J.*, 220, 25 (1978)
51. E. Ullmann, K. Thoma and G. Zelfel. *Pharm. Acta helv.*, 38, 577 (1963)
52. B. Testa and J. C. Etter. *Can. J. pharm. Sci.*, 10, 16 (1975)
53. R. I. Poust and J. C. Colaizzi. *J. pharm. Sci.*, 57, 2119 (1968)
54. J. T. Carstensen, J. Johnson, D. Spera and M. Frank. *ibid.*, 57, 23 (1968)
55. S. S. Kornblum and M. Zoglio. *ibid.*, 56, 1569 (1967)
56. L. Leeson and A. Mattocks. *J. Am. pharm. Ass.*, 47, 329 (1958)
57. J. Tingstad and J. Dudzinski. *J. pharm. Sci.*, 62, 1856 (1973)
58. B. G. Boggiano, R. Drew and R. D. Hancock. *Aust. J. Pharm.*, 51, S14 (1970)
59. M. A. Zoglio, H. V. Maudling, R. M. Haller and S. Briggen. *J. pharm. Sci.*, 57, 1877 (1968)
60. H. V. Maudling, M. A. Zoglio, F. E. Pigois and M. Wagner. *ibid.*, 58, 1359 (1969)

61. H. W. Jun, C. W. Whitworth and L. A. Luzzi. *ibid.*, 61, 1160 (1972)
62. C. W. Whitworth, L. A. Luzzi, B. B. Thompson and H. W. Jun. *ibid.*, 62, 1372 (1973)
63. R. Ekman, L. Liponkoski and P. Kahela. *Acta pharm. suec.*, 19, 241 (1982)
64. N. G. Londi and M. W. Scott. *J. pharm. Sci.*, 54, 531 (1965)
65. A. K. Amirjahed. *ibid.*, 66, 785 (1977)
66. A. R. Rogers. *J. Pharm. Pharmacol.*, 15, 101T (1963)
67. S. P. Eriksen and H. Stelmach. *J. pharm. Sci.*, 54, 1029 (1965)
68. H. V. Maudling and M. A. Zoglio. *ibid.*, 59, 333 (1970)
69. B. W. Madsen, R. A. Anderson, D. Herbison-Evans and W. Sneddon. *ibid.*, 63, 777 (1974)
70. A. I. Kay and T. H. Simon. *ibid.*, 60, 205 (1971)
71. M. A. Zoglio, H. V. Maudling, W. H. Streng and W. C. Vincek. *ibid.*, 64, 1381 (1975)
72. J. V. Swintosky, E. Rosen, M. J. Robinson, R. E. Chamberlain and J. R. Guarini. *J. Am. pharm. Ass.*, 45, 37 (1956)
73. J. Tingstad, J. Dudzinski, L. Lachman and E. Shami. *J. pharm. Sci.*, 62, 1361 (1973)
74. A. R. Rogers. *Pestic. Sci.*, 1, 266 (1970)
75. L. Lachman. *J. pharm. Sci.*, 54, 1519 (1965)

5 Solubility and Partitioning of Drugs

An understanding of the process of solution and the factors governing the solubility of drugs and adjuvants is important in pharmaceutics for several reasons. Drugs must sometimes be formulated as solutions or may be added in powder or solution form to liquids such as infusion fluids where they must be maintained in solution. In whatever way drugs may be presented to the body they must nearly always be in a molecularly dispersed form, that is, they must have dissolved, before absorption can occur across biological membranes. The process of solution has therefore frequently to precede absorption unless the drug is administered in solution; even ingested solutions may form precipitates in the gut contents, and the drug will then have to redissolve. Drugs of low aqueous solubility frequently present problems in relation to their bioavailability.

Aqueous solvents are the most common in pharmaceutical and biological systems so this chapter is concerned mainly with solution in simple and mixed aqueous solvents. The solution of drugs in non-aqueous solvents is also considered because of the many pharmaceutical applications of non-aqueous and heterogeneous systems, and because of the need to understand the process of transport of drugs across biological and artificial membranes. A primary factor in passive membrane transport is the solubility of the drug molecule in the liquid environment of the membrane interior.

A solution is a system in which a solute is molecularly dispersed in a solvent, the solvent usually being the predominant species. When a solution contains a solute at the limit of its solubility at any given temperature and pressure the solution is said to be saturated. If the solubility limit is exceeded, solid particles of solute may be present and the solution phase will be in equilibrium with the solid. Under certain circumstances supersaturated solutions may be prepared.

The over-all solubility of a drug is not only of intrinsic interest but it is of interest as it affects its rate of dissolution, that is, the rate at which the drug enters the solution state from the solid state. As can be seen later, the rate of solution of a substance that dissolves in a solvent without chemical reaction is directly related to the solubility of the drug.

5.1 Expressions of solubility

The solubility of a solute in a solvent can be expressed quantitatively in several ways (see chapter 3, section 3.1). All expressions are readily interconvertible. Other less-specific forms of noting solubility include parts per parts of solvent (for example, parts per million, ppm). The British Pharmacopoeia and other chemical and pharmaceutical compendia frequently use this form and also the expressions insoluble, very slightly soluble, and soluble. These are imprecise, and for quantitative work it is important that more specific concentration terms are used. Most substances have some degree of solubility in water and while they may appear to be 'insoluble' by a qualitative test, their solubility can be measured and quoted precisely. In aqueous media at pH 10 chlorpromazine base has a solubility of 8×10^{-6} mol ℓ^{-1}. It is thus very slightly soluble and perhaps might be considered insoluble if the term was judged by the disappearance of solid placed in a test-tube.

5.2   Prediction of solubility

A diagrammatic representation of the processes involved in the dissolution of a crystalline solute is provided in figure 5.1. The positioning of a solute molecule in the solvent cavity requires greater solute–solvent contact, the larger the solute molecule. If the surface area of the solute molecule is A, the solute–solvent interface increases by $\sigma_{12}A$, where σ_{12} is the interfacial tension between the solvent and the solute. σ is a parameter not readily obtained for interfaces on the molecular scale but reasonable estimates are possible from values obtained at macroscopic interfaces. A predictive theory of solubility based on interfacial area has nevertheless been developed[1-5].

5.2.1   Molecular surface area approach[1-5]

The number of solvent molecules that can pack around the solute molecule is considered in calculations of the thermodynamic properties of the solution. The molecular surface area of the solute is therefore a key parameter and Hermann[4,5] obtained good correlation between solubility in water and this parameter. Most drugs are not simple non-polar hydrocarbons and we must consider in this chapter polar molecules and weak organic electrolytes. The term w_{12} in figure 5.1, which represents solute–solvent interactions, has for these molecules to be further divided to encompass the interaction involving the non-polar part of the molecule and the interactions involving the polar portion. The molecular surface area of each portion should therefore be considered separately.

The free energy change in placing the solute in the solvent cavity is $-\sigma_{12}A$. Indeed it can be shown that the reversible work of solution is $(\tfrac{1}{2}w_{11} + \tfrac{1}{2}w_{22} - w_{12}) \times A$. Implicit in the derivation is that the solution formed be dilute, so that solute–solute interactions in solution do not obtrude.

To correlate calculated solubilities and observed solubilities, the latter values, which refer to compounds in different states and with different boiling points, must be multiplied by the factor $\exp(\Delta H_f T_m - T)/(R T_m T)$ to convert the

table 5.1   **Experimental aqueous solubilities, boiling points, surface areas and predicted solubilities***

Compound	Solubility (molal)	A (nm^2)	Boiling point (°C)	Predicted solubilities (molal)
n-Butanol	1.006	2.721	117.7	0.821
n-Pentanol	2.5×10^{-1}	3.039	137.8	2.09×10^{-1}
n-Hexanol	6.1×10^{-2}	3.357	157	5.32×10^{-2}
n-Heptanol	1.55×10^{-2}	3.675	176.3	1.36×10^{-2}
Cyclohexanol	3.83×10^{-1}	2.905	161	4.3×10^{-1}
1-Nonanol	1×10^{-3}	4.312	213.1	8.8×10^{-4}

*From reference 1

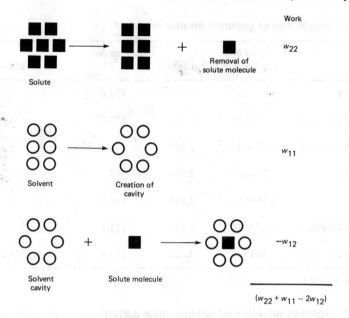

figure 5.1 Diagrammatic representation of the three processes involved in the dissolution of a crystalline solute. The expression for the work involved is $w_{22} + w_{11} - 2w_{12}$. Solute–solvent interaction in the last stage is $-2w_{12}$ as bonds are made with one solute and two solvent molecules.

values into those for 'supercooled' liquids. For 55 compounds (a selection of which are quoted in table 5.1) it was found that

$$\ln S = -4.3A + 11.78 \qquad (5.1)$$

where S is the molal solubility, and A is the total surface area in nm^2.

5.2.2 Structural features of simple molecules and aqueous solubility

Interactions between non-polar groups and water have been discussed in earlier sections, where the importance of both size and shape was indicated. Chain branching of hydrophobic groups influences aqueous solubility, this being readily demonstrated in a series of alcohols (table 5.2). The boiling point correlates with total surface area, and obviously in a large range of compounds we can detect the trend of decreasing aqueous solubility with increasing boiling point. Boiling points of liquids and melting points of solids are indicators of molecular cohesion. These and other empirical correlations are useful. Melting points, even of compounds which form non-ideal solutions, can be used as a guide to order of solubility in a closely related series, as can be seen in the properties of sulphonamide derivatives listed in table 5.3. Boiling points and melting points are related to the cohesion in pure solute liquid and crystalline states and, provided that solute–solvent interactions do not mask the contribution from w_{22}, then these parameters will give a useful guide to sequence of solubility.

table 5.2 **Solubilities of pentanol isomers in water***

	Solubility (molality, m)	A (nm^2)	Boiling point ($^\circ$C)	Structure
n-Pentanol	2.6×10^{-1}	3.039	137.8	
3-Methyl-1-butanol	3.11×10^{-1}	2.914	131.2	
2-Methyl-1-butanol	3.47×10^{-1}	2.894	128.7	
2-Pentanol	5.3×10^{-1}	2.959	119	
3-Pentanol	6.15×10^{-1}	2.935	115.3	
3-Methyl-2-butanol	6.67×10^{-1}	2.843	111.5	
2-Methyl-2-butanol	1.403	2.825	102.0	

*From reference 1

table 5.3 **Aqueous solubility of sulphonamide derivatives**

Compound	Melting point ($^\circ$C)	Solubility
Sulphadiazine	253	1 g in 13 ℓ (0.077 gℓ^{-1})
Sulphamerazine	236	1 g in 5 ℓ (0.20 gℓ^{-1})
Sulphapyridine	192	1 g in 3.5 ℓ (0.29 gℓ^{-1})
Sulphathiazole	174	1 g in 1.7 ℓ (0.59 gℓ^{-1})

The influence of substituents on the solubility of agents in water can be due to their effect on the properties of the pure solute (for example, on molecular cohesion) or to the effect of the substituent on interaction with water molecules. It is not easy to predict what effect a particular substituent will have on crystal properties, but as a guide to the solvent interactions, substituents can be classified as either hydrophobic or hydrophilic, depending on their polarity (see table 5.4). It should be remembered that the position of the substituent on the molecule can influence its effect. For example, the solubilities of *o*-, *m*- and *p*-dihydroxy-benzenes, all, as expected, much greater than the solubility of benzene in water, are not identical, being 4, 9 and 0.6 mol ℓ^{-1}, respectively. The relatively low solubility of the *para* compound is perhaps due to the greater stability of the solid state. The melting points of the derivatives indicate this is so, as they are 105, 111 and 170°C, respectively. The possibility of intramolecular hydrogen bonding in aqueous solutions in the case of the *ortho* derivative, decreasing its ability to interact with water, may explain why its solubility is lower than that of the *meta* analogue.

One can best illustrate the use of the information in table 5.4 by considering the solubility of a series of substituted acetanilides, data for which are provided in table 5.5. The strong hydrophilic characteristics of polar groups capable of

table 5.4 **Substituent group classification in relation to water solubility**

Substituent	Classification
$-CH_3$	Hydrophobic
$-CH_2-$	Hydrophobic
$-Cl, -Br, -F$	Hydrophobic
$-N(CH_3)_2$	Hydrophobic
$-SCH_3$	Hydrophobic
$-OCH_2CH_3$	Hydrophobic
$-OCH_3$	Slightly hydrophilic
$-NO_2$	Slightly hydrophilic
$-CHO$	Hydrophilic
$-COOH$	Slightly hydrophilic
$-COO^-$	Very hydrophilic
$-NH_2$	Hydrophilic
$-NH_3^+$	Very hydrophilic
$-OH$	Very hydrophilic

table 5.5 **The effect of substituents on solubility of acetanilide derivatives in water**

	X	Solubility (mg ℓ^{-1})
NHCOCH₃ (structure)	H	6.38
	Methyl	1.05
	Ethoxyl	0.93
	Hydroxyl	13.9
	Nitro	15.98
	Aceto	9.87

hydrogen bonding with water molecules are evident. The presence of hydroxyl groups can therefore markedly change the solubility characteristics of a compound. Phenol is 100 times more soluble in water than is benzene. In the case of phenol, where there is considerable hydrogen bonding capability, the solute-solvent interaction (w_{12}) outweighs the other factors (w_{22}, w_{11}) in the solution process. But bearing in mind the steric limitations, it is obvious that the position of the substituent on the parent molecule will affect its contribution to solubility.

Steroid solubility

The steroids as a group tend to be poorly soluble in water. This is not surprising in view of their bulk. The complexity of their structure makes prediction of solubility somewhat difficult, but one can generally rationalise the solubility values attributed to members of a series of related steroids. Table 5.6 gives aqueous solubility data for several steroids.

The substitution of an ethinyl group has conferred increased solubility on the oestradiol molecule, as would be expected. Oestradiol benzoate with the 3-OH

table 5.6 **Steroid structure and solubility in water***

Structure	Compound	Solubility ($\mu g\,m\ell^{-1}$)
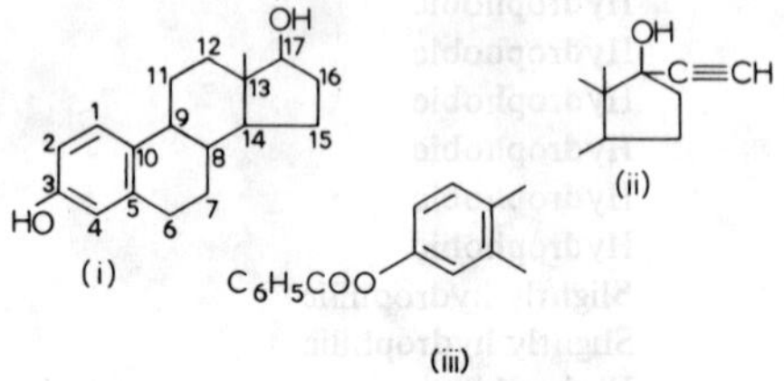 (i), (ii), (iii)	Oestradiol (i)	5
	Ethinyloestradiol (ii)	10
	Oestradiol benzoate (iii)	0.4
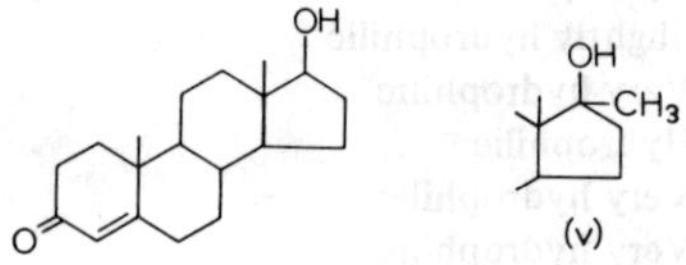 (iv), (v)	Testosterone (iv)	24
	Testosterone propionate	0.4
	Methyltestosterone (v)	32
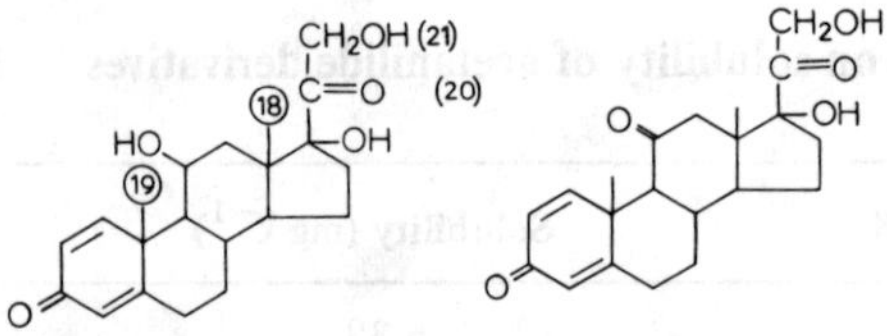 (vi), (vii)	Prednisolone (vi)	215
	Prednisone (vii)	115
	Prednisone acetate	23
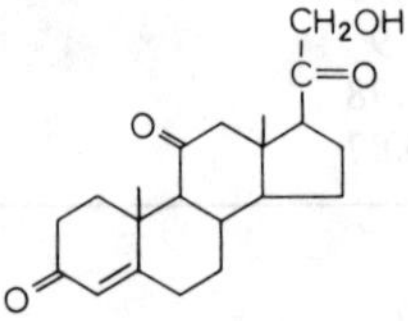 (viii)	Cortisone (viii)	230
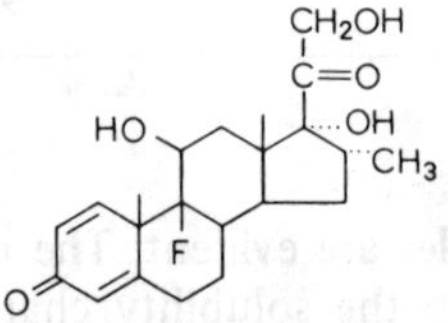 (ix)	Dexamethasone (ix)	84
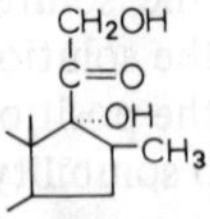 (x)	Betamethasone (x)	58
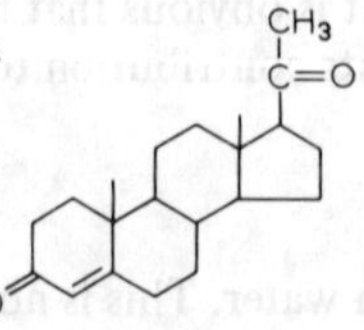 (xi)	Progesterone (xi)	9
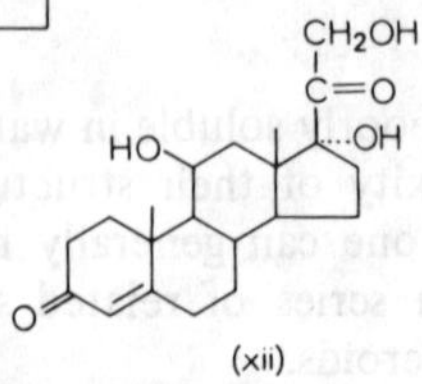 (xii)	Hydrocortisone (xii)	285

*From reference 6

substituted is much less soluble than the parent oestradiol because of the loss of this hydroxyl group and its substitution with a hydrophobic group. The same relationships are seen in testosterone and testosterone propionate. As both oestradiol benzoate and testosterone propionate are oil soluble they are used as solutions in oils such as castor oil and sesame oil for intramuscular and subcutaneous injection (see section 9.4). Methyltestosterone might be expected to be less soluble in water than testosterone but in fact it is not; this demonstrates the importance of crystal properties in determining solubility. The methyl compound is more soluble because of the smaller heat of fusion of this derivative, hence it more readily 'disintegrates' in the solvent.

Dexamethasone and betamethasone are isomeric fluorinated derivatives of methylprednisolone. Solubilities are not identical and this might be a crystal or a solution property. A more simple example of differences in isomeric solubility is that of the *o*-, *m*-, and *p*-dihydroxybenzenes referred to above. A steric argument may be applied to the case of dexamethasone, water molecules being less able to move close to the 17-OH group, when compared with betamethasone.

5.3 Hydration and solvation

5.3.1 Hydration of non-electrolytes

'Solvation' is the general term that describes the process of binding of solvent to solute molecules. If the solvent is water the process is 'hydration'. In solutions of sucrose (I), for example, six water molecules are bound to each sucrose molecule with such avidity that water and sucrose move as a unit in solution. The extent of hydration can indeed be measured by hydrodynamic techniques. The size of the hydrodynamic unit is larger than the crystallographically determined size, there being approximately one water molecule for each hydroxyl group of sucrose.

It has become apparent that chemically very similar molecules such as mannitol (II), sorbitol (III) and inositol show very different hydration behaviour. It may be noted that the solubility of sorbitol is about 3.5 times that of mannitol in water. Most favourable hydration occurs when there is an equatorial −OH group on pyranose sugars[7]. This is thought to be due to the compatibility of the equatorial −OH with the organised structure of water in bulk. Axial hydroxyl groups cannot bond on to the water 'lattice' without causing considerable distortion of the lattice. This may be one explanation of the difference, although difference in the lattice energies of the crystal may also contribute.

(I)Sucrose   (II)Mannitol   (III)Sorbitol

5.3.2 Hydration of ionic species

The study of ionic solvation is complicated, but it is relevant to the present topic because of the effect ions have on the solubility of other species. The forces between cations and water molecules are so strong that the cations may retain a layer of water molecules in their crystals. The effect of ions on water structure is complex and variable. All ions in water, however, possess in the immediate vicinity of the ion a layer of tightly bound water — the water molecules being directionally orientated. Four water molecules are in the bound layer of most monovalent, monatomic ions. The firmly held layer can be regarded as being in a 'frozen' condition around a positive ion. The water molecules could be orientated with all the hydrogens of the water molecules pointing outwards. Because of this and because of steric/size problems, they cannot all participate in the normal tetrahedral arrangements of bulk water. For this to be feasible two of the water molecules must be orientated with the hydrogens of the water molecules pointing in towards the ion. Inevitably with cations and many small anions there tends to be a layer of water around the bound layer which is less ordered than bulk water (figure 5.2). Such ions, which include all the alkali and halide ions except Li^+ and F^-, are called 'structure breakers'. Size of the ion is important, as the surface area of the ion determines the constraints on the polarised water molecules. Many polyvalent ions, for example Al^{3+}, increase the structuredness of water beyond the immediate hydration layer, and are therefore 'structure makers'. Hydration numbers (the number of water molecules in the primary hydration layer) can be determined by various physical techniques (for example, compressibility) and they tend to differ depending on the choice of method. This is because the situation is dynamic — the hydration number can be considered to be the effective number of solvent molecules 'permanently' bound to the ion and to follow it. The over-all total action of the ion on water may be replaced conceptually by a strong binding between ion and some effective number (solvation number) of solvent molecules; this effective number may well be almost zero in the case of large ions such as iodide, caesium and tetraalkylammonium ions. The solvation numbers decrease with increase of ionic radius because the ionic force field diminishes with increasing radius, and consequently water molecules are less inclined to be abstracted from their position in bulk water.

5.3.3 Hydrophobic hydration

Water is associated in a dynamic manner with non-polar groups, but only in rare cases where crystalline clathrates can be formed is this water ever able to be isolated along with the hydrophobic groups. Some authors have spoken of hydrophobic hydration but it is perhaps a term which should not be used because of the possibility of misconception. It has been observed that the motion of water molecules is slowed down in the vicinity of non-polar groups. In the discussion of hydrophobic bonding (section 6.3.1) and non-polar interactions this special relationship between water and hydrocarbon chains is elaborated on.

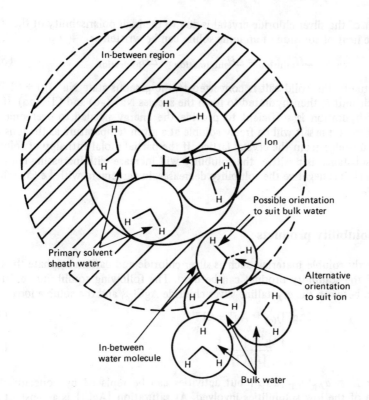

figure 5.2 Schematic diagram to indicate that, in the (hatched) region between the primary solvated ion and bulk water, the orientation of the 'in-between' water molecules must compromise between that which suits the ion (oxygen facing ion) and that which suits the bulk water (hydrogen facing ion). From J. O'M. Bockris and A. K. N. Reddy. *Modern Electrochemistry*, vol. 1, Macdonald, London, 1970 with permission

5.4 The solubility of inorganic materials in water

While few therapeutic agents are inorganic electrolytes, it is nevertheless pertinent to consider the manner of their interaction with water. Electrolytes are, of necessity, components of replacement fluids, injections and eye drops and many other formulations. What determines the solubility of a simple salt such as NaCl and its solubility in relation to another chloride, AgCl? The solubility of sodium chloride is in excess of 5 mol ℓ^{-1} while the solubility of AgCl is 500 000 times less. The heats of solution are 62.8 kJ mol^{-1} for silver chloride and 4.2 kJ mol^{-1} for sodium chloride, suggesting a substantial difference in either the crystal properties or in the interaction of the ions with water. In fact the very great strength of the silver chloride crystal is due to the high polarisability of the silver ion. The heat of solution of an ionic solute can be written as

$$\Delta H_{solution} = \Delta H_{sublimation} - \Delta H_{hydration} \qquad (5.2)$$

Conceptually the solid salt is converted to the gaseous state Na^+ (g) + Cl^- (g) and each unit is then hydrated to form the species Na^+ (aq) and Cl^- (aq). If the heat of hydration is sufficient to provide the energy needed to overcome the lattice forces, the salt will be freely soluble at a given temperature as the ions will readily dislodge from the crystal lattice. If the partial molal enthalpy of solution of the substance is positive, the solubility will increase with increasing temperature, and if it is negative the solubility decreases, in agreement with Le Chatelier's principle.

5.5 Solubility products

For poorly soluble materials such as silver chloride and barium sulphate the concept of the solubility product can be used. The following equilibrium exists in solution between the crystalline silver chloride $AgCl_{(c)}$ and the soluble ions

$$(AgCl)_{(c)} \rightleftharpoons Ag^+ + Cl^- \qquad (5.3)$$

$$K = \frac{[Ag^+]\,[Cl^-]}{[AgCl]} \qquad (5.4)$$

Strictly $K = a_{Ag^+}a_{Cl^-}/a_{AgCl}$ but activities can be replaced by concentrations because of the low solubilities involved. At saturation [AgCl] is a constant and very small and finite. The solubility product, K_{sp}, may therefore be written

$$K_{sp} = [Ag^+]\,[Cl^-] \qquad (5.5)$$

The solubility product is useful for evaluating the influence of other species on the solubility of salts of low aqueous solubility. Some values of solubility products are quoted in table 5.7.

table 5.7 **Solubility products of some inorganic salts**

Compound	K_{sp} (mol^2 ℓ^{-2})
AgCl	1.25×10^{-10}
Al(OH)$_3$	7.7×10^{-13}
BaSO$_4$	1.0×10^{-10}

5.6 The effect of additives on solubility: the solubility coefficient

Additives may either increase or decrease the solubility of a solute in a given solvent. Those salts that increase solubility are said to 'salt in' the solute and those that decrease solubility 'salt out' the solute. The effect of an additive depends very much on the influence it has on the structure of water or its ability to compete with solvent water molecules.

A convenient quantification of the effect of a solute additive on the solubility of another solute may be obtained by the Setschenow equation

$$\log \frac{S_\bullet}{S} = kc_a \tag{5.6}$$

where S is the solubility in the presence of an additive, c_a is the concentration of additive, $S_\bullet$ the solubility in its absence, and k the salting coefficient. The Setschenow equation frequently holds up to additive concentrations of 1 mol ℓ^{-1}. The salting coefficient k is a measure of the sensitivity of the activity coefficient of the solute towards the salt. The sign of k is positive when the activity coefficient is increased; it is negative if the activity coefficient is decreased by the additive. Several salts with large anions or cations which are themselves very soluble in water result in a salting in of non-electrolytes. Sodium benzoate and sodium *p*-toluenesulphonate are good examples of such agents, called 'hydrotropic salts', which increase the solubility of other solutes — a phenomenon known as 'hydrotropy'. Values of k for three salts added to benzoic acid in aqueous solution are 0.17 for NaCl; 0.14 for KCl; and −0.22 for sodium benzoate. That is, NaCl and KCl decrease the solubility of benzoic acid, and sodium benzoate increases solubility.

5.7 Solubility of weak electrolytes

The great majority of drugs are organic electrolytes and there are therefore four parameters which determine their solubility: (i) their degree of ionisation; (ii) their molecular size; (iii) interactions of substituent groups with solvent; and (iv) their crystal properties. In this section consideration is given to the solubility of weak electrolytes and the influence of pH on aqueous solubility.

5.7.1 Acidic drugs

Acidic drugs such as the barbiturates, non-steroidal anti-inflammatories, nitrofurantoin, and phenylbutazone, etc., are less soluble in acidic solutions than in alkaline solutions because the predominant undissociated species cannot interact with water molecules to the same extent as the ionised form which is readily hydrated. If we represent the drug as HA and let the total saturation solubility of the drug be S, with S_0 the solubility of the undissociated species HA, we can write

$$S = S_0 + (\text{concentration of ionised species})$$

The dissociation of the acid in water can be written

$$HA + H_2O \rightleftharpoons H_3O^+ + A^- \tag{5.7}$$

and the dissociation constant K_a is given by

$$K_a = \frac{[H_3O^+]\,[A^-]}{[HA]} \tag{5.8}$$

Rearranging and substituting S_0 for [HA]

$$\frac{K_a}{[H_3O^+]} = \frac{[A^-]}{S_0} \tag{5.9}$$

As $[A^-] = S - S_0$

$$\frac{K_a}{[H_3O^+]} = \frac{S - S_0}{S_0} \tag{5.10}$$

Taking logarithms

$$\boxed{pH - pK_a = \log\left\{\frac{S - S_0}{S_0}\right\}} \tag{5.11}$$

Hence the solubility of the drug at any pH can be calculated provided the pK_a and S_0 are known.

5.7.2 Basic drugs

Basic drugs such as chlorpromazine are, on the contrary, more soluble in acidic solution where the ionised form of the drug is predominant. If S_0 is the solubility of an undissociated base, RNH_2, the expression for the solubility (S) as a function of pH can be obtained similarly as follows.

$$RNH_2 + H_2O \rightleftharpoons RNH_3^+ + OH^- \tag{5.12}$$

$$K_b = \frac{[RNH_3^+][OH^-]}{[RNH_2]} \tag{5.13}$$

That is $\dfrac{K_b}{[OH^-]} = \dfrac{[H^+]}{K_a} = \dfrac{S - S_0}{S_0} \tag{5.14}$

Taking logarithms and writing the equations in terms of K_a we get

$$pK_a - pH = \log\left\{\frac{S - S_0}{S_0}\right\} \tag{5.15}$$

or $\boxed{pH - pK_a = \log\left\{\dfrac{S_0}{S - S_0}\right\}} \tag{5.16}$

Solubility–pH profiles of a number of acidic and basic drugs are plotted in figure 5.3.

5.7.3 Amphoteric drugs

Several drugs are amphoteric, displaying both basic and acidic characteristics. Amino acids are of course amphoteric substances, and probably the most frequently encountered drugs in this category are the sulphonamides and the tetracyclines. If, for simplicity, one were to use a generalised structure

$$R-\underset{\underset{\displaystyle NH_2}{|}}{X}-COOH$$

and write out the solution equilibrium between the species, one would obtain, as before, equations relating solubility to pH. The equilibria for an amphoteric compound in solution are then

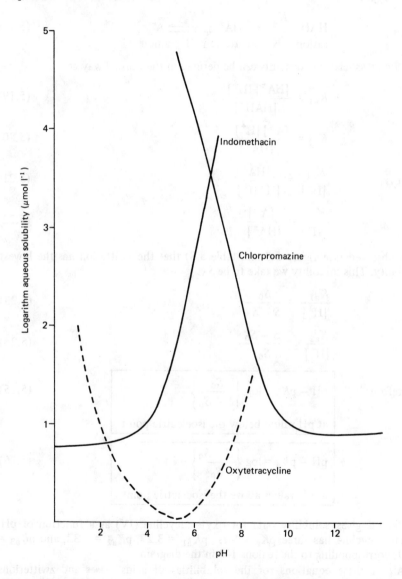

figure 5.3 Solubility of indomethacin, chlorpromazine and oxytetracycline as a function of pH, plotted as logarithm of the solubility in μmol ℓ^{-1}

$$RX\ COOH \xrightleftharpoons[H^+]{K_{a1}} RX\ COO^- \xrightleftharpoons[H^+]{K_{a2}} RX\ COO^- \qquad (5.17)$$

$$\underset{\text{(at low pH)}}{\overset{|}{NH_3^+}} \qquad \overset{|}{NH_3^+} \qquad \underset{\text{(at high pH)}}{\overset{|}{NH_2}}$$

More simply

$$\underset{\text{cation}}{HAH^+} \xrightleftharpoons[H^+]{K_{a1}} \underset{\text{zwitterion}}{HA^{\pm}} \xrightleftharpoons[H^+]{K_{a2}} \underset{\text{anion}}{A^-} \qquad (5.18)$$

The two dissociation constants can be defined in the normal way as

$$K_{a1} = \frac{[HA^{\pm}][H^+]}{[HAH^+]} \qquad (5.19)$$

and

$$K_{a2} = \frac{[A^-][H^+]}{[HA^{\pm}]} \qquad (5.20)$$

That is

$$\frac{K_{a1}}{[H^+]} = \frac{[HA^{\pm}]}{[HAH^+]} \qquad (5.21)$$

and

$$\frac{K_{a2}}{[H^+]} = \frac{[A^-]}{[HA^{\pm}]} \qquad (5.22)$$

It is observed (see figure 5.3 and table 5.8) that the zwitterion has the lowest solubility. This solubility we take to be S_0.

$$\frac{K_{a1}}{[H^+]} = \frac{S_0}{S - S_0} \qquad (5.23)$$

and

$$\frac{K_{a2}}{[H^+]} = \frac{S - S_0}{S_0} \qquad (5.24)$$

Therefore

$$\boxed{pH - pK_a = \log\left\{\frac{S_0}{S - S_0}\right\}} \qquad (5.25)$$

at pH values below the isoelectric point

and

$$\boxed{pH - pK_a = \log\left\{\frac{S - S_0}{S_0}\right\}} \qquad (5.26)$$

at pH values above the isoelectric point

Table 5.8 gives solubility data for oxytetracycline (IV) as a function of pH. Oxytetracycline has three pK_a values: $pK_{a1} = 3.27$, $pK_{a2} = 7.32$, and $pK_{a3} = 9.11$, corresponding to the regions 1–3 in the diagram.

All of these equations for the solubilities of acids, bases and zwitterions (equations 5.11, 5.16, 5.25 and 5.26) can be used to calculate the pH at which drugs will precipitate from solution of a given concentration (or the concentration at which a drug will reach its maximum solubility at a given pH); this is

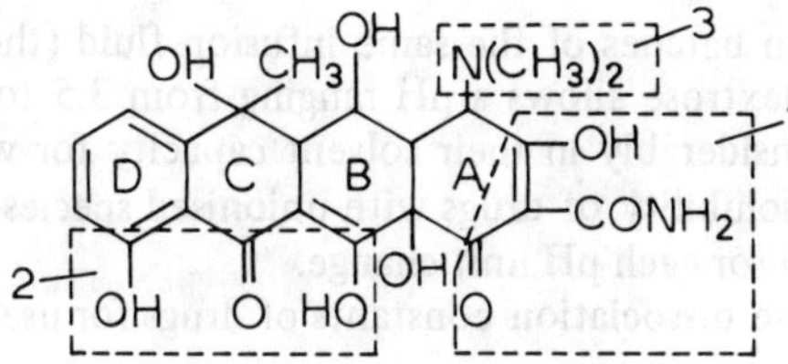

Oxytetracycline

(IV)

table 5.8 **Oxytetracycline: pH dependence of solubility at 20 °C**

pH	Solubility $(g \ \ell^{-1})$
1.2	31.4
2	4.6
3	1.4
4	0.85
5	0.5
6	0.7
7	1.1
8	28.0
9	38.6

*Data from the *United States Dispensatory*, 25th Edition

especially important in determining the maximum allowable levels of a drug in infusion fluids or formulations. Some idea of the range of pH values encountered in common infusion fluids is given in table 5.9. The variation in pH between

table 5.9 **pH of some parenteral solutions***

Solution	pH
5% Dextrose in water (5% D/W)	4.40, 4.70
5% D/W (1 ℓ containing 2 mℓ of vitamins)	4.30, 4.38
5% D/W (1 ℓ containing 100 mg thiamine hydrochloride)	3.90, 3.96
5% D/W (1 ℓ containing 300 mg thiamine hydrochloride)	3.82, 4.00
5% D/W (1 ℓ containing 2 mℓ of vitamins, 200 mg thiamine hydrochloride)	4.15
5% D/W (1 ℓ containing 4 mℓ of vitamins, 300 mg thiamine hydrochloride)	4.28
Normal saline	5.35, 5.40
Lactated Ringer's solution (Ringer)	7.01
Ringer (1 ℓ containing 2 mℓ of vitamins)	5.50
Ringer (1 ℓ containing 2 mℓ of vitamins, 100 mg thiamine hydrochloride)	5.38
Ringer (1 ℓ containing 2 mℓ of vitamins, 200 mg thiamine hydrochloride)	5.16

*From R. L. Tse and M. W. Lee. *J. Am. med. Ass.* 215, 642 (1971)

preparations and within batches of the same infusion fluid (the British Pharma-copoeia standard for dextrose allows a pH ranging from 3.5 to 5.5) means that the fluids will vary considerably in their solvent capacity for weak electrolytes. As a rough guide the solubility of drugs with unionised species of low solubility varies by a factor of 10 for each pH unit change.

A compilation of the dissociation constants of drugs for use in calculations is given in table 3.6.

5.7.4 Calculations

(1) A drug is found to have the following saturation solubilities at room temperature:

pH	S (μmol ℓ^{-1})
7.4	205.0
9.0	10.0
10.0	5.5
12.0	5.0

What type of compound is it likely to be and what is its pK_a?

As solubility decreases with increasing pH the compound is a base. At pH 12 the solubility quoted is likely to be the solubility of the unprotonated species, that is, S_0. Using the figures given we calculate:

From equation 5.15

$$\text{(i)} \ pK_a = pH + \log \frac{S - S_0}{S_0} = 7.4 + \log \frac{200}{5}$$

$$= 7.4 + \log 40 = 7.4 + 1.602 = 9.0$$

$$\text{(ii)} \ pK_a = 9.0 + \log \frac{10 - 5}{5} = 9.0 + \log 1 = 9.0$$

$$\text{(iii)} \ pK_a = 10.0 + \log \frac{5.5 - 5}{5} = 10.0 + \log 0.1 = 9.0$$

The drug has a pK_a value of 9.0 and is thus likely to be an amine.

(2) What is the pH below which sulphadiazine ($pK_a = 6.48$) will begin to precipitate in an infusion fluid, when the initial molar concentration of sulpha-diazine sodium is 4×10^{-2} mol ℓ^{-1} and the solubility of sulphadiazine is 3.07×10^{-4} mol ℓ^{-1}? (D. Preskey and J. B. Kayes. *J. clin. Pharm.*, 1, 39 (1976)).

The pH below which the drug will precipitate is calculated using equation 5.11.

$$pH = 6.48 + \log \frac{(4.00 \times 10^{-2}) - (3.07 \times 10^{-4})}{3.07 \times 10^{-4}} = 8.60$$

(3) What is the solubility of penicillin G at a pH sufficiently low to allow only the non-dissociated form of the drug to be present? The pK_a of penicillin G is 2.76 and the solubility of the drug at pH 8.0 is 0.174 mol ℓ^{-1}. (From R. E. Notari, *Biopharmaceutics and Pharmacokinetics*, Marcel Dekker, New York, 2nd Edition.)

If only the undissociated form is present at low pH then we need to find S_0. This can be obtained from the information given using equation 5.11.

$$pH - pK_a = \log \left\{ \frac{S - S_0}{S_0} \right\}$$

$$8.0 - 2.76 = \log \left\{ \frac{0.174 - S_0}{S_0} \right\}$$

Therefore $\quad 5.24 = \log \left\{ \frac{0.174 - S_0}{S_0} \right\}$

Therefore $\quad S_0 = 1 \times 10^{-6}$ mol ℓ^{-1}

That is, at pH values well below the pK_a the saturation solubility will equal S_0 (= 1×10^{-6} mol ℓ^{-1}).

(4) Tryptophan has two pK_a values, 2.4 and 9.4, respectively. Calculate the solubility of tryptophan at pH 10 and pH 2, given that the solubility of the compound in neutral solution is 2×10^{-2} mol ℓ^{-1}.

$S_0 = 2 \times 10^{-2}$ mol ℓ^{-1}. We must use equations 5.25 and 5.26. At pH 2.0

$$pH - pK_a = \log \left\{ \frac{S_0}{S - S_0} \right\}$$

$$2 - 2.4 = \log \left\{ \frac{2 \times 10^{-2}}{S - (2 \times 10^{-2})} \right\}$$

$$\log \left\{ \frac{S - (2 \times 10^{-2})}{2 \times 10^{-2}} \right\} = 0.4$$

That is $\quad \left\{ \frac{S - (2 \times 10^{-2})}{2 \times 10^{-2}} \right\} = 2.5118$

Therefore $\quad S = (5.02 \times 10^{-2}) + (2 \times 10^{-2}) = 7.02 \times 10^{-2}$ mol ℓ^{-1}

At pH 10 (using equation 5.26)

$$10 - 9.4 = \log \left\{ \frac{S - (2 \times 10^{-2})}{2 \times 10^{-2}} \right\}$$

That is $\quad \left\{ \frac{S - (2 \times 10^{-2})}{2 \times 10^{-2}} \right\} = 3.981$

Therefore $\quad S = (7.96 \times 10^{-2}) + (2 \times 10^{-2}) = 9.96 \times 10^{-2}$ mol ℓ^{-1}

(5) Calculate the pH at which the following drugs will precipitate from solution given the information supplied.

Drug	pK_a	Solubility of unionised species	Concentration of solution
(i) Thioridazine HCl (mol. wt. 407)	9.5	1.5×10^{-6} mol ℓ^{-1}	0.407% w/v
(ii) Oxytetracycline HCl	3.3, 7.3 and 9.1	0.5 g ℓ^{-1}	1.4 mg mℓ^{-1}

(i) We use equation 5.15 or 5.16 to calculate the pH above which thioridazine will precipitate:

$$pH = pK_a + \log \left\{ \frac{S_0}{S - S_0} \right\}$$

The concentration of solution is the saturation solubility at the point of precipitation. 0.407% w/v = 1×10^{-2} mol ℓ^{-1} = S. $S_0 = 1.5 \times 10^{-6}$ mol ℓ^{-1}.

$$pH = 9.5 + \log \left\{ \frac{1.5 \times 10^{-6}}{(1 \times 10^{-2}) - (1.5 \times 10^{-6})} \right\}$$

$$\approx 9.5 + \log \left\{ \frac{1.5 \times 10^{-6}}{1.0 \times 10^{-2}} \right\}$$

$$= 9.5 - 3.824$$

$$= 5.68$$

(ii) The concentration of solution is 1.4 mg mℓ^{-1} which is 1.4 g ℓ^{-1}. $S_0 = 0.5$ g ℓ^{-1}.

At pH values below 7 the pH at which S is the maximum solubility is given by

$$pH = pK_a + \log \left\{ \frac{S_0}{S - S_0} \right\}$$

$$= 3.3 + \log (0.556)$$

$$= 3.3 - 0.255$$

$$= 3.05$$

At pH values above 7 the pH at which S is the maximum solubility is given by

$$pH = pK_a + \log \left\{ \frac{S - S_0}{S_0} \right\}$$

$$= 7.3 + \log (1.8)$$

$$= 7.3 + 0.255$$

$$= 7.56$$

Thus at pH values between 3.05 and 7.56 the solution containing 1.4 mg mℓ^{-1} will precipitate.

5.8 Determination of solubility of organic electrolytes in aqueous solution

A simple turbidity method for the determination of the solubility of acids and bases in buffers of different pH has been described by Green[8]. Solutions of the hydrochloride (or other salt) of a basic drug, or the soluble salt of an acidic compound, are prepared in water over a range of concentrations. Portions of each solution are added to buffers of known pH and the turbidity of the solutions determined in the visible region. Typical results are shown in figure 5.4. Below the solubility limit there is no turbidity. As the solubility limit is progressively exceeded the turbidity rises. The solubility can be determined by extrapolation as shown in figure 5.4. Green obtained the results shown in table 5.10 for phenothiazines and tricyclic antidepressant compounds. Determination of the solubility of weak electrolytes at several pH values allows one method of obtaining the dissociation constant of the drug substance. For the basic drugs that Green studied, equation 5.16 is rearranged to give

$$pH = pK_a + \log \left\{ \frac{S_0}{S - S_0} \right\}$$

S_0, the solubility of the undissociated species (base), is determined at high pH, and S is determined at several different lower pH values.

A plot of $\log [S_0/(S - S_0)]$ versus pH will have the pK_a as the intercept on the pH axis. Alternatively, S may be plotted against $[H^+]$ as in figure 5.4*b*. Equation 5.14 may be written in the form

$$[H^+] = \left(\frac{K_a S}{S_0} \right) - K_a \tag{5.27}$$

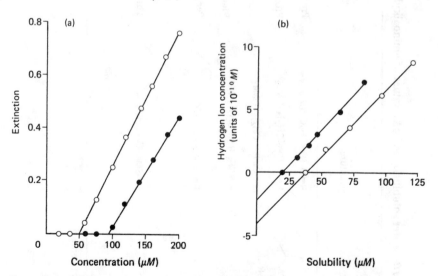

figure 5.4 (a) Plot of extinction against concentration of amitriptyline hydrochloride at pH 9.78 (open circles) and pH 9.20 (filled circles). (b) Relationship between hydrogen ion concentration and solubility of pecazine (filled circles) and amitriptyline (open circles). From reference 8

table 5.10   **Water solubilities and pK_a values of aminoalkylphenothiazines and related compounds***

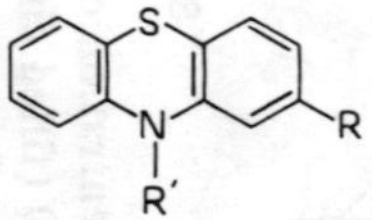

Structure	Trivial or approved name	pK_a		Solubility (μmol ℓ^{-1})	Calculated relative solubility at pH 7.4
		Solubility method	Chatten and Harris (1962)		
R=H; R'=CH$_2$CH(Me)·NMe$_2$	Promethazine	9.1	9.1	55	4.5
R=H, R'=[CH$_2$]$_3$·NMe$_2$	Promazine	9.4	—	50	8.0
R=Cl, R'=[CH$_2$]$_3$·NMe$_2$	Chlorpromazine	9.3	9.2	8	1.0
R=CF$_3$; R'=[CH$_2$]$_3$·NMe$_2$	Fluopromazine	9.2	9.4	5	0.4
R=H; R'=CH$_2$— 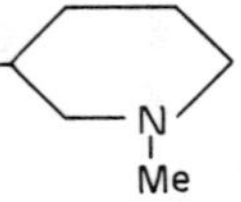 (N—Me)	Pecazine	9.7	—	18	5.0
R=SMe; R'=[CH$_2$]$_2$— 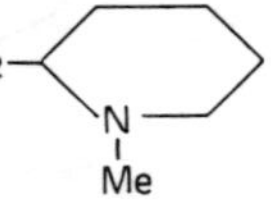 (N—Me)	Thioridazine	9.5	9.2	1.5	0.3

*From reference 8

Plotting data as in figure 5.4b yields S_0 when the line crosses the x-axis as $[H^+] = 0$ (and $S = S_0$). The intercept on the y-axis gives K_a and the slope of the line is K_a/S_0.

5.9 The solubility parameter

Regular solution theory characterises non-polar solvents in terms of solubility parameter, δ_1, defined as

$$\delta_1 = \left(\frac{\Delta U}{V}\right)^{1/2} = \left(\frac{\Delta H - RT}{V}\right)^{1/2} \tag{5.28}$$

where ΔU is the molar energy and ΔH the molar heat of vaporisation of the solvent. ΔH is determined by calorimetry at temperatures below the boiling point at constant volume. V is the molar volume of the solvent. The solubility parameter is a measure of the intermolecular forces within the solvent and therefore must give us some information on the ability of the liquid to act as a solvent. Table 5.11 gives the solubility parameters of some common solvents calculated using equation 5.28.

table 5.11 **Solubility parameters of common solvents**

Solvent	δ_1 (cal$^{1/2}$ cm$^{-3/2}$)*
1. Methanol	14.50
2. Ethanol	12.74
3. 1-Propanol	11.94
4. 2-Propanol	11.56
5. 1-Butanol	11.40
6. 1-Octanol	10.24
7. Ethyl acetate	8.58
8. Isoamyl acetates	8.07
9. Hexane	7.3
10. Hexadecane	8
11. Carbon disulphide	10
12. Membrane lipid** (erythrocytes)	10.3 ± 0.40
13. Cyclohexane	8.2
14. Benzene	9.2

*The solubility parameter is commonly expressed in hildebrand units: 1 hildebrand = 1 (cal cm^{-3})$^{1/2}$ [1 cal = 4.18 J]

Compounds 1–8: calculated from $\delta_1 = \left(\dfrac{\Delta H_{25} - RT}{V}\right)^{1/2}$ (from reference 9)

Note that $\Delta U = \Delta H - P\Delta V = \Delta H - RT$.

**Bennett and Miller, reference 12. See also J. H. Hildebrand, J. M. Prausnitz and R. L. Scott. *Regular and Related Solutions*, Van Nostrand, New York, 1970

The energy of a molecule in its own liquid may be obtained from the energy required to remove it from the liquid. A measure of this attraction is the energy of vaporisation per unit volume, which Scatchard has called the liquid's 'cohesive energy density', $\Delta U/V$. As cavities have to be formed in a solvent to accommodate a solute molecule, by separating other solvent molecules, the solubility parameter δ_1 enables predictions of solubility to be made in a semiquantitative manner, especially in relation to the solubility parameter of the solute, δ_2.

By itself the solubility parameter can explain the behaviour of only a relatively small group of solvents — those with little or no polarity and those unable to participate in hydrogen-bonding interactions. The difference between the solubility parameters expressed as $(\delta_1 - \delta_2)$ will give an indication of solubility relationships. For solid solutes a hypothetical value of δ_2 can be calculated from $(U/V)^{1/2}$, where U is the lattice energy of the crystal[9]. In a study of solubility of ion pairs in organic solvents it has been found that the logarithm of the solubility $(\log S)$ correlated well with $(\delta_2 - \delta_1)^2$.

5.9.1 Solubility parameters and biological processes

The solubility of small molecules in biological membranes is of importance from pharmacological, physiological and toxicological viewpoints. Biological membranes are not simple solvents — the bilayer has an interior core of hydrocarbon chains about 2.5–3.5 nm thick — and therefore one would not expect simple solution theory to hold[10]. However, a correlation is observed between oil solubility and general anaesthetic potency of gases (see chapter 1) and the permeability of membranes to non-ionised solutes has also been related to olive-oil solubility[11]. Bennett and Miller[12] have attempted to apply regular solution theory to biomembranes to obtain a value of δ for the membrane system. From experimental solubility data for the gases in erythrocyte ghosts a mean empirical solubility parameter of 10.3 ± 0.40 for the whole membrane and 8.7 ± 1.03 for membrane lipid was calculated. The values compare with the solubility parameter of 7.3 for hexane and 8.0 for hexadecane. The value for the whole membrane (10.3) is very close to the solubility parameter of 1-octanol (10.2). Octanol is a solvent which is used widely in partition coefficient work to simulate biological lipid phases.

Solubility parameters of drugs have been correlated with membrane absorption rates in model systems[13]. A reasonable relationship was obtained between δ_2 and a logarithmic absorption term, thus providing one predictive index of absorption in this system. Scott[14] has said of solubility parameters and equations employing them that the 'theory offers a useful initial approach to a very wide area of solutions. Like a small-scale map for a very broad long-distance view of a sub-continent they are unlikely to prove highly accurate when a small area is examined carefully, but they are equally unlikely to prove completely absurd'.

5.10   Solubility in mixed solvents

The device of using mixed solvents is resorted to when solubility in one solvent is limited or when the stability characteristics of soluble salts forbid their use. Many pharmaceutical preparations are complex systems. As can be imagined, the addition of another component complicates any system and explanations of the often complex solubility patterns are not easy. Only recently has there been any attempt to predict theoretically solubility in mixed solvents, although the solubility parameters of the mixed solvent systems have been used for this purpose for some time. Common water-miscible solvents used in pharmaceutical formulations include glycerol, propylene glycol, ethyl alcohol and polyoxyethylene glycols. Toxicity considerations are a constraint on choice of solvent for products for administration by any route. Figure 5.5 shows the solubility of phenobarbitone in glycerol-water, alcohol-water and alcohol-glycerol mixtures. In water phenobarbitone dissolves up to 0.12% w/v at 25 °C. Glycerol, even in high concentrations, does not significantly increase the solubility of the drug. Ethanol is a much more efficient co-solvent than glycerol as it is a less polar solvent. Solubility is at a maximum at 90 per cent ethanol in ethanol-water mixtures, and at 80 per cent ethanol in ethanol-glycerol mixtures. It is naive to assume that the drug dissolves in 'pockets' of the co-solvent (for example, ethanol in ethanol-water mixtures), although obviously the affinity of co-solvent for the solute is of importance.

Additives will influence solute–solvent interfacial energies or dissociation of electrolytes through dielectric constant changes. A reduction in ionisation

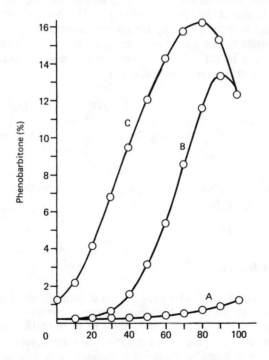

figure 5.5 The solubility of phenobarbitone in glycerol–water, ethanol–water and absolute ethanol–glycerol mixtures as a function of the percentage composition of the mixtures. Abscissa indicates percentage of: A, glycerol in water; B, ethanol in water; and C, absolute ethanol in glycerol. From G. M. Krause and J. M. Cross. *J. Am. pharm. Ass.*, 40, 137 (1951)

through a decrease in dielectric constant will favour decreased solubility but this effect may be counterbalanced by the greater affinity of the undissociated species in the presence of the co-solvent.

5.11 Problems in formulation

5.11.1 Mixtures of acidic and basic compounds

Sometimes a combination formulation requires the admixture of acidic and basic drugs. One example is discussed here[15] (Septrin infusion).

Because sulphamethoxazole (V) is a weakly acidic substance and trimethoprim (VI) a weakly basic one, for optimal solubility basic and acidic solutions, respectively, are required. In consequence, in an ordinary aqueous solution sulphamethoxazole and trimethoprim demonstrate a high degree of incompatibility and on mixing mutual precipitation occurs. To optimise mutual dissolution an aqueous solution which includes 40 per cent propylene glycol is utilised in the formulation of the infusion. This solution, which has a pH between 9.5 and 11.0, allows adequate amounts of both substances to co-exist in solution to give the correct ratio of concentration for antibacterial action. On dilution the infusion becomes less stable and at the recommended 1 in 25 dilution stability is about seven hours. Owing to incompatibility of the two constituents, their degrees of solubility are sensitive to changes in ionic composition, pH, and any drug additives. If an imbalance in pH or ionic composition occurs then precipitation of one or other of the components may well occur.

Sulphamethoxazole
pK_a 6.03
(V)

Trimethoprim
pK_a 7.05
(VI)

5.11.2 Choice of drug salt

The choice of a particular salt of a drug for use in formulations may depend on several factors. The solubility of the drug in aqueous media may be markedly dependent on the salt form. Deliberate choice of an insoluble form for use in suspensions is an obvious ploy; the formation of water-soluble entities from poorly soluble acids or bases by the use of hydrophilic counterions is frequently attempted to produce injectable solutions of the drug. Table 5.12 gives some indication of the range of solubilities that can be obtained through the use of different salt forms, in this case of an experimental antimalarial drug (VII). The chemical stability rather than the solubility may be the criterion and in many cases this is dependent on the choice of salt, sometimes through a pH effect.

(VII)

table 5.12 **Solubilities of salts of an antimalarial drug (VII)***

Salt	Melting point (°C)	Solubility (mg ml^{-1})	Saturated solution pH
Free base	215	7–8	–
Hydrochloride	331	32–15	5.8
dl-Lactate	172 dec	1800	3.8
l-Lactate	193 dec	900	–
2-Hydroxy-1-sulphonate	250 dec	620	2.4
Methane sulphonate	290 dec	300	5.1
Sulphate	270 dec	20	–

*From S. Agharkar, S. Lindenbaum and T. Higuchi. *J. pharm. Sci.*, 65, 747 (1976)

The large hydrophobic compound VII, even as its hydrochloride salt, is poorly soluble and this is presumably the reason for its poor oral bioavailability. Similar conclusions were drawn several years ago for novobiocin. The acid salt administered at 12.5 mg kg^{-1} to dogs was not absorbed, but the monosodium salt, which is about 300 times as soluble in water, produced plasma levels of 22 μg ml^{-1} after three hours. Unfortunately, the sodium salt is unstable in solution. An amorphous form of the acid produced even higher levels of drug than the sodium salt, illustrating the fact that choice of salt and crystalline form of a drug substance may be of critical importance.

Some of the solubility differences obviously arise from differences in the pH of the salt solutions, which in the case of compound VII ranged from 2.4 to 5.8 pH units. This is not atypical. The pH of solutions of salts of a 3-oxyl-1,4-benzodiazepine derivative at 5 mg ml^{-1} ranged from 2.3 for its dihydrochloride, to 4.3 for the maleate, to 4.8 for the methane sulphonate[16].

Further examples of the solubility range in drug salts and derivatives are shown in table 5.13. The increase in the solubility on the formation of the hydrochloride is readily attributable in the case of tetracycline to a lowering of the solution pH (see figure 5.3) by the hydrochloride. The common ion effect can, however, produce an unexpected trend in the solubilities of bases in the presence

table 5.13 **Aqueous solubilities of tetracycline salts and erythromycin salts**

Compound	Solubility in water (mg mℓ^{-1})
Tetracycline	1.7
Tetracycline hydrochloride	10.9
Tetracycline phosphate	15.9
Erythromycin	2.1
Erythromycin estolate*	0.16
Erythromycin stearate	0.33
Erythromycin lactobionate	20

*Lauryl sulphate ester of erythromycin propionate

of high concentrations of hydrochloric acid. Increase in Cl$^-$ concentrations will cause the equilibrium between solid and solution forms

$$BH^+Cl^-_{(c)} \overset{K_{sp}}{\rightleftharpoons} (BH^+)_{aq} + (Cl^-)_{aq}$$

to be pushed to the left-hand side, with a resultant decrease in solubility. The solubility of VII as the hydrochloride is 24×10^{-5} mol ℓ^{-1} in 1.3 mmol ℓ^{-1} chloride ion, 10.9×10^{-5} mol ℓ^{-1} in 5 mmol ℓ^{-1} chloride, and 3×10^{-5} mol ℓ^{-1} in 40 mmol ℓ^{-1} chloride ion concentrations. It should be noted that the stomach contents are rich in chloride ions. The common ion effect will be apparent in many infusion fluids to which drugs may be added, and therefore pH as well as electrolyte concentrations must be noted.

Consideration of table 5.13 suggests that the hydrochloride salts of tetracyclines are always more readily dissolved than the base. However, the situation is more complex than at first appears. In dilute HCl at pH 1.2 the free base dissolves more than the hydrochloride, probably due to the differences in crystallinity. The amount of compound derived from the base in solution decreases with time as the drug is converted to the hydrochloride. At pH 1.6 the rate of solution of the two forms is identical, and at pH 2.1 the hydrochloride has a higher solubility due to its effect on local pH around the dissolving particles.

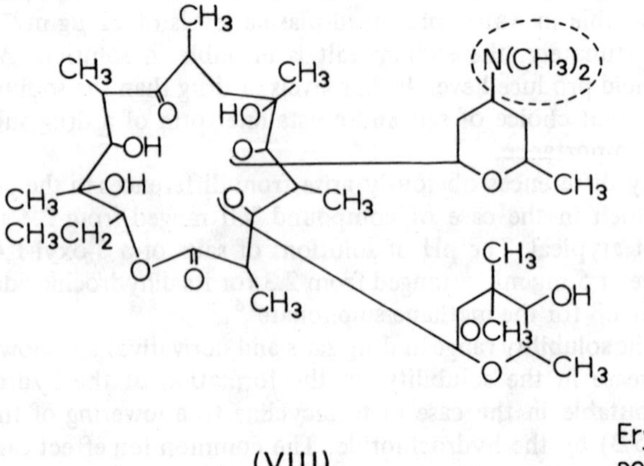

(VIII)

Erythromycin, showing position of amine group

Erythromycin (VIII) is labile at pH values below pH 4, and hence is unstable in the stomach contents. Erythromycin stearate — the salt of the tertiary aliphatic amine and stearic acid — being less soluble, is not as susceptible to degradation. The salt dissociates in the intestine yielding the free base, which is absorbed. There are differences in the absorption behaviour of the erythromycin salts and differences in toxicity, which may be related to their aqueous solubilities. Erythromycin ethylsuccinate was originally developed for paediatric use because its low water solubility and relative tastelessness were suited to paediatric formulations. The soluble lactobionate is used in intravenous infusions.

5.12   Drug solubility and biological activity

There should be a broad correlation between aqueous solubility and indices of biological activity. On the one hand, as drug solubility in aqueous media is inversely related to the solubility of the agent in biological lipid phases, there will be some relationship between pharmacodynamic activity and drug solubility. On the other, we should expect that drug or drug salt solubility might influence pharmacokinetic behaviour; drugs of very low aqueous solubility will dissolve slowly in the gastro-intestinal tract and in many cases the rate of dissolution is the rate-controlling step in the absorption phase of drug action. It is with drugs of low aqueous solubility such as digoxin, chlorpropamide, indomethacin, griseofulvin, and the steroids, that variation in the physical properties of the drug can cause variability in the biological properties. At some stage in a drug's development pharmacological and toxicological tests are frequently carried out on extemporaneously prepared solutions and suspensions whose physical characteristics are not well defined. Boyd[17] has shown that the toxicity of some drugs given by gavage to rats is dependent on the solubility of the drug species (table 5.14). This has been shown to be true with polymorphic forms of the same drug, but in the cases discussed in table 5.14 different salts of the drugs were used.

Other examples are given by Stenlake[18] in which aqueous solubility acts as a rough and ready guide to absorption characteristics. Of the cardiotonic glycosides, digitoxin, digoxin and ouabain, the least water-soluble, being the most lipid-soluble, is best absorbed. But because of the lipophilicity of digitoxin and digoxin, the rate-limiting step must be the rate of solution which is influenced directly by the solubility of the compounds. High molecular weight quaternary

table 5.14   **The effect of solubility in water on the toxicity of drugs given by gavage to albino rats***

Drug	Salt	Solubility	$LD_{50} \pm$ s.e.
Benzylpenicillin	Ammonium	<20 mg ml^{-1}	8.4 ± 0.13
Benzylpenicillin	Potassium	>20 mg ml^{-1}	6.7 ± 0.1
Iron	Free metal	insoluble	98.6 ± 26.7
Iron	Ferrous sulphate	soluble	0.78
Spiramycin	Free base	poorly soluble	9.4 ± 0.8
Spiramycin	Adipate	soluble	4.9 ± 0.2

*Adapted from reference 17

salts such as bephenium hydroxynaphthoate (IX) and viprynium embonate (X), being quaternary, have low lipid solubility but also have low aqueous solubility. They are virtually unabsorbed from the gut and indeed are used in the treatment of worm infestation of the lower bowel.

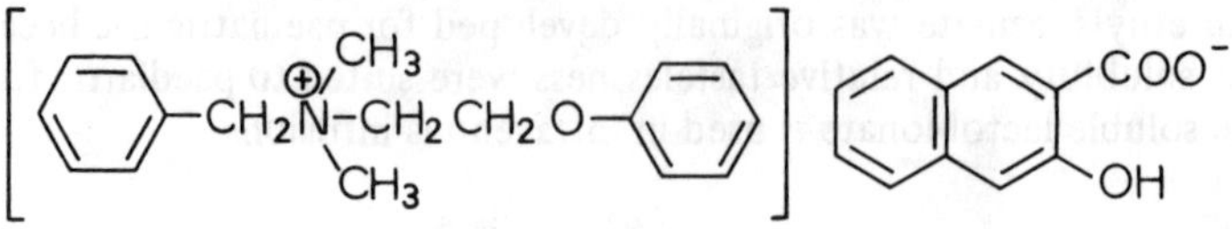

Bephenium hydroxynaphthoate

(IX)

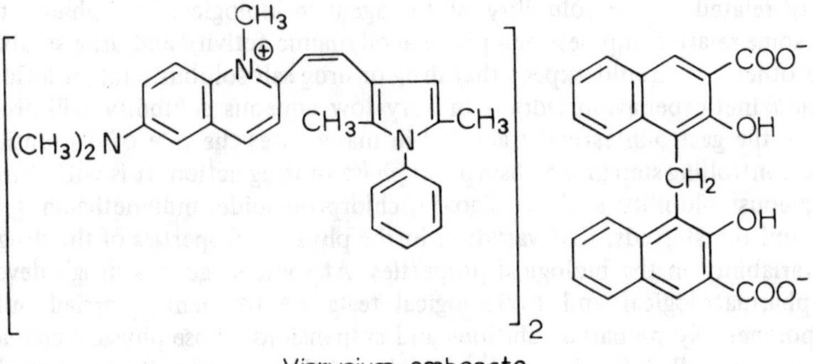

Viprynium embonate

(X)

5.13 Partitioning phenomena*

The movement of molecules from one phase to another is called partitioning. Drugs partition between aqueous phases and lipid biophases. Preservative molecules in emulsions partition between the water and oil phases. Antibiotics partition into micro-organisms and solutes — either drug or preservative or other adjuvant can partition into plastic giving sets and rubber stoppers. It is therefore important that the process is quantified and understood.

If two immiscible phases are placed in contact, one containing a solute soluble in both phases, the solute will distribute itself so that when equilibrium is attained no further net transfer of solute takes place. At equilibrium the chemical potential of the solute in one phase is equal to its chemical potential in the other phase. If we are considering an aqueous (w) and an organic (o) phase we can write according to equations 3.27 and 3.31

$$\mu_w^{\ominus} + RT \ln a_w = \mu_o^{\ominus} + RT \ln a_o \qquad (5.29)$$

Rearranging equation 5.29 we obtain

*An excellent review on partitioning has been written by Leo et al.[19].

$$\frac{\mu_o^\ominus - \mu_w^\ominus}{RT} = \ln \frac{a_w}{a_o} \tag{5.30}$$

The term on the left-hand side of equation 5.30 is constant at a given temperature and pressure, so it follows that

$$\frac{a_w}{a_o} = \text{constant}$$

or $\qquad \dfrac{a_o}{a_w} = \text{constant} \tag{5.31}$

These constants are the partition coefficients or distribution coefficients, P. If the solute forms an ideal solution in both solvents, activities can be replaced by concentrations; so

$$\boxed{P = \frac{C_o}{C_w}} \tag{5.32}$$

In this book reference to P, unless otherwise stated, implies that it has been calculated according to the convention in equation 5.32 where the concentration in the organic oily phase is divided by the concentration in the aqueous phase. The greater the value of P, the higher the lipid solubility of the solute. It has been shown for several systems that the partition coefficient can be approximated by the solubility of the solute in the organic phase divided by its solubility in the aqueous phase. Thus P is a measure of the relative affinities of the solute for an aqueous and a non-aqueous or lipid phase.

In many systems the ionisation of the solute in one or both phases or the association of the solute in one of the solvents complicates the consideration of partition. As early as 1891, Nernst stressed the fact that the partition coefficient would be constant only if a single molecular species was involved in the partition.

table 5.15 **Distribution of two acids between immiscible phases†**

$C_1{}^*$ (mol ℓ^{-1})	$C_2{}^{**}$ (mol ℓ^{-1})	C_1/C_2	$C_1/\sqrt{C_2}$
Succinic acid			
0.191	0.0248	7.69	
0.370	0.0488	7.58	
0.547	0.0736	7.43	
0.749	0.101	7.41	
Benzoic acid			
4.88×10^{-2}	3.64×10^{-1}	0.134	0.0256
8.00	8.59	0.093	0.0273
16.00	33.8	0.047	0.0275
23.7	75.3	0.030	0.0273

†From S. Glasstone and D. Lewis. *Elements of Physical Chemistry*, 2nd Edition, Macmillan, 1964
*Aqueous phase concentration ⁣ ⁣ ⁣ ⁣ ⁣ ⁣**Non-aqueous phase concentration

If the solute associates, partitioning can be considered as follows. Consider the following equilibrium between the phases 1 and 2

$$2A \quad \rightleftharpoons \quad A\text{-}A$$
$$\text{phase 1} \qquad \text{phase 2}$$

$$K = \frac{[A]_{\text{phase 2}}}{[A]^2_{\text{phase 1}}}$$

Hence
$$K' = \frac{\sqrt{C_2}}{C_1} \tag{5.33}$$

K' is a constant combining the partition coefficient and the association constant. Table 5.15 illustrates the use of equations 5.32 and 5.33.

5.13.1   Partitioning of weak electrolytes

As the site of action of many biologically active species is in lipid regions such as membranes, correlations between partition coefficients and biological activity were found by early investigators of structure–action relationships.

Many drugs are weak electrolytes and will ionise in at least one phase, generally the aqueous phase. If the partition coefficient refers to the distribution of one species then the assay figures for the concentration of the solute in the phases have to be corrected for ionisation. We use the convention that C_a is the total concentration of all species in the aqueous phase, C_o is the concentration in the organic phase and C_{ion} is the concentration of ions in the aqueous phase. If we consider the dissociation of a weak acid

$$\text{HA} \qquad \rightleftharpoons \quad \text{H}^+ \quad + \quad \text{A}^-$$
$$(C_a - C_{\text{ion}}) \qquad (C_{\text{ion}}) \quad (C_{\text{ion}})$$

the concentration of the species may be written as shown, and therefore the dissociation constant in the aqueous phase, K_a is given by

$$K_a = \frac{(C_{\text{ion}})^2}{(C_a - C_{\text{ion}})} \tag{5.34}$$

If dimerisation occurs in the organic phase and if K_D is the dissociation constant of dimers into single molecules, we can consider the process in the organic phase to be

$$(\text{HA})_2 \quad \rightleftharpoons \quad 2\text{HA}$$
$$\text{organic} \qquad \text{aqueous}$$

and
$$K_D = \frac{2[P(C_a - C_{\text{ion}})]^2}{C_o - P(C_a - C_{\text{ion}})} = \frac{2(PN)^2}{C_o - PN} \tag{5.35}$$

where N is $(C_a - C_{\text{ion}})$, that is, the concentration of unionised molecules in water, the species which will distribute into the non-aqueous phase. It is generally accepted that only the non-ionised species partition from the aqueous phase into the non-aqueous phase. Ionised species, being hydrated and highly soluble in the aqueous phase, disfavour the organic phase. Transfer of such a hydrated species would involve its dehydration. In addition, organic solvents of low polarity do not favour the existence of free ions.

Equation 5.35 can be rearranged to give

$$K_D (C_o - PN) = 2 (PN)^2 \qquad (5.36)$$

Multiplying by $1/K_D N^2$ and rearranging we obtain

$$C_o/N^2 = P (1/N) + \text{constant} \qquad (5.37)$$

A plot of (C_o/N^2) *vs* $1/N$ will yield a straight line with slope $= P$. If ionisation and its consequences are neglected, an apparent partition coefficient, P_{app}, is obtained by assay of both phases. The relationship between the true P and P_{app} is given by the following equations:

for acids
$$\boxed{\log P = \log P_{app} - \log \left[\frac{1}{1 + 10^{pH - pK_a}} \right]} \qquad (5.38)$$

for bases
$$\boxed{\log P = \log P_{app} - \log \left[\frac{1}{1 + 10^{pK_a - pH}} \right]} \qquad (5.39)$$

The standard free energy of transfer of solute between the phases is given by

$$\Delta G^{\ominus}_{trans} = \mu_w^{\ominus} - \mu_o^{\ominus} = RT \ln P \qquad (5.40)$$

In a homologous series, P and therefore $\ln P$ can be measured and the increase in partition coefficient observed for each substituent group (for example, $-CH_2-$). As the chain length of non-polar aliphatic compounds increases it has been found that P increases by a factor of two to four per methylene group. Additivity of the substituent contributions to P are observed and a substituent constant, π, may be defined as

$$\pi_x = \log P_x - \log P_H \qquad (5.41)$$

where P_x is the partition coefficient of the derivative of the parent compound whose partition coefficient is P_H. π_x is the logarithm of the partition coefficient of the function x; for example, π_{Cl} can be obtained by subtracting from log $P_{chlorobenzene}$ the value log $P_{benzene}$. Octanol has frequently been used as the non-aqueous phase to obtain partitioning data for comparing the lipophilicity of drugs. The polarity of this solvent means that water is solubilised to some extent in the octanol and thus the partition process is more complex than with an anhydrous solvent. Perhaps its usefulness stems from the fact that the biological membrane is no simple lipid phase. However, other non-aqueous solvents have been used. For example, isobutanol has been used to show that the binding of many drugs to serum protein follows the relationship

$$\log K = 0.9 \log P_{isobutanol} + \text{constant} \qquad (5.42)$$

where K is an equilibrium constant measuring the binding of solute to protein. Transfer of a hydrophobic drug from an aqueous phase to a protein is, of course, a complex type of partitioning.

Work on the correlation of lipophilicity and biological activity frequently produces equations of the type[16]

$$\log \frac{1}{C} = A \log P + \text{constant} \qquad (5.43)$$

where C is the concentration required to produce a given pharmacological response.

5.13.2 Biological activity and partition coefficient — thermodynamic activity and Ferguson's principle

A wide range of simple organic compounds can exert qualitatively identical depressant actions on many organisms. The noticeable lack of any chemical specificity in the compounds tested led several workers to suggest that the physical, rather than the chemical, properties governed the activity of the compounds. Early work by Meyer and Overton related narcotic potency to the oil/water partition coefficient, and Meyer and Hemmi[20] in a later re-interpretation of the data concluded that narcosis commences when any chemically non-specific substance has attained a certain molar concentration in the lipids of the cells.

Ferguson[21] placed the Overton–Meyer theory on a more quantitative basis by applying a thermodynamic analysis to the problem of narcotic action. By expressing the potency in terms of the thermodynamic activity, rather than the concenration, he avoided the problem of the various distribution coefficients between the numerous different phases within the cell, any of which might be the phase on which the drug exerts its pharmacological effects (the biophase). The fact that the narcotic action of a drug remains at a constant level whilst a critical concentration of drug is supplied and decreases rapidly when administration of the drug is stopped, suggests that an equilibrium exists between some external phase and the biophase.

According to equation 3.27 the chemical potentials in two phases at equilibrium are equal. Thus, from equation 3.35

$$\mu_A^{\ominus} + RT \ln a_A = \mu_B^{\ominus} + RT \ln a_B \qquad (5.44)$$

If the standard states are identical, $\mu_A^{\ominus} = \mu_B^{\ominus}$, and consequently the activities will be equal in the two phases. The activity of a substance in the biophase at equilibrium is thus identical to the readily determined value in an external phase.

For narcotic agents applied as a vapour, the standard state was taken to be the saturated vapour, and thus activity $a = p_t/p_s$, where p_t is the partial pressure of the vapour and p_s is the saturated vapour pressure at the same temperature. When the narcotic agent was applied in solution and was a substance of limited solubility, the activity was equated with the ratio S_t/S_0 where S_t is the molar concentration of the narcotic solution and S_0 its solubility; the ratio S_t/S_0 therefore represents proportional saturation. This is in contrast to the normal procedure of taking the standard state as an infinitely dilute solution.

From recalculations of the data of earlier workers and also from measurements of the potency of many different compounds, Ferguson concluded that, within reasonable limits, substances present at approximately the same proportional saturation (that is, activity) in a given medium have the same biological potency.

table 5.16 **Bactericidal concentrations and activities of organic substances in solution***

Substance	Bactericidal concentration (mol ℓ^{-1}) (S_t)	Solubility at 25 °C (mol ℓ^{-1}) (S_0)	Activity or S_t/S_0
Phenol	0.097	0.90	0.11
o-Cresol	0.039	0.23	0.17
Propaldehyde	1.08	2.88	0.37
Thymol	0.0022	0.0057	0.38
Acetone	3.89	∞	0.40
Methyl ethyl ketone	1.25	3.13	0.40
Aniline	0.17	0.40	0.44
Cyclohexanol	0.18	0.38	0.47
Resorcinol	3.09	6.08	0.54
Methyl propyl ketone	0.39	0.70	0.56
Butyraldehyde	0.39	0.51	0.76

*From reference 20

Table 5.16, for example, shows the activities of various organic substances equitoxic to *B. typhosus*. Although the bactericidal concentrations vary widely (0.0022–3.89 mol ℓ^{-1}), the activities are constant within a relatively restricted range. Other workers have reported on the applicability of Ferguson's principle to many organic compounds acting on a variety of physiological processes.

5.14 Uses of log *P*

As the examples of the uses of log *P* are legion, only a selection of applications can be considered here. The interest in partition coefficients derives from the fact that the relatively simple *in vitro* measurement (of *P*) can give an accurate prediction of activity in a complex biological system provided that the obvious limitations of the simple system are recognised and that the activity of the drug depends on its lipophilic nature. The value of *P* or of log *P* is mainly found in homologous series or series of closely related compounds where the influence of substituent groups can be examined. As the range of information on the relationship of biological activity to log *P* is so wide, the following section deals only with selected systems to indicate the scope of the subject.

The relationship between lipophilicity and pharmacological behaviour of tetracyclines has been studied by Barga *et al.*[22] and by Hoeptrich and Warshauer[23]. Four tetracyclines (minocycline, doxycycline, tetracycline and oxytetracycline) were investigated; lipid solubility correlated inversely with the mean concentration of antibiotic in plasma and with renal uptake and excretion. Only the more lipophilic congeners (minocycline and doxycycline) passed the blood–brain and blood–ocular barriers in detectable concentrations. Table 5.17 gives some of the characteristics of the tetracyclines studied. These analogues of tetracycline, while active *in vitro* against meningococci, are not all of value in clinical use; oxytetracycline and doxycycline fail to change the state of 'carriers' of the disease,

table 5.17 **Some characteristics of four tetracyclines**

	P_{app} (chloroform/water)*	P_{app} (octanol/water)**	Serum protein binding (per cent)
Minocycline	30	1.1	76–83
Doxycycline	0.48	0.60	25–93
Tetracycline	0.09	0.036	20–67
Oxytetracycline	0.007	0.025	10–40

*pH 7.4   **pH 7.5

whereas minocycline has a significant effect. It is thought that the ability to enter the saliva and tears influences the clinical activity, for although saliva does not usually wet the nasopharynx, tears pass into the nasopharynx as the normal route of drainage from the conjunctival sac.

The pH dependence of partition coefficients of tetracyclines is more complex than for most drugs, as the tetracyclines are amphoteric. A partition mechanism for tetracycline has been proposed by Terada and Inagi[24] whose studies confirm that the neutral form of tetracycline is hydrophobic but that the concentration of the zwitterionic species in the aqueous phase is very small.

For slightly simpler amphoteric compounds, such as *p*-aminobenzoic acid and sulphonamides, the apparent partition coefficient is maximal at the isoelectric point. When pK_{a1} is much smaller than pK_{a2} the value of the apparent partition coefficient coincides with the true partition coefficient (defined as before as the ratio of activities of the neutral form of the acid in the two phases). The true partition coefficient changes linearly with slopes of +1 and −1 in the regions $pK_{a1} \gg pH$ and $pK_{a2} \ll pH$, respectively. Figure 5.6 illustrations the variation of $\log P_{app}$ for *p*-aminobenzoic acid and for two sulphonamides[25].

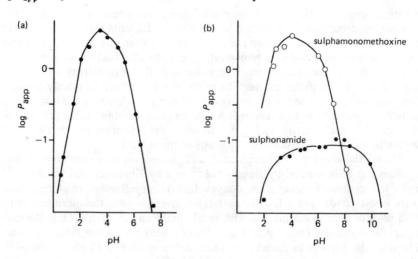

figure 5.6 (a) Variation of $\log P_{app}$ with pH for *p*-aminobenzoic acid. (b) Variation of $\log P_{app}$ with pH for sulphamonomethoxine and sulphonamide. From reference 25

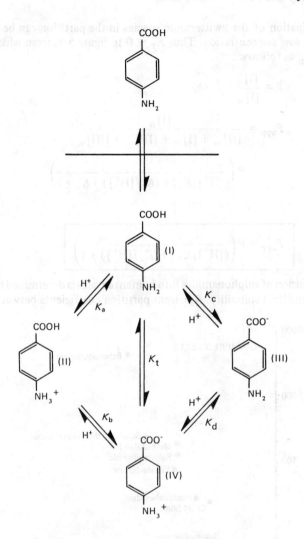

figure 5.7 Partition of *p*-aminobenzoic acid (from reference 25). K_a, K_b, K_c and K_d are microdissociation constants for each equilibrium (see section 3.4.4) and the relation between them is

$$K_1 = K_a + K_b$$

$$\frac{1}{K_2} = \frac{1}{K_c} + \frac{1}{K_d}$$

$$K_t = \frac{[\text{zwitterion}]}{[\text{neutral form}]} = \frac{K_b}{K_a} = \frac{K_c}{K_d}$$

where K_1 and K_2 are composite or macroscopic acid dissociation constants and K_t is the tautomeric constant between the zwitterionic and neutral forms

The participation of the zwitterionic species in the partition can be excluded because of its low concentration. Thus $K_t \to 0$ in figure 5.7, from which we obtain P and P_{app} as follows:

$$P = \frac{[I]_o}{[I]_w} \qquad (5.45)$$

and

$$P_{app} = \frac{[I]_o}{[II]_w + [I]_w + [IV]_w + [III]_w}$$

$$= P \left(\frac{1}{([H^+]/K_a) + (K_c/[H^+]) + K_t + 1} \right) \qquad (5.46)$$

If $K_t \to 0$

$$\boxed{P_{app} = P \left(\frac{1}{([H^+]/K_1) + (K_2/[H^+]) + 1} \right)} \qquad (5.47)$$

Passive diffusion of sulphonamides into human red cells is determined by plasma drug binding and lipid solubility. Apparent partition coefficients between chloro-

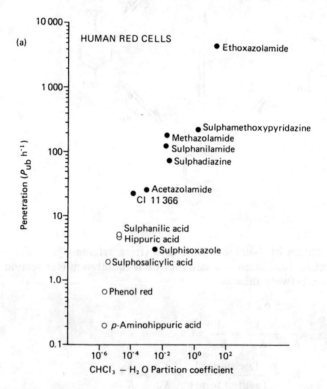

figure 5.8 (a) Penetration rates of drugs into human red cells plotted as a function of chloroform/water partition coefficients. Data from reference 26, and Schanker. *J. Pharm. exp. Ther.*, 133, 325 (1961)

form and water at pH 7.4 show an almost linear relation with penetration constant for sulphonamides and a number of other acids. Penetration rates of sulphonamides into the aqueous humour and cerebrospinal fluid also correlate with partition coefficients[26]. The data in figure 5.8, which are derived from several sources, illustrate this fact.

Antibacterial effects of fatty acids and esters towards *B. subtilis* correlate with octanol/water partition coefficients[27] (figure 5.9).

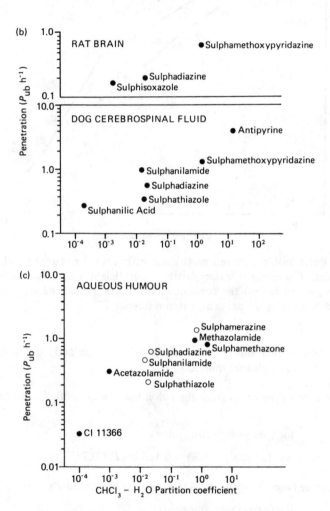

figure 5.8 (b) Penetration rates of sulphonamides from plasma into rat brain and into canine cerebrospinal fluid. Rat brain data from T. F. Muller; dog cerebrospinal fluid data from D. P. Rall. *J. Pharm. exp. Ther.*, 125, 185 (1959). (c) Penetration rates of sulphonamides from plasma into aqueous humour for the rabbit (o) and for the rat (•) against partition coefficients (chloroform/water). Data from P. J. Wistrand. *Acta pharmacol. toxicol.*, 17, 337 (1960) and A. Sorsby. *Br. J. Ophthalmol.*, 33, 347 (1949)

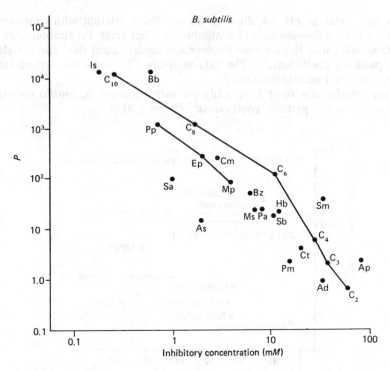

figure 5.9 Relationship between partition coefficient (P) and growth inhibitory concentration. The octanol/water partition coefficient of undissociated compounds was plotted against the concentrations of compounds needed to inhibit growth of *B. subtilis* by 50 per cent. From reference 27

The quantitative relations between physiological action and $\log P$ or $\log P_{app}$ are legion. A few examples are quoted.

Absorption of acidic drugs from the colon may be quantified according to P and pK_a:

$$\log (\text{percentage absorption})$$
$$= 0.156 \, (pK_a - 6.8) - 0.366 \log P + 0.755 \qquad (5.48)$$

Absorption of bases from the small intestine has been similarly treated:

$$\log (\text{percentage absorbed})$$
$$= -0.131 \, (\log P)^2 + 0.362 \log P \qquad (5.49)$$

Local anaesthetic action on peripheral nerve is also proportional to P since the unionised form must diffuse across the continuous cell layer of the perineurium[28]. Once across the perineurium the molecules ionise and they combine with the receptors in the nerve membrane in their ionised form.

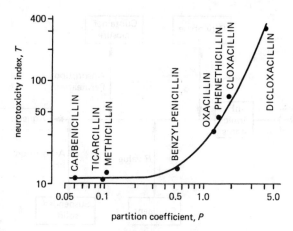

figure 5.10 Relation of neurotoxicity to the hydrophobic character of various penicillins. The hydrophobic character is measured by the partition coefficient *P* and toxicity by the neurotoxicity index *T*. From reference 30

The toxicity of some agents has been related to their lipophilicity. Rates of entry into the brain of X-ray contrast agents used in cerebral angiography are proportional to *P*, and *P* is furthermore correlated with clinical neurotoxicity[29]. The toxicity of the penicillins has also been connected with partition coeffic-ient[30], as shown in figure 5.10.

Figure 5.11 summarises the physicochemical problems in the use of preserva-tive molecules in formulation. Solubility and partition coefficients are determined, as we have seen, by the pH and ionic strength of the system. Partitioning may occur to the oily phase of an emulsion, to the micellar phase of a surfactant (emulgent), or to a closure. Adsorption may also occur onto container closures and suspended solid particles.

Permeation of antimicrobial agents into rubber stoppers and other closures is, then, another example of partitioning. Although the rubber is an amorphous solid, partitioning between the aqueous phase and the rubber depends, as in liquid systems, on the relative affinities of the solute for each phase. Glyceryl trinitrate, a volatile drug with a chloroform/water partition coefficient of 109, diffuses from simple tablet bases into the walls of plastic bottles and into plastic liners used in packaging tablets. This partitioning can be prevented if there is included in the tablet formulation an agent which complexes with the drug substance, thereby increasing its affinity for the 'tablet phase'; in other words, reducing its escaping tendency or fugacity[31]. Significant losses of glyceryl trinitrate have been detected when the drug was given as an infusion through plastic giving sets from a plastic reservoir, the absorption (of as much as 50 per cent of the drug) resulting in the necessity to use unusually high doses of the drug[32].

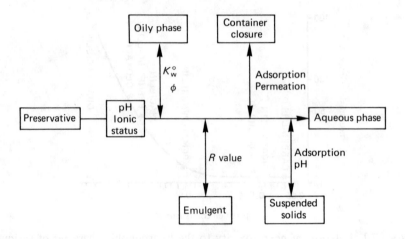

figure 5.11 The potential fates of preservative molecules. From M. S. Parker. In *Bentley's Textbook of Pharmaceutics*, Baillière, London, in press

table 5.18 Sorption into PVC and partitioning of drugs*

Compound	Initial rate of sorption $(h^{-1}) (\times 10^2)$	Extent of sorption at equilibrium (per cent)	Log P_{app}	Log P (hexane/water)
Medazepam	51	85	1.7	2.9
Diazepam	27	90	1.9	0.9
Warfarin (pH 2–4)	22	90	1.9	0.2
Nitroglycerin	20	83	1.6	0.2
Thiopental (pH 4)	6	73	1.4	0.1
Oxazepam	4.6	46	0.9	−0.1
Nitrazepam	4.1	47	0.9	−0.1
Hydrocortisone acetate	0	0	–	−1.2
Pentobarbitone (pH 4)	0	0	–	−1.3

*From reference 33

Some relation between the rate of sorption by PVC bags of a series of drugs and their hexane/water partition coefficients has been found[33]. Table 5.18 shows the data for sorption to 100 mℓ PVC infusion bags (equivalent to 11 g of PVC). In the table, P_{app} values have been calculated from

$$P_{app} = \left(\frac{1 - F_\infty}{F_\infty}\right) \frac{W_s}{W_p} \qquad (5.50)$$

where W_s is the weight of solution in contact with a given weight W_p of plastic and F_∞ is the equilibrium fraction of drug remaining in solution. Only the unionised form of the drug is sorbed; the sorption kinetics can be accounted for in considering the diffusion of the molecules in the plastic matrix.

References

1. G. L. Amidon, S. H. Yalkowsky and S. Leung. *J. pharm. Sci.*, 63, 1858 (1974)
2. G. L. Amidon. *ibid.*, 63, 1520 (1974)
3. G. L. Amidon, S. H. Yalkowsky, A. T. Anik and S. C. Valvani. *J. phys. Chem.*, 79, 2239 (1975)
4. R. B. Hermann. *ibid.*, 76, 2754 (1972)
5. R. B. Hermann. *ibid.*, 75, 363 (1971)
6. P. Kabasakalian, E. Britt and M. D. Yudis. *J. pharm. Sci.*, 55, 642 (1966)
7. F. Franks. In *Water Relations of Foods* (ed. R. D. Duckworth), Academic Press, London, 1975
8. A. L. Green. *J. Pharm. Pharmacol.*, 19, 10 (1976)
9. T. Takamatsu. *Bull. chem. Soc. Japan*, 47, 2647 (1974)
10. R. Fettiplace, S. M. Andrews and D. A. Haydon. *J. membr. Biol.*, 5, 277 (1971)
11. R. Collander. *Acta physiol. Scand.*, 13, 363 (1947)
12. L. J. Bennett and K. W. Miller. *J. med. Chem.*, 17, 1124 (1974)
13. S. A. Khalil, O. Y. Abdallah and M. A. Moustafa. *Can. J. pharm. Sci.*, 11, 26 (1976)
14. R. L. Scott. *Annu. Rev. phys. Chem.*, 7, 43 (1956)
15. Burroughs Wellcome. *Septrin for Infusions*
16. A. Nudelman, R. J. McCaully and S. C. Bell. *J. pharm. Sci.*, 63, 1880 (1974)
17. E. Boyd. *Predictive Toxicometrics*, Scientechnica, Bristol, 1972
18. J. B. Stenlake. *Pharm. J.*, 215, 533 (1975)
19. A. J. Leo, C. Hansch and D. Elkins. *Chem. Rev.*, 71, 525 (1971)
20. K. H. Meyer and H. Hemmi, *Biochem. Z.*, 277, 39 (1935)
21. J. Ferguson. *Proc. R. Soc. B*, 127, 387 (1939)
22. M. Barga *et al. Antimicrob. Agents Chemother.*, 8, 713 (1975)
23. P. D. Hoeptrich and D. M. Warshauer. *ibid.*, 5, 330 (1974)
24. H. Terada and T. Inagi. *Chem. pharm. Bull.*, 23, 1960 (1975)
25. H. Terada. *ibid.*, 20, 765 (1972)
26. L. B. Holder and S. L. Hayes. *Mol. Pharmacol.*, 1, 266 (1965)
27. G. W. Sheu *et al. Antimicrob. Agents Chemother.*, 7, 349 (1975)
28. S. I. Rapoport. *Blood Brain Barrier in Physiology and Medicine*, Raven Press, New York, 1976
29. S. I. Rapoport and H. Levitan. *Am. J. Roentgenol. radium Ther. nucl. Med.*, 122, 1986 (1974)
30. T. R. Weihrauch *et al. Arch Pharmacol. (NS)*, 289, 55 (1975)
31. B. A. Edelman, A. V. Contractor and R. F. Shangraw. *J. Am. pharm. Ass.*, 60, 30 (1971)
32. P. A. Cossum, A. J. Galbraith, M. S. Roberts and G. W. Boyd. *Lancet*, ii, 349 (1978)
33. L. Illum and H. Bundgaard, *Int. J. Pharmaceut.*, 10, 339 (1982)

Bibliography

On drug salts: P. Gould, *Int. J. Pharmaceut.*, 33, 201 (1986)

6 Surfactants

Certain compounds, because of their chemical structure, have a tendency to accumulate at the boundary between two phases. Such compounds are termed amphiphiles, surface-active agents, or surfactants. The adsorption at the various interfaces between solids, liquids and gases results in changes in the nature of the interface which are of considerable importance in pharmacy. Thus, the lowering of the interfacial tension between oil and water phases facilitates emulsion formation, the adsorption of surfactants on the insoluble particles enables these particles to be dispersed in the form of a suspension, and the incorporation of insoluble compounds within micelles of the surfactant can lead to the production of clear solutions.

6.1 Amphipathic compounds

Surface-active compounds are characterised by the possession of two distinct regions in their chemical structure, termed hydrophilic (water-liking) and hydrophobic (water-hating) regions. The existence of two such moieties in a molecule is referred to as amphipathy and the molecules are consequently often referred to as amphipathic molecules.

The hydrophobic portions are usually saturated or unsaturated hydrocarbon chains or, less commonly, heterocyclic or aromatic ring systems. The hydrophilic regions can be anionic, cationic or non-ionic. Surfactants are generally classified according to the nature of the hydrophilic group. Typical examples are given in table 6.1.

In addition to the types listed in table 6.1, some surfactants possess both positively and negatively charged groups; that is, they are ampholytic compounds. This type can exist as either an anionic or cationic surfactant depending on the pH of the solution. A typical example is N-dodecyl-N,N-dimethylbetaine ($C_{12} H_{25} N^+ (CH_3)_2 CH_2 COO^-$).

The dual structure of amphipathic molecules is the unique feature which is responsible for the characteristic behaviour of this type of compound. Thus the surface activity arises from adsorption at the solution–air interface – the means by which the hydrophobic region of the molecule 'escapes' from the hostile aqueous environment by protruding into the vapour phase above. Similarly, adsorption at the interface between non-aqueous and aqueous solutions occurs in such a way that the hydrophobic group is in solution in the non-aqueous phase, leaving the hydrophilic group in contact with the aqueous solution. Adsorption on hydrophobic solutes such as carbon again represents a means of removal of hydrophobic groups as far as possible from the aqueous environment and allows the consequent attainment of a minimum energy state. Perhaps the most striking consequence of the dual structure is micellisation – the formation in solution of aggregates in which the component molecules are generally arranged in a spheroidal structure with the hydrophobic cores shielded from the water by a mantle of hydrophilic groups.

table 6.1 **Classification of surfactants**

Anionic	Hydrophobic	Hydrophilic
Sodium stearate	$CH_3(CH_2)_{16}$	$COO^- \ Na^+$
Sodium dodecyl sulphate	$CH_3(CH_2)_{11}$	$SO_4^- \ Na^+$
Sodium dodecyl benzene sulphonate	$CH_3(CH_2)_{11}C_6H_4$	$SO_3^- \ Na^+$
Sodium cholate	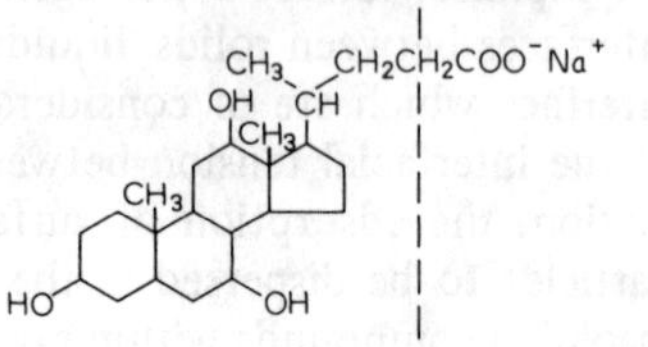	$CH_2CH_2COO^- Na^+$

Cationic		
Hexadecyltrimethyl ammonium bromide	$CH_3(CH_2)_{15}$	$N^+CH_3 \ Br^-$
Dodecylpyridinium chloride	$C_{12}H_{25}$	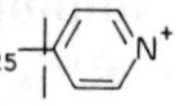 $N^+ \ Cl^-$

Non-ionic		
Heptaoxyethylene monohexadecyl ether	$CH_3(CH_2)_{15}$	$(OCH_2CH_2)_7OH$

In the following sections we shall examine in more detail the various character-istic properties of surfactants which arise as a consequence of their amphipathic nature.

6.2 Surface and interfacial properties of surfactants

6.2.1 Effects of amphiphiles on surface and interfacial tension

The molecules at the surface of a liquid are not completely surrounded by other like molecules as they are in the bulk of the liquid. As a result there is a net in-ward force of attraction exerted on a molecule at the surface from the molecules in the bulk solution, which results in a tendency for the surface to contract.

The contraction of the surface is spontaneous; that is, it is accompanied by a decrease in free energy. The contracted surface thus represents a minimum free energy state and any attempt to expand the surface must involve an increase in the free energy. The surface free energy of a liquid is defined as the work, w, required to increase the surface area A by 1 m^2

$$w = \gamma \Delta A \tag{6.1}$$

where ΔA is the increase in surface area. γ is also referred to as surface tension and in this context is defined as the force acting at right-angles to a line 1 m in

table 6.2 **Surface tensions and interfacial tensions against water at 20 °C**

Substance	Surface tension (mN m^{-1})	Interfacial tension against H$_2$O (mN m^{-1})
Water	72	–
Glycerol	63	–
Oleic acid	33	16
Benzene	29	35
Chloroform	27	33
n-Octanol	27	8.5
Carbon tetrachloride	27	45
Castor oil	39	–
Olive oil	36	33
Cottonseed oil	35	–
n-Octane	22	51
Ethyl ether	17	11

length along the surface. Surface free energy and surface tension are numerically equal and both have SI units of N m^{-1}. It is usual, however, to quote values of surface tension in mN m^{-1} which is numerically equivalent to the cgs unit, dyne cm^{-1}.

A similar imbalance of attractive forces exists at the interface between two immiscible liquids. Table 6.2 lists surface tensions of various liquids and also interfacial tensions at the liquid–water interface. The value of the interfacial tension is generally between those of the surface tensions of the two liquids involved except where there is interaction between them. Table 6.2 includes several such examples. The interfacial tension at the octanol–water interface is considerably lower than the surface tension of octanol due to hydrogen bonding between these two liquids.

Amphiphilic molecules in aqueous solution have a tendency to seek out the surface and to orientate themselves in such a way as to remove the hydrophobic group from the aqueous environment and hence achieve a minimum free energy state (see figure 6.1).

A consequence of the intrusion of surfactant molecules into the surface or interfacial layer is that some of the water molecules are effectively replaced by hydrocarbon or other non-polar groups. Since the forces of intermolecular attraction between water molecules and non-polar groups are less than those existing between two water molecules, the contracting power of the surface is reduced and so therefore is the surface tension. In some cases the interfacial tension between two liquids may be reduced to such a low level (10^{-3} mN m^{-1}) that spontaneous emulsification of the two immiscible liquids is observed. These very low interfacial tensions are of relevance in an understanding of the formation and stabilisation of emulsions and will be dealt with in more detail in chapter 7.

6.2.2 Gibbs adsorption equation

It is important to remember that there is an equilibrium established between the surfactant molecules at the surface or interface and those remaining in the bulk

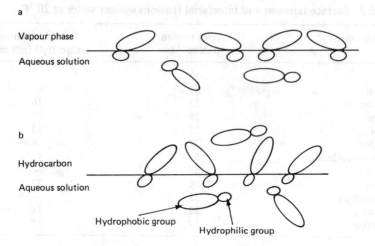

figure 6.1 Orientation of amphiphiles at (a) solution–vapour interface, and (b) hydrocarbon–solution interface

of the solution. This equilibrium is expressed in terms of the Gibbs equation. In developing this expression it is necessary to imagine a definite boundary between the bulk of the solution and the interfacial layer (see figure 6.2). The real system containing the interfacial layer is then compared with this reference system in which it is assumed that the properties of the two bulk phases remain unchanged up to the dividing surface.

We can treat the thermodynamics of the surface layer in a similar way to the bulk of the solution. The energy change, dU, accompanying an infinitesimal, reversible change in the system is given by

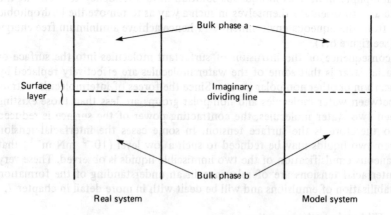

figure 6.2 Diagrammatic representation of an interface between two bulk phases in the presence of an adsorbed layer

$$dU = q_{rev} - w$$

or $\qquad\qquad dU = TdS - w \qquad\qquad\qquad (6.2)$

For an open system (one in which there is transference of material between phases) equation 6.2 must be written

$$dU = TdS - w + \sum \mu_i \, dn_i \qquad\qquad (6.3)$$

where μ_i and n_i are the chemical potential and number of moles of the i-th component, respectively.

When applying equation 6.3 to the surface layer, the work is that required to increase the area of the surface by an infinitesimal amount, dA, at constant T, P and n. This work is done against the surface tension and is given by equation 6.1 as $w = \gamma dA$.

Thus, equation 6.3 becomes

$$dU^s = T \, dS^s + \gamma dA + \sum \mu_i \, dn_i^s \qquad\qquad (6.4)$$

where the superscript, s, denotes the surface layer.

If the energy, entropy and number of moles of component are allowed to increase from zero to some finite value, equation 6.4 becomes

$$U^s = TS^s + \gamma A + \sum \mu_i n_i^s \qquad\qquad (6.5)$$

Differentiating equation 6.5 generally, gives

$$dU^s = TdS^s + S^s dT + \sum \mu_i \, dn_i^s$$
$$+ \sum n_i^s \, d\mu_i + \gamma dA + A \, d\gamma \qquad\qquad (6.6)$$

Comparison with equation 6.4 gives

$$0 = S^s dT + \sum n_i^s \, d\mu_i + A \, d\gamma \qquad\qquad (6.7)$$

At constant temperature equation 6.7 becomes

$$d\gamma = - \sum \Gamma_i \, d\mu_i \qquad\qquad (6.8)$$

where $\Gamma_i = n_i^s / A$ and is termed the *surface excess concentration*. Γ_i is the amount of the i-th component in the surface phase s, in excess of that which there would have been had the bulk phases a and b extended to the dividing surface without change in composition.

For a two-component system at constant temperature equation 6.8 reduces to

$$d\gamma = -\Gamma_1 \, d\mu_1 - \Gamma_2 \, d\mu_2 \qquad\qquad (6.9)$$

The surface excess concentrations are defined relative to an arbitrarily chosen dividing surface. A convenient choice of location of this surface is that at which the surface excess concentration of the solvent, Γ_1, is zero. Indeed, this is the most realistic position since we are now considering the surface layer of adsorbed solute. Equation 6.9 then becomes

$$d\gamma = -\Gamma_2 \, d\mu_2 \qquad\qquad (6.10)$$

The chemical potential of the solute is given by equation 3.31 as

$$\mu_2 = \mu_2^{\ominus} + RT \ln a_2$$

$$d\mu_2 = RT \, d \ln a_2$$

Therefore
$$\Gamma_2 = \frac{-1}{RT} \frac{d\gamma}{d \ln a_2} = \frac{-a_2}{RT} \frac{d\gamma}{da_2} \qquad (6.11)$$

For dilute solutions we may substitute concentration for activity

$$\boxed{\Gamma_2 = \frac{-1}{RT} \frac{d\gamma}{d \ln c} = \frac{-c}{RT} \frac{d\gamma}{dc}} \qquad (6.12)$$

Equation 6.12 is the usual form of the Gibbs equation and is applicable to the adsorption of non-ionic surfactants. For ionic surfactants the derivation becomes more complex since consideration must be taken of the adsorption of both surfactant ion and counterion. The general form of the Gibbs equation is written[1]

$$\boxed{\Gamma_2 = \frac{-1}{xRT} \frac{d\gamma}{d \ln c} = \frac{-1}{xRT} \frac{d\gamma}{2.303 \, d \log c}} \qquad (6.13)$$

where x has a numerical value varying from 1 (for ionic surfactants in dilute solution) to 2 (in concentrated solution). In the presence of excess electrolyte the normal form (that is, $x = 1$) of the Gibbs equation may be applied.

6.2.3   Application of the Gibbs equation to surfactant solutions

Figure 6.3 shows typical plots of surface tension against log of concentration for solutions of non-ionic surfactants with the general formula $CH_3(CH_2)_{15}$ $(OCH_2 CH_2)_n$ OH. Appreciable lowering of surface tension is evident even at low concentrations. As the surfactant concentration is increased the surface layer becomes saturated with surfactant molecules. An alternative means of shielding the hydrophobic portion of the amphiphile from the aqueous environment occurs at higher concentrations when micelles are formed within the bulk of the solution. The surface excess concentration of surfactant molecules remains approximately constant in the presence of micelles and hence the γ-log concentration plot becomes almost horizontal. The discontinuity in the plot is identified with the critical micelle concentration (cmc). The slope of the plot reaches a limiting value at concentrations just below the cmc. The surfactant molecules are closely packed in the surface over this narrow concentration range and the surface area A occupied per molecule may be determined from

$$\boxed{A = 1/N_A \Gamma_2} \qquad (6.14)$$

where Γ_2 is the value of surface excess concentration calculated from the Gibbs equation using the limiting $d\gamma/d \ln c$ value and N_A is Avogadro's constant.

Example 6.1

The limiting slope of a plot of γ against $\log c$ for the antihistamine diphenhydramine hydrochloride (see figure 6.5) is -0.0115 N m^{-1} at 30 °C. Calculate the area per molecule of this drug at the air–solution interface.

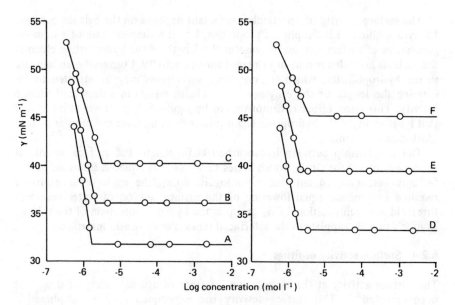

figure 6.3 Surface tension versus log concentration plots for non-ionic surfactants with the general formula $CH_3(CH_2)_{15}(OCH_2CH_2)_nOH$: A, $n = 6$; B, $n = 9$; C, $n = 15$; D, $n = 7$; E, $n = 12$; F, $n = 21$. From P. H. Elworthy and C. B. Macfarlane. *J. Pharm. Pharmacol.*, 14, 100T (1962) with permission

The surface excess concentration, Γ_2, may be calculated from equation 6.13, assuming a value of $x = 1$

$$\Gamma_2 = \frac{(0.0115)}{8.314 \times 303 \times 2.303}$$

$$= 1.982 \times 10^{-6} \text{ mol m}^{-2}$$

Substituting in equation 6.14

$$A = \frac{1}{6.023 \times 10^{23} \times 1.982 \times 10^{-6}}$$

$$= 83.8 \times 10^{-20} \text{ m}^2 \text{ molecule}$$

The area per molecule of diphenhydramine = 0.84 nm^2. ∎

An interesting effect arises when the surfactant is contaminated with surface-active impurities. A pronounced minimum is observed at the cmc which would seem to be an apparent violation of the Gibbs equation, suggesting a desorption (positive $d\gamma/d \log c$ value) in the vicinity of the cmc. The minimum in fact arises because of the release below the cmc of the surface-active impurities on the breakup of the surfactant micelles in which they were solubilised.

The surface activity of a particular surfactant depends on the balance between its hydrophilic and hydrophobic properties. For the simplest case of a homologous series of surfactants, an increase in the length of the hydrocarbon chain as the series is ascended results in increased surface activity. Conversely, an increase in the hydrophilicity, which for non-ionic surfactants may be effected by increasing the length of the ethylene oxide chain, results in a decreased surface activity. This latter effect is demonstrated by figure 6.3, from which it is noted that lengthening of the hydrophilic chain results in an increase in both the surface tension and the cmc.

The relationship between hydrocarbon chain length and surface activity is expressed by Traube's rule, which states that 'in dilute aqueous solutions of surfactants belonging to any one homologous series, the molar concentrations required to produce equal lowering of the surface tension of water decreases threefold for each additional CH_2 group in the hydrocarbon chain of the solute'. Traube's rule also applies to the interfacial tension at oil–water interfaces.

6.2.4   Surface activity of drugs

The surface activity at the air–solution interface of a wide variety of drugs has been reported[2,3]. This surface activity is a consequence of the amphipathic nature of the drugs. The hydrophobic portions of the drug molecules are in general more complex than those of typical surfactants, being composed of aromatic or heterocyclic ring systems. Examples of the types of drug that exhibit surface activity are illustrated in figure 6.4. They include the phenothiazine tranquillisers such as promazine, chlorpromazine, promethazine, and the antidepressants, for example, imipramine, amitriptyline, nortriptyline, which have tricyclic hydrophobic moieties; the antihistamines (for example, chlorcyclizine and diphenhydramine) and the antiacetylcholine drugs such as orphenadrine and lachesine, which are based on a diphenylmethane hydrophobic group; the local anaesthetics (tetracaine, for example) and several antihistamines, such as brompheniramine and mepyramine, which have a hydrophobic group consisting of a single phenyl ring.

As with typical surfactants, the surface activity depends on the nature of the hydrophobic and hydrophilic portions of the drug molecule. The presence of any substituents on the aromatic ring systems can have an appreciable effect on surface activity. Figure 6.5 shows the increased hydrophobicity that is imparted by a Cl or Br substituent on a phenyl ring of several antihistamines. Similarly, substitution on the phenothiazine ring systems was found[4] to increase surface activity in the order $CF_3 \gg Cl > H$.

Several attempts have been made to correlate the surface activity of a drug and its biochemical and pharmacological actions. In a few cases such correlations have been established; for example, the surface activity of a series of neuroleptic phenothiazines has been related to their clinical potencies[5].

6.2.5   Insoluble monolayers

In section 6.2.3 we examined the case where the surface of a solution containing an amphiphile became covered with a monomolecular film as a result of adsorption from solution. The molecules in such films are in equilibrium with those in

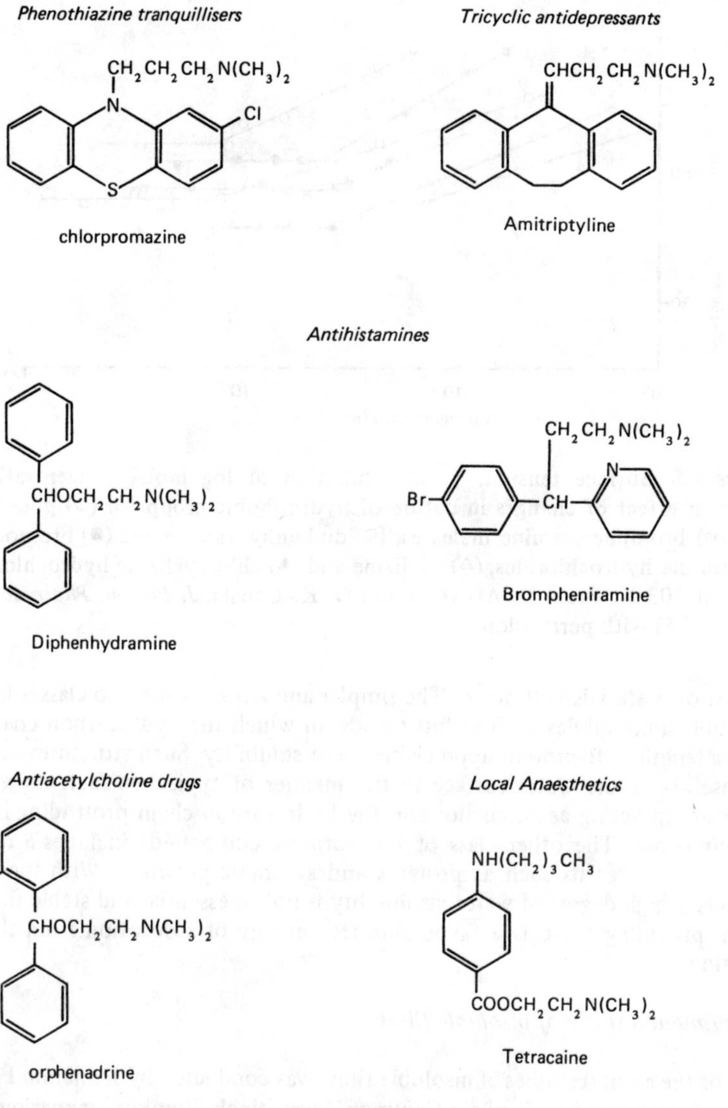

Phenothiazine tranquillisers

CH$_2$CH$_2$CH$_2$N(CH$_3$)$_2$

chlorpromazine

Tricyclic antidepressants

CHCH$_2$CH$_2$N(CH$_3$)$_2$

Amitriptyline

Antihistamines

CHOCH$_2$CH$_2$N(CH$_3$)$_2$

Diphenhydramine

CH$_2$CH$_2$N(CH$_3$)$_2$

Br—⟨ ⟩—CH—

Brompheniramine

Antiacetylcholine drugs

CH$_3$

CHOCH$_2$CH$_2$N(CH$_3$)$_2$

orphenadrine

Local Anaesthetics

NH(CH$_2$)$_3$CH$_3$

COOCH$_2$CH$_2$N(CH$_3$)$_2$

Tetracaine

figure 6.4   Examples of surface-active drugs

the bulk of the solution. In contrast, films of certain insoluble substances may be formed on the surface of a substrate by dissolving the substance in a suitable volatile solvent and carefully injecting the solution on to the surface. The film so obtained is not in equilibrium with the bulk solution. All of the spread film molecules remain on the surface and hence the number of molecules per unit area of surface is generally known directly.

Although the films are called insoluble films, this is not meant to imply that any insoluble substance will form a stable monolayer and, in fact, only two

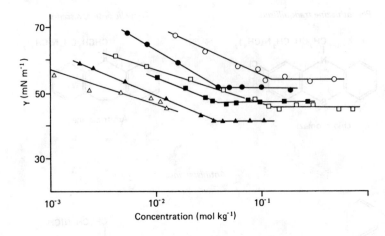

figure 6.5 Surface tension, γ, as a function of log molal concentration, m, showing effect of changes in nature of hydrophobic group for ($\circ$) pheniramine and ($\bullet$) brompheniramine maleates, ($\square$) diphenhydramine and ($\blacksquare$) bromodiphenhydramine hydrochlorides, ($\triangle$) cyclizine and ($\blacktriangle$) chlorcyclizine hydrochlorides in H_2O at 303 K. From D. Attwood and O. K. Udeala. *J. Pharm. Pharmacol.*, 27, 754 (1975) with permission

classes of materials will do so. The simpler and larger of the two classes includes insoluble amphiphiles such as fatty acids, in which the hydrocarbon chain is of such a length as to prevent appreciable water solubility. Such structures orientate themselves at the water surface in the manner of typical surfactants with the polar group acting as an anchor and the hydrocarbon chain protruding into the vapour phase. The other class of film-forming compounds includes a range of polymeric materials such as proteins and synthetic polymers. With these compounds a high degree of water insolubility is not so essential and stable films will form, providing there is a favourable free energy of adsorption from the bulk solution.

Experimental study of insoluble films

One of the earliest studies of insoluble films was conducted by Benjamin Franklin in 1765 on a pond in Clapham Common. Surprisingly Franklin's experiment was sufficiently controlled to establish that olive oil formed a film of monolayer thickness (quoted as one ten millionth of an inch: approximately 2.5 nm). Nowadays experiments are conducted on a less grand scale. Figure 6.6 illustrates a commonly used apparatus for this purpose.

The apparatus consists essentially of a trough with waxed or Teflon sides (non-wetting), along which a non-wetting barrier may be mechanically moved. In use, the trough is filled completely so as to build up a meniscus above the level of the sides. The surface is swept clean with the moveable barrier and any surface impurities are sucked away using a water pump. The film-forming material is dissolved in a suitable volatile solvent and an accurately measured amount, usually about 0.01 mℓ, of this solution is carefully distributed on to the surface. The

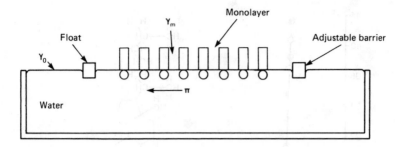

figure 6.6   Langmuir trough for monolayer studies

solvent evaporates and leaves a uniformly spread film which can now be compressed using the moveable barrier. For each setting of the barrier, a force is
applied to a torsion wire attached to the float to maintain the float at a fixed
position. This force is a direct measure of the surface pressure, π, of the film;
that is, the difference between the surface tension of the clean surface γ_0 and
that of the film-covered surface, γ_m.

$$\pi = \gamma_0 - \gamma_m \qquad (6.15)$$

The results are generally presented as graphs of π against the surface area per
molecule, A (readily calculated from the number of molecules added to the surface and the area enclosed between the float and the barrier).

Monolayer states

The surface film acts as a two-dimensional analogue to normal matter in that it
may exist in different physical states, which in some ways resemble solids, liquids
and gases. In this section we shall consider the monolayer states of simple
amphiphiles.

(A) Solid or condensed state

Figure 6.7 shows the π–A curve for cholesterol which produces a typical condensed film on an aqueous substrate. The film pressure remains very low at high
film areas and rises abruptly when the molecules become tightly packed on
compression. Simultaneous electron micrographs of the film-covered surface have
shown cholesterol clusters or islands which gradually pack more tightly at greater
pressures. The film becomes continuous as the pressure is further increased and
at such high pressures the molecules are in contact and orientated as depicted in
figure 6.7. The extrapolated limiting surface area of 0.39 nm^2 is very close to
the cross-sectional area of a cholesterol ring system calculated from molecular
models.

Similar films are formed by long-chain fatty acids such as stearic and palmitic
acid, for which a limiting surface area of about 0.20 nm^2 is found. This value is
very close to the cross-sectional area of the compounds in the bulk crystal as
determined by X-ray diffraction.

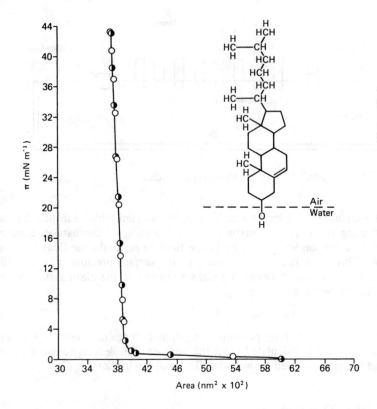

figure 6.7 Surface pressure π versus area per molecule for cholesterol, which forms a typical condensed monolayer, and a schematic drawing of the orientated molecule. From H. E. Reiss, *et. al. J. colloid interface Sci.*, 57, 396 (1976) with permission

(B) Gaseous monolayers
These films represent the other extreme in behaviour to the condensed film. They resemble the gaseous state of three-dimensional matter in that the molecules move around in the film, remaining a sufficiently large distance apart so as to exert very little force on each other. Upon compression the surface pressure approaches zero asymptotically, in marked contrast to the behaviour of solid films. It is thought that the molecules in these types of monolayer lie along the surface and this is certainly so with those dibasic esters with terminal polar groups which anchor the molecules flat on the surface. Those steroids in which the polar groups are distributed about the molecule tend to form gaseous films for similar reasons.

(C) Expanded monolayers
Variously named liquid-expanded, expanded or liquid, these monolayers represent intermediate states between gaseous and condensed films. The $\pi-A$ plots are quite steeply curved and extrapolation to a limiting surface area yields a value which is usually several times greater than the cross-sectional area from molecular

models. Films of this type tend to be formed by molecules in which close packing into condensed films is prohibited by bulky side chains or, as in the case of oleyl alcohol (figure 6.8), by a *cis* configuration of the molecule.

It was suggested by Langmuir that the hydrophobic regions, rather than being arranged in a regular manner, are in fact randomly orientated or intertwined as in a liquid.

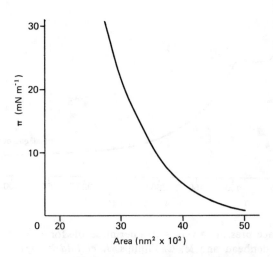

figure 6.8   Surface pressure, π versus area per molecule for oleyl alcohol which forms a typical expanded monolayer. From reference 6 with permission

Transition between monolayer states

Many simple molecules, rather than exhibiting behaviour exclusively characteristic of one monolayer state, show transitions between one state and another as the film is compressed. Oestradiol diacetate, for example (figure 6.9), shows typical gaseous behaviour at a large area per molecule, and in this state the molecules are thought to be lying along the surface, as might be expected from the location of the hydrophilic groups on the molecule. As compression is applied, the molecules are gradually pressed closer together until at a molecular surface area of approximately $0.96 \ nm^2$ the molecules begin to stand upright. The film now undergoes a gradual transition to a condensed film as the proportion of upright molecules increases with further compression, until at approximately $0.38 \ nm^2$ the film is totally in the condensed form.

In some compounds, notably myristic acid, the extent of the gaseous, expanded and condensed regions varies with temperature (figure 6.10). There is an analogy between the π–A curves of such compounds and the PV isotherms of three-dimensional gases.

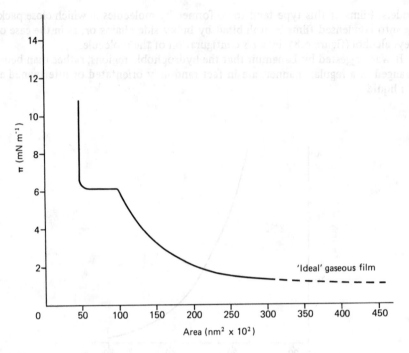

figure 6.9   Surface pressure π versus area per molecule for β-oestradiol diacetate. From D. A. Cadenhead and M. C. Phillips. *J. colloid interface Sci.*, 24, 491 (1967)

Polymer monolayers

Monolayers of polymers and proteins lack the characteristic features described in the previous section. Most produce smooth curves, typical of those for gaseous monolayers of amphiphiles, although attempts at classification into condensed and expanded classes have been made[6].

6.2.6   Pharmaceutical applications of surface film studies

In a study of polymers used as packaging materials and film coatings

Packaging materials must protect the drug without altering in any way the composition of the product. One of the problems which is discussed in section 6.2.7 is that of adsorption of constituents, for example, preservatives, from the drug product. The permeability of the packing material to gases or liquids should also be considered, since this may result in deterioration of the product due to oxidation, hydrolysis or loss of volatile ingredients. Monolayers are useful models by which the properties of polymers used as packaging materials can be investigated.

Several methods have been employed in the determination of the resistance of monolayers to evaporation. The evaporation rate may be determined from the increase in mass of a desiccant suspended over the monolayer, or from the loss

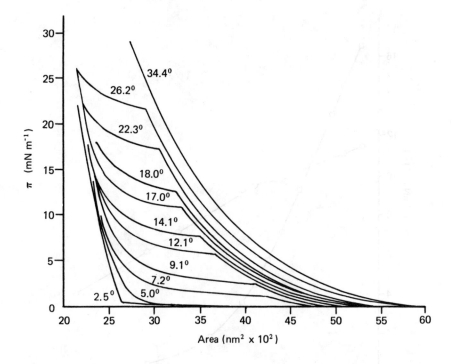

figure 6.10 Surface pressure π versus area per molecule for myristic acid spread on HCl (0.01 mol ℓ^{-1}) at various temperatures (°C). From N. K. Adam and G. Jessop. *Proc. R. Soc.* A, 112, 364 (1926) with permission

of weight of a petri dish containing solution and spread monolayer, under carefully controlled conditions. Such experiments are useful in determining the effect on permeability of incorporation of a plasticiser into the polymer structure[7].

Polymer monolayers have been used as models to assess the suitability of new polymers and of polymer mixtures as potential enteric and film coatings for solid dosage forms. Zatz and Knowles[8] reported the effects of substrate pH on the properties of three esters of cellulose, namely, cellulose acetate phthalate (CAP), cellulose acetate butyrate (CAB) and cellulose acetate stearate (CAS). Monolayers of CAB and CAS were virtually unaffected by changes of pH of the substrate from 3 to 6.5. Condensed films were formed at both pH values, indicating that disintegration in either the stomach or small intestine would be prevented. Neither of these cellulose esters would therefore be of use as enteric coatings. CAP, on the other hand, formed a much more condensed monolayer at pH 3 than at pH 6.5 (see figure 6.11). These conformational changes of CAP suggested its suitability as an enteric coating: the more tightly packed film at low pH would restrict dissolution in the stomach, whereas the more expanded film at higher pH would allow penetration of water and tablet disintegration in the small intestine where the environmental pH is approximately 6.

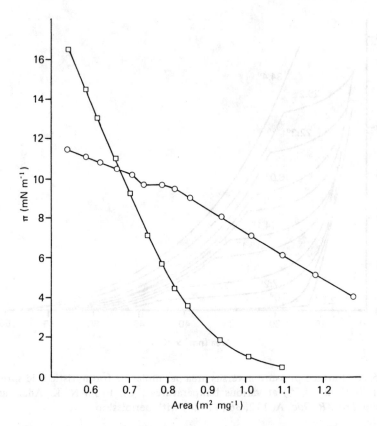

figure 6.11 Surface pressure π versus area for cellulose acetate phthalate at two subphase pH values: (□), pH 3.1, and (○), pH 6.5. From reference 8 with permission

As cell membrane models

Insoluble films of combinations of phospholipids and proteins have been used in attempts to provide information on the structure and functions of cell membranes (see reviews by Colacicco[9] and Arnold and Pak[10]). A number of studies of the interaction of drugs and other agents with phospholipid or cholesterol monolayers have been reported[3].

Figure 6.12 shows the interaction of a series of antihistamines with lecithin monolayers. In all cases, an increase in surface pressure was noted for the film spread on the surface of drug solution compared with that on a clean water surface. This effect is indicative of penetration of the film by the drug molecules. The magnitude of the surface pressure increase is related to the hydrophobicity of the drug. Thus the diphenylmethane derivatives, chlorcyclizine, bromodiphenhydramine and diphenhydramine, produced a greater rise of surface pressure than the less hydrophobic antihistamines such as tripelennamine, mepyramine and pheniramine, in which the hydrophobic moiety is composed of a single phenyl

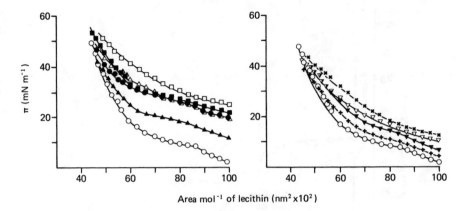

figure 6.12 Plot of surface pressure π, against area per molecule of dipalmitoyl lecithin for lecithin films on (O) H_2O and on 5×10^{-3} mol kg^{-1} solutions of (□) dimenhydrinate, (■) chlorcyclizine HCl, (●) bromodiphenhydramine HCl, (△) cyclizine HCl, (▲) diphenhydramine HCl, (✕) tripelennamine HCl, (▽) mepyramine maleate, (▼) thenyldiamine HCl, and (✛) pheniramine maleate. From D. Attwood and O. K. Udeala. *J. Pharm. Pharmacol.*, 27, 806 (1975) with permission

ring. Similar results were reported for several phenothiazine tranquillisers[11,12]. Excellent correlations between film penetration studies and the ability of a number of local anaesthetics to block nerve impulse conduction have been reported[13,14]. Similarly, the ability of a group of polyene antibiotics to produce membrane damage showed good agreement with their interactions with lipid monolayers[15].

A more sophisticated model of the cell membrane is provided by monolayers that are mixtures of molecules which are horizontally orientated at the interface, thus resembling proteins, and those which are vertically orientated resembling lipids. Figure 6.13 shows the surface pressure–area isotherms for equimolar mixtures of valinomycin (a cyclic dodecadepsipeptide) which orientates horizontally at the air–solution interface to give an expanded film, and cholesterol which orientates vertically giving a solid film. The shape of the mixed isotherm at low and intermediate pressures is similar to that of valinomycin, whilst the behaviour at high surface pressures is similar to that of cholesterol, suggesting that the valinomycin has been squeezed out of the mixed film. The position of the mixed curve to the left of the calculated average curve suggests some form of interaction between the components which condenses the mixed film. Such interactions are quite common in mixed films.

6.2.7 Adsorption at the solid–liquid interface

The term adsorption is used to describe the process of accumulation at an interface. Adsorption is essentially a surface effect and should be distinguished from *absorption*, which implies the penetration of one component throughout the body of a second. The distinction between the two processes is not always clear

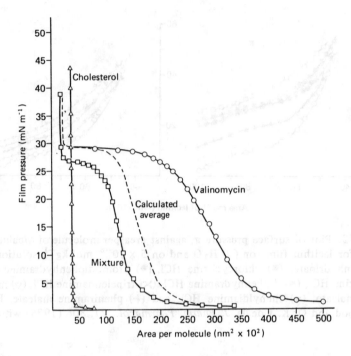

figure 6.13 Pressure–area isotherms for cholesterol, valinomycin and an equi-molar mixture of the two. From H. E. Reiss and H. S. Swift. *J. colloid interface Sci.*, 64, 111 (1978) with permission

cut, however, and in such cases the non-commital word 'sorption' is sometimes used.

There are two general types of adsorption: physical adsorption, in which the adsorbate is bound to the surface through the weak van der Waals forces, and chemical adsorption (chemisorption), which involves the stronger valence forces. Of the two processes, chemisorption is the more specific, and usually involves an ion-exchange process. Frequently both physical and chemical adsorption may be involved in a particular adsorption process. This is the case with the adsorption of toxins in the stomach by attapulgite and kaolin: there is chemisorption involv-ing cation exchange with the basic groups of the toxins and also physical adsorp-tion of the remainder of the molecule.

Adsorption isotherms

The study of adsorption from solution is experimentally straightforward. A known mass of the adsorbent material is shaken with a solution of known con-centration at a fixed temperature. The concentration of the supernatant solution is determined by either physical or chemical means and the experiment is con-tinued until no further change in the concentration of the supernatant is observed; that is, until equilibrium conditions have been established. Equations originally derived for the adsorption of gases on solids are generally used in the interpretation of the data, the Langmuir and Freundlich equations being the most commonly used. When applied to adsorption from solution, the Langmuir equation becomes

$$x/m = abc/(1 + bc) \qquad (6.16)$$

where x is the amount of solute adsorbed by a weight, m, of adsorbent, c is the concentration of solution at equilibrium, b is a constant related to the enthalpy of adsorption, and a is related to the surface area of the solid. Figure 6.14 shows a typical Langmuir isotherm for the adsorption of the antidepressant drug, amitriptyline, on carbon black.

Equation 6.16 can be arranged into the linear form

$$\boxed{c/(x/m) = 1/ab + c/a} \qquad (6.17)$$

Values of a and b may be determined from the intercept and slope of plots of $c/(x/m)$ against concentration.

Example 6.2

Calculate the Langmuir constants for the adsorption of amitriptyline on carbon black using the data from figure 6.14.

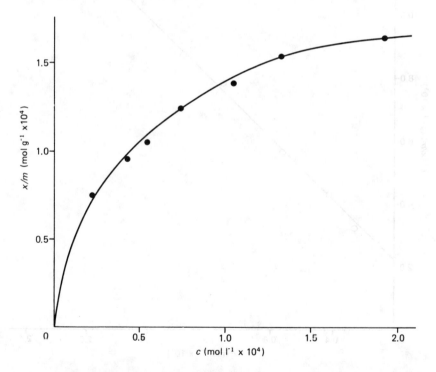

figure 6.14 Langmuir adsorption isotherm of amitriptyline on carbon black from aqueous solution at 30 °C. From N. Nambu, S. Sakurai and T. Nagai. *Chem. pharm. Bull.*, 23, 1404 (1975) with permission

From figure 6.14

x/m (mol g^{-1} × 10^3)	0.75	0.95	1.10	1.25	1.40	1.55	1.65
c (mol ℓ$^{-1}$ × 10^4)	0.25	0.40	0.60	0.70	1.10	1.35	1.95
$c/(x/m)$ (g ℓ$^{-1}$ × 10^2)	3.33	4.21	5.45	5.60	7.86	8.71	11.82

Slope of plot of $c/(x/m)$ against c (see figure 6.15) = 4.88 × 10^2 g mol^{-1} = 1/a

Therefore $a = 1/\text{slope} = 2.05 \times 10^{-3}$ mol g^{-1}

Intercept = $2.35 \times 10^{-2} = 1/ab$

Therefore $b = \dfrac{1}{2.35 \times 10^{-2} \times 2.05 \times 10^{-3}} = 2.07 \times 10^4$ ℓ mol^{-1}. ∎

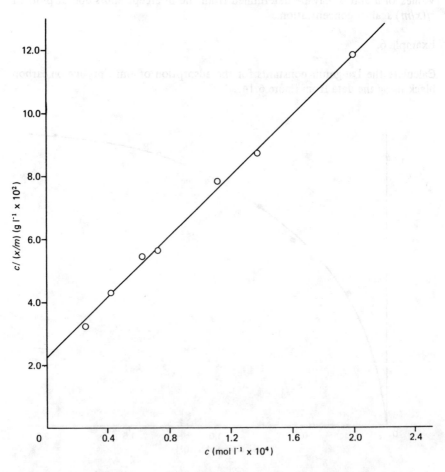

figure 6.15 Adsorption of amitriptyline by carbon black plotted according to equation 6.17 using the data from figure 6.14

The value of a is the measure of the adsorptive capacity of the adsorbent for the particular adsorbate under examination. Table 6.3 gives the adsorptive capacity of carbon black for a series of antidepressant and phenothiazine drugs, arranged in order of decreasing degree of adsorption.

Deviations from the typical Langumuir plot can occur at high concentrations and are then usually attributed to the formation of multilayers.

table 6.3 **Langmuir constants a and b in the adsorption of antidepressants and phenothiazines by carbon black***

Drugs	a $(\times 10^3 \text{ mol g}^{-1})$	b $(\times 10^{-4} \text{ } \ell \text{ mol}^{-1})$
Antidepressants		
Amitriptyline	2.05	2.07
Imipramine	1.80	1.48
Carpipramine	1.54	2.36
Opipramol	1.51	1.77
Desipramine	1.36	4.70
Dipiperon	1.15	0.17
Phenothiazines		
Promazine	1.70	3.36
Chlorpromazine	1.70	4.37
Isothipendyl	1.30	2.23
Chlorpromazine sulphoxide	1.13	3.18

*From N. Nambu, S. Sakurai and T. Nagai. *Chem. pharm. Bull.*, 23, 1404 (1975)

The second equation, the Freundlich equation, is generally written in the form:

$$x/m = ac^{1/n} \tag{6.18}$$

where a and n are constants, the form $1/n$ being used to emphasise that c is raised to a power less than unity. Equation 6.18 can be written in a linear form by taking logarithms of both sides, giving

$$\boxed{\log (x/m) = \log a + (1/n) \log c} \tag{6.19}$$

A plot of $\log (x/m)$ against $\log c$ should be linear, with an intercept of $\log a$ and slope of n^{-1}. It is generally assumed that, for systems which obey this equation, adsorption results in the formation of multilayers rather than a single monolayer. Figure 6.16 shows Freundlich isotherms for the anticholinergic drug oxyphencyclimine on various adsorbents.

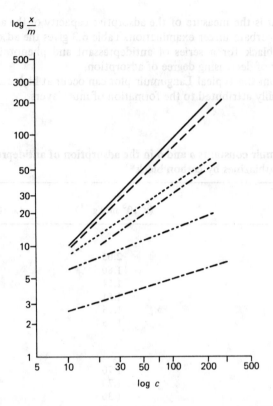

$\log \dfrac{x}{m}$

$\log c$

figure 6.16 Freundlich adsorption isotherms for oxyphencyclidine hydro-chloride on various antacids. (———) Magnesium trisilicate; (— — —) magnesium oxide; (------) calcium carbonate; (—··—) aluminium hydroxide; (—···—) kaolin; (— — ···— —) bismuth oxycarbonate. From S. A. Khalil and M. A. Moustafa. *Pharmazie*, 28, 116 (1973) with permission

Factors affecting adsorption

(A) Solubility of the adsorbate

Solubility is an important factor affecting adsorption. In general the extent of adsorption of a solute is inversely proportional to its solubility in the solvent from which adsorption occurs. This empirical rule is termed Lundelius' rule. There are numerous examples of the applicability of this rule; for example, in Lundelius' original work it was noted that the adsorption of iodine onto carbon from CCl_4, $CHCl_3$ and CS_2 was 1:2:4.5, respectively. These ratios are close to the inverse ratios for the solubilities of iodine in the respective solvents. The effect of solubility on adsorption might be expected since, in order for adsorption to occur, solute–solvent bonds must first be broken. The greater the solubility, the stronger are these bonds and hence the smaller the extent of adsorption.

For compounds with hydrocarbon chains, adsorption from solution increases as the homologous series is ascended and the molecules become more hydrophobic. This effect is analogous to that noted for the dependence of surface tension on hydrocarbon chain length (Traube's rule). Thus Traube's rule is a special case of Lundelius' rule.

The amount of adsorption of barbiturates on carbon black was found to increase regularly with increase in molar volume (see figure 6.17). All the barbituric acid derivatives examined had the same hydrophilic moiety and thus the gradual increase in the amount adsorbed was directly related to an increase in the size of the hydrophobic group, as would be expected from Traube's rule.

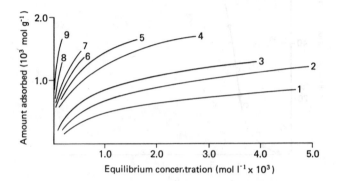

figure 6.17 Adsorption isotherms of barbituric and thiobarbituric acid derivatives from aqueous solution by carbon black at 40 °C. 1, Barbitone; 2, probarbitone; 3, allobarbitone; 4, cyclobarbitone and phenobarbitone; 5, pentobarbitone and amobarbitone; 6, heptabarbitone; 7, secobarbitone; 8, thiopentone; and 9, thiamylal. From H. Nogami, T. Nagai and H. Uchida. *Chem. pharm. Bull*, 17, 168 (1969) with permission

(B) pH

pH affects adsorption for a variety of reasons, the most important from a pharmaceutical viewpoint being its effect on the ionisation and solubility of the adsorbate drug molecule. In general, for simple molecules adsorption increases as the ionisation of the drug is suppressed, the extent of adsorption reaching a maximum when the drug is completely unionised. Figure 6.18 shows that the pH profile for the sorption (this is not a true adsorption process) of benzocaine by nylon 6 powder was indeed almost superimposable on the drug dissociation curve. For amphoteric compounds, adsorption is at a maximum at the isoelectric point; that is, when the compound bears a net charge of zero.

In general, pH and solubility effects act in concert, since the unionised form of most drugs in aqueous solution has a low solubility. Of the two effects, the solubility effect is usually the stronger. Thus, in the adsorption of hyoscine and atropine on magnesium trisilicate it was noted[16] that hyoscine, although in its completely unionised form, was less strongly adsorbed than atropine, which at the pH of the experiment was 50 per cent ionised. The reason for this apparently anomalous result is clear when the solubilities of the two bases are considered. Hyoscine base is freely soluble (1 in 9.5 parts of water at 15 °C) compared with

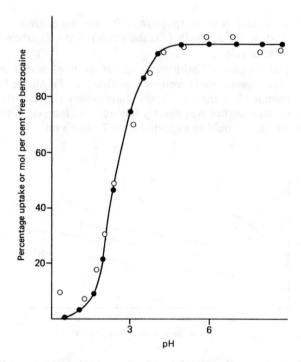

figure 6.18 pH profile for the sorption of benzocaine by nylon 6 powder from buffered solutions at 30 °C and ionic strength 0.5 mol ℓ^{-1} (○) and the corresponding drug dissociation curve (●). From N. E. Richards, and B. J. Meakin. *J. Pharm. Pharmacol.*, 26, 166 (1974) with permission

atropine base (1 in 400 at 20 °C). Even when 50 per cent ionised, atropine is less soluble than hyoscine and consequently more strongly adsorbed.

(C) Nature of the adsorbent
The physicochemical nature of the adsorbent can have profound effects on the rate and capacity for adsorption. The most important property affecting adsorption is the surface area of the adsorbent; the extent of adsorption is proportional to the specific surface area. Thus the more finely divided or the more porous the solid, the greater will be its adsorption capacity. Indeed, adsorption studies are frequently used to calculate the surface area of a solid.

Adsorbent–adsorbate interactions are of a complex nature and beyond the scope of this book. Particular adsorbents have affinities for particular adsorbates for a wide variety of reasons. The adsorbent clays such as bentonite, attapulgite and kaolin have cation-exchange sites on the surface. Such clays have strong affinities for protonated compounds which they adsorb by an ion-exchange process. In many cases, different parts of the surface of the same adsorbent have different affinities for different types of adsorbents. There is evidence, for example, that anionic materials are adsorbed on the cationic edge of kaolin particles whilst cationics are adsorbed on the cleavage surface of the particles which are negatively charged. An example of the differing affinities of a series of adsorbents

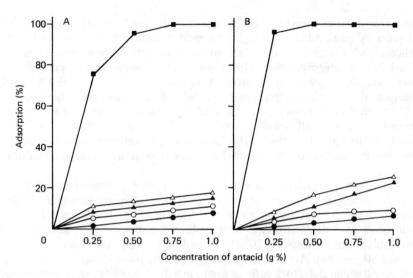

figure 6.19 Adsorption of (A) digoxin and (B) digitoxin by some antacids at 37 ± 0.1 °C. (■) Magnesium trisilicate, (△) aluminium hydroxide gel BP.*, (▲) light magnesium oxide, (○) light magnesium carbonate, (●) calcium carbonate. Initial concentration of the glycoside: 0.25 mg%. *Aludrox was used in the concentration range 2.5–10% v/v. From S. A. H. Khalil. *J. Pharm. Pharmacol.*, 26, 961 (1974) with permission

used an antacids is shown in figure 6.19. The adsorptive capacity of a particular adsorbent often depends on the source from which it was prepared and also on the pretreatment received.

(D) Temperature
Since adsorption is generally an exothermic process, an increase in temperature normally leads to a decrease in the amount adsorbed. The changes in enthalpy of adsorption are usually of the order of those for condensation or crystallisation. Thus small variations in temperature tend not to alter the adsorption process to a significant extent.

Medical and pharmaceutical applications and consequences of adsorption

The important application of adsorption in preparative and analytical chromatography is too wide a subject to be covered here and the reader is referred to reviews on the subject[17]. Similarly, the role of adsorption in heterogeneous catalysis, water purification and solvent recovery has been adequately reviewed[18]. We shall be concerned with examples of the involvement of adsorption in more medical and pharmaceutical situations.

(A) Adsorption of noxious substances from the alimentary tract
The 'universal antidote' for use in reducing the effects of poisoning by the oral route is composed of activated charcoal, magnesium oxide and tannic acid. A

more recent use of adsorbents has been in dialysis to reduce toxic concentrations of drugs by passing blood through a haemodialysis membrane over charcoal and other adsorbents. Several drugs are adsorbed effectively by activated charcoal. Drugs in this category include chlorpheniramine, propoxyphene hydrochloride, colchicine, diphenylhydantoin and acetylsalicylic acid. Some of these are easily recognisable as surface-active molecules (chlorpheniramine, propoxyphene) and will be expected to adsorb onto solids. Highly ionised substances of low molecular weight are not well adsorbed, neither are drugs such as tolbutamide that are poorly soluble in acidic media. The formation of a monolayer of drug molecules covering the surface of the charcoal particles through non-polar interactions is indicated.

The direct application of *in vitro* data for estimating doses of activated charcoal for antidotal purposes may lead to use of inadequate amounts of adsorbent[19]. In an animal study, charcoal/drug ratios of 1:1, 2:1, 4:1 and 8:1 reduced absorption of drugs as follows: pentobarbitone sodium, 7, 38, 62 and 89 per cent; chloroquine phosphate 20, 30, 70 and 96 per cent; and isoniazid 1.2, 7.2, 35 and 80 per cent. Activated charcoal, of course, is not effective in binding all poisons. Biological factors such as gastro-intestinal motility, secretions and pH may influence charcoal adsorption. While 5 g of activated charcoal have been said to be capable of binding 8 g of aspirin *in vitro*[20], 30 g of charcoal *in vivo* were reported to inhibit the gastro-intestinal absorption of 3 g of aspirin by only 50 per cent[21]. The surface area of the charcoal is a factor in its effectiveness; charcoal tablets have been found to be approximately half as effective as powdered material.

The purposeful adsorption of drugs such as diazepam onto solid substrates should be mentioned, where the object is to minimise taste problems. Desorption of the drug *in vivo* is essential but should not occur during the shelf-life of the preparation. Desorption may be a rate-limiting step in absorption. Diazepam adsorbed onto an inorganic colloidal magnesium aluminium silicate (Veegum) had the same potency in experimental animals as a solution of the drug, but adsorbed onto microcrystalline cellulose (Avicel) its efficacy was much reduced. Flocculation of the cellulose in the acidic environment of the stomach probably retards the desorption process.

An extra-corporeal method of treating cases of severe drug overdoses was introduced by Yatzidis[22]. The technique is termed carbon haemoperfusion and originally involved perfusion of the blood directly over charcoal granules. Although activated charcoal granules were very effective in adsorbing many toxic materials, they were found to give off embolising particles and also to lead to removal of blood platelets. Micro-encapsulation of activated charcoal granules by coating with biocompatible membranes such as acrylic hydrogels was found to be a successful means of eliminating charcoal embolism and to lead to a much reduced effect on platelet count. *In vitro* tests showed that the coated granules had a reduced adsorption rate although the adsorptive capacity was unchanged[23]. A large proportion of drug overdoses in Great Britain involve barbiturates, and the applicability of carbon haemoperfusion in the treatment of such cases has been demonstrated[24]. Many other drugs taken as overdoses are also present in the plasma at sufficiently high concentration to allow removal by this technique. The use of micro-encapsulated granules containing enzymes, ion-exchange resins and activated charcoal for the removal of toxins and waste metabolites in artificial kidneys and livers has been investigated[25]

(B) Adsorption problems in drug formulation

Problems arising from the adsorption of medicaments by adsorbents such as ant-acids which may be taken simultaneously by the patient or, in some cases, may be present in the same formulation are discussed in chapter 10. Problems also arise from the adsorption of medicaments on to the container walls. Containers for medicaments, whether glass or plastic, may adsorb a significant quantity of the drug, bacteriostatic or fungistatic agent present in the formulation and there-by affect the potency and possibly the stability of the product. The problem is particularly significant where the drug is highly surface active and present in low concentration. With plastic containers the process is often referred to as sorption rather than adsorption since it often involves significant penetration of the drug into the polymer matrix. Plastics are a large and varied group of materials and their properties are often modified by various additives, such as plasticisers, fillers and stabilisers (see chapter 8). Such additives may have a pronounced effect on the sorption characteristics of the plastics. Saski[26] studied the sorption of the fungistatic agent, sorbic acid, from aqueous solution by plastic cellulose acetate and cellulose triacetate. An appreciable pH dependence of amount sorbed was noted, the sorption declining to zero in the vicinity of the point of maximum ionisation of the sorbic acid.

A study of the sorption of local anaesthetics by polyamide and polyethylene was reported by Bauer and Ullmann[27]. The amount sorbed depended on the kind of plastic, the reaction conditions and the chemical structure of the drugs. As with sorbic acid, significant sorption was only observed when the drugs were in their unionised forms.

6.3 Micellisation

As the concentration of aqueous solutions of many amphipathic substances in-creases, a pronounced deviation from ideality occurs, which is generally consider-ably larger than that exhibited by strong electrolytes. This onset of non-ideality is readily detected by a change of slope when certain physical properties such as surface tension, conductivity and light scattering are plotted as a function of concentration (see figure 6.20) and is attributable to the self-association of the amphiphile into micelles. The concentration at which the change of slope occurs is called the critical micelle concentration (cmc).

Although the concept of micellisation has now gained universal acceptance, it should be realised that the idea that molecules should come together at a critical concentration to form aggregates in solution was quite novel when first proposed by McBain in 1913. The micelles are in dynamic equilibrium with free molecules (monomers) in solution; that is, the micelles are continuously breaking down and reforming. It is this fact that distinguishes micellar solutions from other types of colloidal solution and this difference is emphasised by referring to micelle-forming compounds as *association* colloids.

The primary reason for micelle formation is the attainment of a state of mini-mum free energy. At low concentration, amphiphiles can achieve an adequate decrease in the over-all free energy of the system by accumulation at the surface or interface, in such a way as to remove the hydrophobic group from the aqueous environment. As the concentration is increased, this method of free energy reduc-

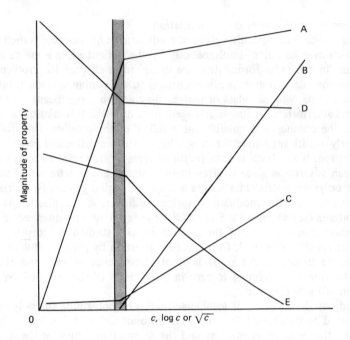

Magnitude of property

0   c, log c or $\sqrt{c}$

figure 6.20 Solution properties of an ionic surfactant as a function of concentration, c. A, Osmotic pressure (against c); B, solubility of a water-insoluble solubilisate (against c); C, intensity of light scattered by the solution (against c); D, surface tension (against log c); E, molar conductivity (against $\sqrt{c}$)

tion becomes inadequate and the monomers form into micelles. The hydrophobic groups form the core of the micelle and so are shielded from the water.

The free energy change of a system is dependent on changes in both the entropy and enthalpy; that is, $\Delta G = \Delta H - T\Delta S$. For a micellar system at normal temperatures the entropy term is by far the most important in determining the free energy changes ($T\Delta S$ constitutes approximately 90–95 per cent of the ΔG value). Micelle formation entails the transfer of a hydrocarbon chain from an aqueous to a non-aqueous environment (the interior of the micelle). To understand the changes in enthalpy and entropy which accompany this process, we must first consider the structure of water itself.

6.3.1 Water structure and hydrophobic bonding

Water possesses many unique features which distinguish it from other liquids. These arise from the unusual structure of the molecule in which the O and H atoms are arranged at the apices of a triangle, as shown in figure 6.21.

The covalent bonds between the H atoms and the O atom of the water molecule involve the pairing of the electron of each H atom with an electron in the oxygen atom's outer shell of six electrons. This pairing leaves two lone pairs of electrons in the outer shell with their orbitals extending tetrahedrally. The result-

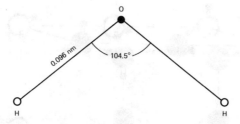

figure 6.21   Diagram of water molecule showing bond angle and length

ing tetrahedral structure has two positively charged sites at one side and two negatively charged sites at the other. It will readily attach itself by hydrogen bonds to four neighbouring molecules, two at the negatively charged sites and two at the positively charged sites. In its usual form, ice demonstrates an almost perfect tetrahedral arrangement of bonds with a distance of about 0.276 nm between neighbouring oxygen atoms. There is much unfilled space in the crystal which accounts for the low density of ice (see figure 6.22).

When ice melts there is still a high degree of hydrogen bonding in the resulting liquid. In spite of extensive investigation by a variety of techniques such as X-ray diffraction, thermochemical determination, infrared and Raman spectroscopy, the problem of the structural nature of liquid water is still to be completely resolved. There are, broadly speaking, two distinct types of model: those which involve distortion but not breaking of hydrogen bonds, and a second type in which unbonded detached water molecules exist in addition to the hydrogen-bonded structures. Of the former type, the model which is considered to be the most acceptable is one in which all the water molecules continue to be hydrogen-bonded to their four neighbours, but the intermolecular links are bent or stretched to give an irregular framework. Such distorted networks are known to exist in some of the denser forms of ice.

There have been many proposed structures for water which involve mixtures of structured material and free water molecules. One of the most highly develop-ed theories encompasses the so-called 'flickering cluster' concept of water structure. The model is based on the co-operative nature of hydrogen bonding. The formation of one hydrogen bond on a water molecule leaves the molecule more susceptible to further hydrogen bonding and similarly when one bond breaks there is a tendency for large groups of bonds to break. As a result clusters of ice-like hydrogen-bonded material are imagined to be suspended in a fluid of unbonded water (figure 6.23). Because of the continual formation and rupture of hydrogen bonds throughout the liquid, these clusters have only a temporary existence, and are aptly described by the term 'flickering'.

Most of the models proposed for the structure of water, only two of which have been considered here, can account for some, but not all of the physical and thermodynamic anomalies which have been observed with water.

The flickering cluster model can be used to describe possible structural changes which occur when non-polar and polar solutes are dissolved in water. A non-polar molecule or portion of a molecule tends to seek out the more ice-like regions

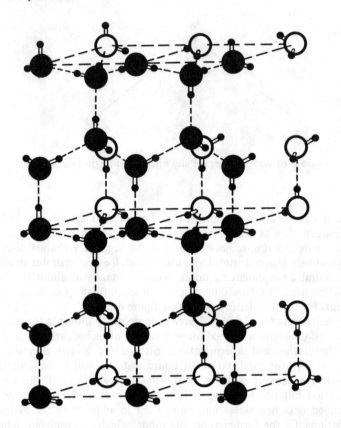

figure 6.22 The structure of ordinary ice. Large spheres represent oxygen atoms, the small spheres hydrogen atoms. Reprinted with permission from Linus Pauling *The Nature of the Chemical Bond*, Cornell University Press, 1960

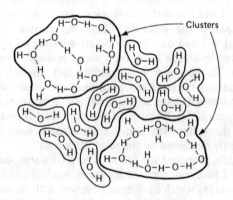

figure 6.23 Water clusters with unassociated water molecules around them. From Némethy and Scheraga *J. Chem. Phys.*, 36, 3382 (1962)

within the water. Such regions, as we have seen, contain open structures into which the non-polar molecules may fit without breaking hydrogen bonds or otherwise disturbing the surrounding ice-like material. In solution, therefore, hydrophobic molecules tend always to be surrounded by structured water. This fact is important in discussing interactions between non-polar molecules in aqueous solution, such as those which occur in micelle formation. The interaction of hydrocarbons in aqueous solution was first thought to arise simply as a consequence of the van der Waals forces between the hydrocarbon molecules. However, it was later realised that changes in the water structure around the non-polar groups must play an important role in the formation of bonds between the non-polar molecules — the so-called 'hydrophobic bonds'. In fact the contribution from the van der Waals forces is only about 45 per cent of the total free energy of formation of a hydrophobic bond. When the non-polar groups approach each other until they are in contact, there will be a decrease in the total number of water molecules in contact with the non-polar groups. The formation of the hydrophobic bond in this way is thus equivalent to the partial removal of hydrocarbon from an aqueous environment and a consequent loss of the ice-like structuring which always surrounds the hydrophobic molecules. The increase in entropy and decrease in free energy which accompany the loss of structuring make the formation of the hydrophobic bond an energetically favourable process.

There is much experimental evidence to support this explanation for the decrease in ΔG. Thus the enthalpy of micelle formation becomes more negative as the temperature is increased; a fact which was attributed to a reduction in water structure as temperature is increased. Nuclear magnetic resonance (n.m.r.) measurements indicate an increase in the mobility of water protons at the onset of micellisation. The addition of urea, a water-structure breaking compound, to surfactant solutions leads to an increase of cmc, again stressing the role of water structure in the micellisation process. An alternative explanation of the free energy decrease emphasises the increase in internal freedom of the hydrocarbon chains which occurs when these chains are transferred from the aqueous environment, where their motion is restrained by the hydrogen-bonded water molecules, to the interior of the micelle. It has been suggested that the increased mobility of the hydrocarbon chains, and of course their mutual attraction, constitute the principal hydrophobic factor in micellisation.

6.3.2 Theories of micelle formation

Two general approaches have been employed in attempting to describe the process of micellisation. In one of these, the phase separation model, the cmc is assumed to represent the saturation concentration of the unassociated polymer and the micelles are regarded as a distinct phase which separates out at the cmc. In the alternative approach, the micelles and associated monomers are assumed to be in an association–dissociation equilibrium to which the law of mass action may be applied. Neither of these models is rigorously correct, although the mass action approach seems to give a more realistic description of micellisation, and thus will be considered in more detail.

The aggregation process may in its simplest form be described by

$$ND^+ + (N - p)X^- \rightleftarrows M^{p+} \tag{6.20}$$

Equation 6.20 represents the formation of a cationic micelle M^{p+} from N surfactant ions D^+ and $N - p$ firmly held counterions X^-. Whenever the thermodynamics of a process is under consideration, it is important to define the standard states of the species. In this example, the standard states are such that the mole fractions of the ionic species are unity and the solution properties are those of the infinitely dilute solutions. The equilibrium constant K_m may be written in the usual way

$$K_m = \frac{[M^{p+}]}{[D^+]^N [X^-]^{N-p}} \tag{6.21}$$

where activity coefficients have been neglected. The analogous equation for non-ionic micelles is of a simpler form since counterion terms and charges need not be considered.

$$K_m = \frac{[M]}{[D]^N} \tag{6.22}$$

Equations 6.21 and 6.22 are important in that they can be used to predict the variation of both monomers and micelles with total solution concentration. Figure 6.24 shows the result of such a calculation for a model system.

Figure 6.24 illustrates several important points about the micellisation process. According to the mass action treatment the monomer concentration decreases very slightly above the cmc. It should be emphasised that this is a very small effect (although it can be detected experimentally from surface-tension measurements) and for most purposes it is reasonable to assume that the monomer concentration remains constant at the cmc value. A second point of interest illustrated by the mass action treatment concerns the predicted sharpness of the cmc. It is readily shown by calculations that combinations of low values of N and K_m lead

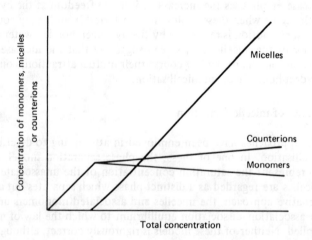

figure 6.24 Concentration of micelles, monomers and counterions against total concentration (arbitrary units) calculated from equation 6.21 for an aggregation number (N) of 100, micellar equilibrium constant (K_m) of 1 and with 85 per cent of the counterions bound to the micelle

to gradual changes of slope of the cmc region whilst larger values for both of these parameters give sharp inflections. The cmc, rather than being an exact concentration, is often a region of concentration over which the solution properties exhibit a gradual change and hence is often difficult to locate exactly.

6.3.3 Micellar structure

(A) Ionic micelles
Although the exact shape of ionic micelles is currently the subject of much discussion, it is generally assumed for the purposes of interpretation of experimental data that charged micelles of low aggregation number adopt a spherical or near spherical shape at concentrations not too far removed from the cmc. The hydrophobic part of the amphiphile is located in the core of the micelle. Around this core is a concentric shell of hydrophilic head groups together with $(1 - \alpha)N$ counterions. This compact region is termed the Stern layer (see figure 6.25). For most ionic micelles the degree of ionisation α is between 0.2 and 0.3; that is, 70-80 per cent of the counterions may be considered to be bound to the micelles.

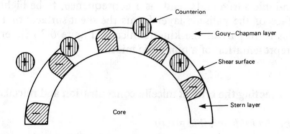

figure 6.25 Partial cross-section of an anionic micelle

The outer surface of the Stern layer is the shear surface of the micelle. The core and the Stern layer together constitute what is termed the 'kinetic micelle'. Surrounding the Stern layer is a diffuse layer called the Gouy–Chapman electrical double layer which contains the αN counterions required to neutralise the charge on the kinetic micelle. The thickness of the double layer is dependent on the ionic strength of the solution and is greatly compressed in the presence of electrolyte.

In highly concentrated solution, a gradual change in micellar shape is thought to occur with many ionic systems, the micelles elongating to form cylindrical or lamellar structures (see figure 6.26). In non-aqueous media 'inverted micelles' may form in which the hydrophilic charge groups form the micellar core shielded from the non-aqueous environment by the hydrophobic chains.

(B) Non-ionic micelles
In general, non-ionic surfactants form larger micelles than their ionic counterparts. The reason for this is clearly attributable to the removal of electrical work which must be done when a monomer of an ionic surfactant is added to an existing charged micelle. As a consequence of the larger size, the non-ionic micelles are frequently

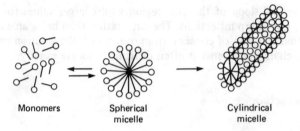

Monomers   Spherical   Cylindrical
               micelle   micelle

figure 6.26 Schematic representation of possible changes in micellar structure with increase in concentration

asymmetric. The micelles of Cetomacrogol 1000 ($C_{16}H_{33}(OCH_2CH_2)_{21}OH$) are thought[28] to be ellipsoidal with an axial ratio not exceeding 2:1.

Non-ionic micelles have a hydrophobic core surrounded by a shell of oxyethylene chains which is often termed the palisade layer. As well as the water molecules which are hydrogen bonded to the oxyethylene chains, this layer is also capable of mechanically entrapping a considerable number of water molecules. Micelles of non-ionic surfactants tend, as a consequence, to be highly hydrated. The outer surface of the palisade layer forms the shear surface; that is, the hydrating molecules form part of the kinetic micelle. Figure 6.33 (later) includes a diagrammatic representation of a non-ionic micelle.

6.3.4 Factors affecting the critical micelle concentration and micellar size

(A) Structure of the hydrophobic group
The hydrophobic group plays an important role in determining the type of association of the amphiphile. Compounds with rigid aromatic or hetero-aromatic ring structures (many dyes, purines and pyrimidines, for example) associate by a non-micellar process involving the face-to-face stacking of molecules one on top of the other, rather than by micellisation[29]. Such systems do not exhibit cmcs. Association usually commences at very low concentrations and growth of aggregates may occur by the stepwise addition of monomers. Consequently the aggregates continuously increase in size rather than attain an equilibrium size as in micellisation. Some drugs are thought to associate in this manner[3].

The most common type of micellar amphiphiles have hydrophobic groups constructed from hydrocarbon chains. Increase in length of this chain results in a decrease in cmc, and for compounds with identical polar head groups this relationship is expressed by the linear equation

$$\log [cmc] = A - Bm \qquad (6.23)$$

where m is the number of carbon atoms in the chain and A and B are constants for a homologous series. A corresponding increase in micellar size with increase in hydrocarbon chain length is also noted.

Many drugs are surface active and form small micelles in aqueous solution. For these and other amphiphiles with more complex hydrophobic regions, the effect of substituents on hydrophobicity can be roughly estimated from table

table 6.4 **Effect of substituents on the micellar properties of some diphenylmethane drugs***

Drug	R_1	R_2	R_3	cmc $(mol\ kg^{-1})$	Micellar aggregation number
Diphenhydramine HCl	H	H	H	0.132	3
Orphenadrine HCl	H	CH_3	H	0.096	7
Bromodiphenhydramine HCl	Br	H	H	0.053	11
Chlorphenoxamine HCl	Cl	H	CH_3	0.045	13

*From D. Attwood. *J. Pharm. Pharmacol.*, 24, 751 (1972); 28, 407 (1976)

5.4. In the series of micellar diphenylmethane drugs shown in table 6.4 there is an increased hydrophobicity (as evidenced by a decrease in cmc and increase in aggregation number) following the introduction of $-CH_3$, $-Br$ and $-Cl$ substituents to the hydrophobic ring systems.

(B) Nature of the hydrophilic group
The most important point to be noted here is the pronounced difference in properties between amphiphiles with ionic hydrophilic groups and those in which this group is uncharged. In general, non-ionic surfactants have very much lower cmc values and higher aggregation numbers than their ionic counterparts with similar hydrocarbon chains, mainly because the micellisation process for such compounds does not involve any electrical work.

The properties of the polyoxyethylated non-ionic surfactants show a pronounced dependence on the length of the polyoxyethylene chain. An increase in the chain length confers a greater hydrophilicity to the molecule and the cmc increases, as shown in table 6.5.

table 6.5 **Values of cmc and micellar weights of hexadecyl polyoxyethylene ethers $CH_3(CH_2)_{15}(OCH_2CH_2)_n OH$***

$n =$	6	7	9	12	15	21
10^6 cmc $(mol\ \ell^{-1})$	1.7	1.7	2.1	2.3	3.1	3.9
10^{-5} Micellar weight	12.3	3.27	1.4	1.17	–	0.82
Aggregation number	2430	590	220	150	–	70

*From P. H. Elworthy and C. B. Macfarlane, *J. Chem. Soc.*, 907 (1963); 537 (1962)

(C) Nature of the counterion
The counterion associated with the charged group of ionic surfactants has a significant effect on the micellar properties. There is an increase in micellar size for a particular cationic surfactant as the counterion is changed according to the series $Cl^- < Br^- < I^-$, and for a particular anionic surfactant according to $Na^+ < K^+ < Cs^+$. Generally, the more weakly hydrated a counterion, the larger the micelles formed by the surfactant. This is because the weakly hydrated ions can be absorbed more readily in the micellar surface and so decrease the charge repulsion between the polar groups. A greater depression of cmc and a greater increase in micellar size is noted with organic counterions such as maleates, than with inorganic ions.

(D) Addition of electrolytes
Addition of electrolytes to ionic surfactants decreases the cmc and increases the micellar size. The effect is simply explained in terms of a reduction in the magnitude of the forces of repulsion between the charged head groups in the micelle and a consequent decrease in the electrical work of micellisation. Table 6.6 shows the effect of sodium chloride addition on the micellar properties of the cationic surfactant, dodecyltrimethylammonium bromide. The micellar properties of non-ionic surfactants, in contrast, are little affected by electrolyte addition.

table 6.6   **Effect of electrolyte on the micellar properties of dodecyltrimethylammonium bromide***

NaCl conc. $(mol \, \ell^{-1})$	cmc $(mol \, \ell^{-1})$	Aggregation number
0.000	0.0146	61
0.100	0.00428	74
0.502	0.00171	90

*From E. W. Anacker. *Cationic Surfactants* (ed. E. Jungermann), Marcel Dekker, New York, 1970

(E) Effect of temperature
If aqueous solutions of many non-ionic surfactants are heated, they become turbid at a characteristic temperature called the cloud point. Other non-ionic surfactants have cloud points above $100\,^{\circ}C$. The process is reversible; that is, cooling the solution restores its clarity. The turbidity at the cloud point is due to separation of the solution into two phases. At temperatures up to the cloud point an increase in micellar size and a corresponding decrease in cmc is noted for many non-ionic surfactants. The cloud point is very sensitive to additives in the system, which can increase or decrease the clouding temperature.

Temperature has a comparatively small effect on the micellar properties of ionic surfactants. The temperature dependence of the cmc of sodium lauryl (dodecyl) sulphate shown in figure 6.27 is typical of the effect observed.

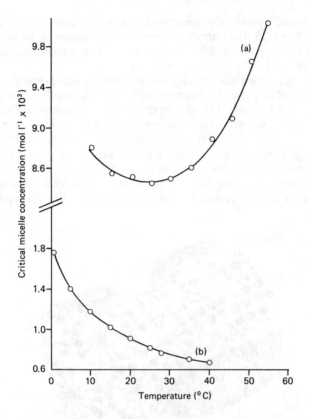

figure 6.27 Variation of cmc with temperature for: (a) sodium dodecyl sulphate; (b) $CH_3(CH_2)_9(OCH_2CH_2)_5OH$. After Goddard and Benson. *Can. J. Chem.*, 35, 986 (1957) with permission

6.4 Liquid crystals and surfactant vesicles

Surfactant solutions at concentrations close to the cmc are clear, isotropic solutions, i.e. the magnitudes of such physical properties as viscosity and refractive index do not depend on the direction in which these properties are measured. As the concentration is increased there is frequently a transition from the typical spherical micellar structure to a more elongated or rod-like micelle. Further increase in concentration may cause the orientation and close packing of the elongated micelles into hexagonal arrays. A new phase containing these ordered arrays separates out from the remainder of the solution which contains randomly orientated rods but remains in equilibrium with it. This new phase is termed the 'middle phase'. With some surfactants, further increase of concentration results in the separation of a second transition phase, the 'neat phase' which has a lamellar structure. Finally in all systems surfactant separates out of solution. Both the middle and neat phases are liquid crystalline states and are referred to as *lyotropic* liquid crystals. The structure of these two states is shown diagrammatically in figure 6.28. The transition from micellar solution to liquid crystalline phase and finally to pure amphiphile may be shown on a phase diagram (see for example figure 6.32).

The liquid crystals are anisotropic; that is, their physical properties vary with direction. The middle phase, for example, will flow only in a direction parallel to the long axis of the arrays. It is rigid in the other two directions. On the other hand, the neat phase is more fluid and behaves as a solid only in the direction perpendicular to that of the layers. Similarly, plane-polarised light is rotated when travelling along any axis except the long axis in the middle phase and a direction perpendicular to the layers in the neat phase. Because of this ability to rotate polarised light the liquid crystals are visible when placed between crossed polarisers and this provides a useful means of detecting the liquid crystalline state.

(a)

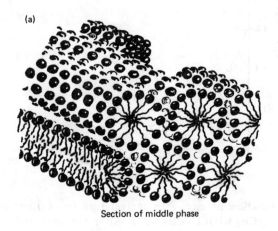

Section of middle phase

(b)

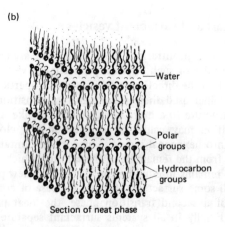

Section of neat phase

figure 6.28 Diagrammatic representation of forms of lyotropic liquid crystals

A second category of liquid crystals is the type produced when certain substances, notably the esters of cholesterol, are heated. These systems are referred to as *thermotropic* liquid crystals. The formation of a cloudy liquid at temperatures between 145 and 179 °C when cholesteryl benzoate is heated was first noted in 1888 by the Austrian botanist, Reinitzer. The name 'liquid crystal' was applied to this cloudy intermediate phase because of the presence of areas with crystal-like molecular structure within this solution.

Although the compounds that form thermotropic liquid crystalline phases are of a variety of chemical types such as anils, azo compounds, azoxy compounds or esters, the molecular geometries of the molecule have some characteristic features in that they are generally elongated, flat and rigid along their axes. The presence of easily polarisable groups often enhances liquid crystal formation.

The arrangement of the elongated molecules in thermotropic liquid crystals is generally recognisable as one of three principal types: namely, smectic (soap-like), nematic (thread-like) and cholesteric. The molecular arrays are illustrated diagrammatically in figure 6.29.

In the nematic liquid crystalline state, groups of molecules orientate spontaneously with their long axes parallel, but they are not ordered into layers. Because the molecules have freedom of rotation about their long axis, the nematic liquid crystals are quite mobile and are readily orientated by electric or magnetic fields. Nematic liquid crystals are formed when *p*-azoxyanisole is heated.

The molecules in smectic liquid crystals are more ordered than the nematic since, not only are they arranged with their long axes parallel, but they are also arranged into distinct layers. As a result of this two-dimensional order the smectic liquid crystals are viscous and are not orientated by magnetic fields. Examples of compounds forming smectic liquid crystals are octyl *p*-azoxycinnamate and ethyl *p*-azoxybenzoate.

The cholesteric phase, which is produced by several cholesteryl esters, resembles the smectic phase in that the molecules are arranged in layers. Within each layer, however, the elongated molecules lie parallel to each other in the plane of the layer, producing very thin layers. The orientation of the long axes in each layer is displaced from that in the adjacent layer. This displacement is cumulative through successive layers so that the over-all displacement traces out a helical path through the layers. The helical path causes very pronounced rotation of polarised light which can be as much as 50 rotations per millimetre. The pitch of the helix is very sensitive to small changes in temperature and pressure and dramatic colour changes can result from variations in these properties. When non-polarised light is passed through the cholesteric material, the light is separated into two components, one with the electric vector rotating clockwise and the other with the electric vector rotating anticlockwise. One of these components is transmitted and the other reflected, depending on the material involved. This process is called circular dichroism and it gives the cholesteric phase a characteristic iridescent appearance when illuminated by white light.

Liposomes, niosomes and surfactant vesicles

Phospholipid vesicles (liposomes) are liquid crystals of the smectic type which are formed when naturally occurring phospholipids such as lecithin are equilibrated with excess water or aqueous salt solution. These liquid crystals when

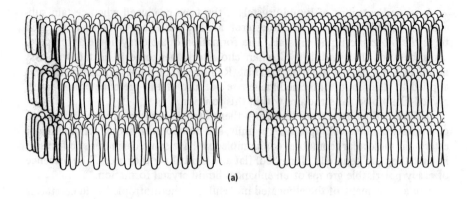

(a)

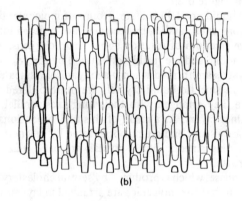

(b)

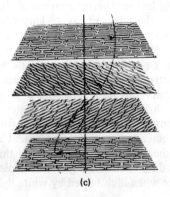

(c)

figure 6.29 Diagrammatic representation of forms of thermotropic liquid crystals. (a) Smectic, (b) nematic, and (c) cholesteric liquid crystals. From J. S. Ferguson: Liquid crystals, *Sci. Am*., 211, 76 (1964)

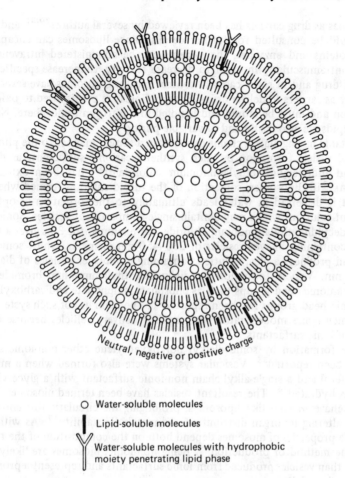

figure 6.30 Diagrammatic representation of a liposome in which three bilayers of polar phospholipids alternate with aqueous compartments. Water-soluble and lipid-soluble substances may be accommodated in the aqueous and lipid phases, respectively. Certain macromolecules can insert their hydrophobic regions into the lipid bilayers with the hydrophilic portions extending into water. From G. Gregoriadis. *New Engl. J. Med.*, 295, 704 (1976) with permission

first formed are usually composed of several bimolecular lipid lamellae separated by aqueous layers (multilamellar liposomes). Figure 6.30 shows a diagrammatic representation of a multilamellar liposome with three phospholipid bilayers. Sonication of these units can give rise to unilamellar liposomes. The net charge of the liposome can be varied by incorporation of, for example, a long-chain amine such as stearylamine (to give positively charged vesicles) or dicetyl phosphate (to give negatively charged species). Water-soluble drugs can be entrapped in liposomes by intercalation in the aqueous layers while lipid-soluble drugs can be solubilised within the hydrocarbon interiors of the lipid bilayers (see figure 6.30). The use of liposomes as drug carriers has been reviewed by several authors[30,31] and these texts should be consulted for further details. Since liposomes can encapsulate drugs, proteins and enzymes, the systems can be administered intravenously, orally or intramuscularly in order to decrease toxicity, to increase specificity of uptake of drug and in some cases to control release. Liposomes have several disadvantages as carriers to deliver drugs. Phospholipids are predisposed to oxidative degradation and must be stored and handled in a nitrogen atmosphere. Natural phospholipids are also costly and of variable purity.

Surfactants having dialkyl chains can pack in a similar manner to the phospholipids, and vesicle formation by such synthetic cationic surfactants as dioctadecyl- and didodecyldimethylammonium chloride[32,33] has been extensively studied. As with liposomes, sonication of the turbid solution formed when the surfactant is dispersed in water leads ultimately to the formation of optically transparent solutions which may contain single-compartment vesicles. Sonication of dioctadecyldimethylammonium chloride, for example, for 30 s gives a turbid solution containing bilayer vesicles of 250–450 nm diameter, whilst sonication for 15 min produces a clear solution containing monolayer vesicles of diameter 100–150 nm. Vesicle formation by dialkyldimethylammonium bromide with modified anionic head groups such as phosphate, sulphonate and carboxylate or zwitterionic head groups has been reported[34]. The main use of such systems has been as membrane models rather than as drug delivery vehicles because of the toxicity of ionic surfactants.

Vesicle formation by some dialkyl polyoxyethylene ether non-ionic surfactants has been reported[35]. Vesicular systems were also formed when a mixture of cholesterol and a single-alkyl-chain non-ionic surfactant with a glyceryl head group was hydrated[36]. The resultant vesicles have been termed niosomes. These vesicles behave *in vivo* like liposomes, prolonging the circulation of entrapped drug and altering its organ distribution and metabolic stability[37]. As with liposomes, the properties of niosomes depend both on the composition of the bilayer and on the method of production[38]. Being non-ionic, niosomes are likely to be less toxic than vesicles produced from ionic surfactants and represent a promising vehicle for drug delivery.

6.5   Properties of some commonly used surfactants

6.5.1   Anionic surfactants

Sodium lauryl sulphate BP is a mixture of sodium alkyl sulphates, the chief of which is sodium dodecyl sulphate, $C_{12}H_{25}SO_4^-$ Na^+. It is very soluble in water giving a turbid solution. It is used pharmaceutically as a pre-operative skin cleaner, having bacteriostatic action against gram-positive bacteria, and also in medicated shampoos.

Sodium dodecyl sulphate has been studied in depth. The cmc at 25 °C is 8.2×10^{-3} mol ℓ^{-1} (0.23% w/v). Figure 6.27 showed the effect of temperature on the cmc.

6.5.2 Cationic surfactants

The quaternary ammonium and pyridinium cationic surfactants are important pharmaceutically because of their bactericidal activity against a wide range of gram-positive and some gram-negative organisms. They may be used on the skin especially in the cleaning of wounds. Aqueous solutions are used for cleaning contaminated utensils.

Cetrimide BP consists mainly of tetradecyltrimethylammonium bromide together with smaller amounts of dodecyl- and hexadecyltrimethylammonium bromides. The properties of the individual components have been studied in detail and are summarised in table 6.7. Solutions containing 0.1–1 per cent of cetrimide are used for cleaning the skin, wounds and burns, for cleansing contaminated vessels, for storage of sterilised surgical instruments, and for cleansing polythene tubing and catheters. Solutions of cetrimide are also used in shampoos to remove scales in seborrhoea. In the form of cetrimide emulsifying wax, it is used as an emulsifying agent for producing oil-in-water creams suitable for the incorporation of cationic and non-ionic medicaments (anionic medicaments would, of course, be incompatible with this cationic surfactant).

Benzalkonium chloride is a mixture of alkylbenzyldimethylammonium chlorides of the general formula $[C_6H_5CH_2N(CH_3)_2R]Cl$, where R represents a

table 6.7 **Micellar properties of a commercial sample of cetrimide and its main constituents at 25 °C***

Constituent	Percentage (calculated on dry weight basis)	cmc (mmol ℓ^{-1})	Micellar molecular weight ($\times 10^{-4}$)
Tetradecyltrimethylammonium bromide	68	3.3	2.7
Dodecyltrimethylammonium bromide	22	5.3	2.1
Hexadecyltrimethylammonium bromide	7	0.82	3.3
Cetrimide	–	2.9	2.5

*Data from B. W. Barry *et. al. J. colloid interface Sci.*, 33, 554 (1970); *ibid*, 40, 174 (1972)

mixture of the alkyls from C_8H_{17} to $C_{18}H_{37}$. In dilute solution (1 in 1000 to 1 in 2000) it may be used for the pre-operative disinfection of skin and mucous membranes, for application to burns and wounds, and for cleansing polythene and nylon tubing and catheters. Benzalkonium chloride is also used as a preservative for eye-drops and a permitted vehicle for the preparation of certain eye-drops.

6.5.3 Non-ionic surfactants

The amphiphilic nature of non-ionic surfactants is often expressed in terms of the balance between the hydrophobic and hydrophilic portions of the molecule.

An empirical scale of HLB (hydrophile–lipophile balance) numbers has been devised (section 7.3.2). The lower the HLB number, the more lipophilic is the compound and vice versa. HLB values are quoted in tables 6.8 and 6.9 for a series of commercial non-ionic surfactants. The choice of surfactant for medicinal use involves a consideration of the toxicity of the substance which may be ingested in large amounts. The following surfactants are widely used in pharmaceutical formulations.

Sorbitan esters

The commercial products are mixtures of the partial esters of sorbitol and its mono- and di-anhydrides with oleic acid. The formula of a representative component is shown below. They are generally insoluble in water and are used as water-in-oil emulsifiers and as wetting agents. The main sorbitan esters are listed in table 6.8.

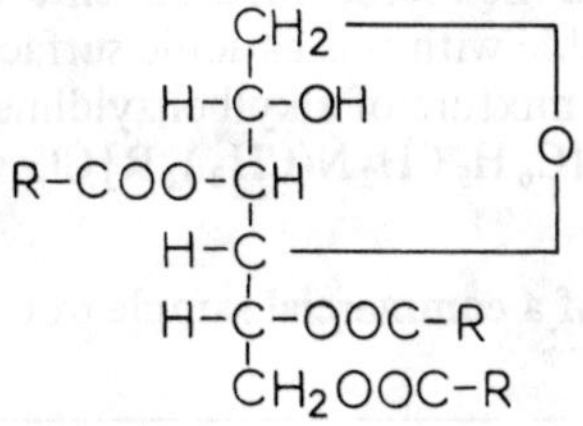

where R is H or an alkyl chain

table 6.8 **HLB values of sorbitan esters**

Chemical name	Commercial name	HLB
Sorbitan monolaurate	Span 20	8.6
Sorbitan monopalmitate	Span 40	6.7
Sorbitan monostearate	Span 60	4.7
Sorbitan tristearate	Span 65	2.1
Sorbitan monooleate	Span 80	4.3
Sorbitan trioleate	Span 85	1.8

Polysorbates

Commercial products are complex mixtures of partial esters of sorbitol and its mono- and di-anhydrides condensed with an approximate number of moles of ethylene oxide. The formula of a representative component is shown above table 6.9. The Polysorbates are miscible with water, as reflected in their higher HLB values (see table 6.9), and are used as emulsifying agents for oil-in-water emulsions.

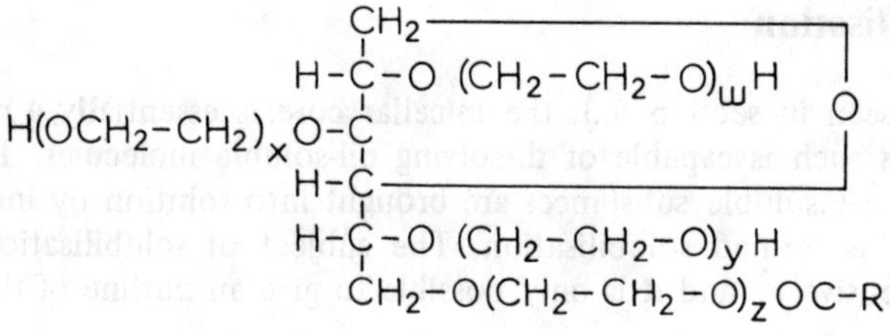

$$
\begin{array}{c}
CH_2 \longrightarrow \\
H-C-O\ (CH_2-CH_2-O)_w H \\
H(OCH_2-CH_2)_x O-C \qquad\qquad\qquad O \\
H-C \\
H-C-O\ (CH_2-CH_2-O)_y H \\
CH_2-O(CH_2-CH_2-O)_z OC-R
\end{array}
$$

where R is an alkyl chain

table 6.9 **HLB and cmc values of Polysorbates**

Chemical name	Commercial name	HLB	cmc* $(g\ \ell^{-1})$
Polyoxyethylene (20) sorbitan monolaurate	Polysorbate (Tween) 20	16.7	0.060
Polyoxyethylene (20) sorbitan monopalmitate	Polysorbate (Tween) 40	15.6	0.031
Polyoxyethylene (20) sorbitan monostearate	Polysorbate (Tween) 60	14.9	0.028
Polyoxyethylene (20) sorbitan tristearate	Polysorbate (Tween) 65	10.5	0.050
Polyoxyethylene (20) sorbitan monooleate	Polysorbate (Tween) 80	15.0	0.014
Polyoxyethylene (20) sorbitan trioleate	Polysorbate (Tween) 85	11.0	0.023

*Data from L. S. Wan and P. F. S. Lee. *J. pharm. Sci.*, 63, 136 (1974)

Cetomacrogol 1000 BP and other macrogol ethers

Cetomacrogol is a water-soluble substance with the general structure $CH_3(CH_2)_m$ $(OCH_2CH_2)_n OH$, where m may be 15 or 17 and the number of oxyethylene groups, n, is between 20 and 24. It is used in the form of cetomacrogol emulsifying wax in the preparation of oil-in-water emulsions and also as a solubilising agent for volatile oils. The cmc and micellar molecular weight in aqueous solution are $6 \times 10^{-2}\ g\ \ell^{-1}$ and 1.01×10^5, respectively.

Other macrogol ethers — for example, polyoxyethylene lauryl ether (Brij 30), polyoxyethylene stearyl ether (Brij 72), and polyoxyethylene oleyl ether (Brij 92) — are commercially available.

Cremophor EL is a polyoxyethylated castor oil containing approximately 40 oxyethylene groups to each triglyceride unit. It is used as a solubilising agent in the preparation of intravenous anaesthetics.

6.6   Solubilisation

As we have seen in section 6.3, the micellar core is essentially a paraffin-like region and as such is capable of dissolving oil-soluble molecules. This process whereby water-insoluble substances are brought into solution by incorporation into micelles is termed solubilisation. The subject of solubilisation has been reviewed extensively[3] and it is only possible to give an outline of this phenomenon.

6.6.1   Determination of maximum additive concentration

The maximum amount of solubilisate that can be incorporated into a given system at a fixed concentration is termed the maximum additive concentration (MAC). The simplest method of determining the MAC is to prepare a series of vials containing surfactant solution of known concentration. Increasing concentrations of solubilisate are added and the vials are then sealed and agitated until equilibrium conditions are established. The maximum concentration of solubilisate forming a clear solution can be detemined by visual inspection or from extinction or turbidity measurements on the solutions.

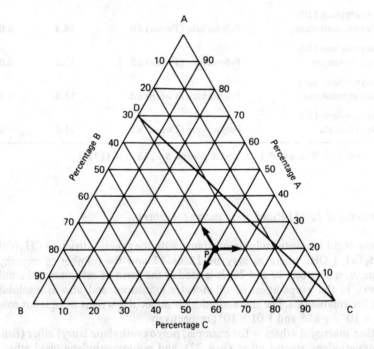

figure 6.31   Three-component phase diagram. A point P represents a system of composition 20 per cent A, 30 per cent B, and 50 per cent C. Line CD represents the dilution of a mixture, originally containing 70 per cent A and 30 per cent B with component C

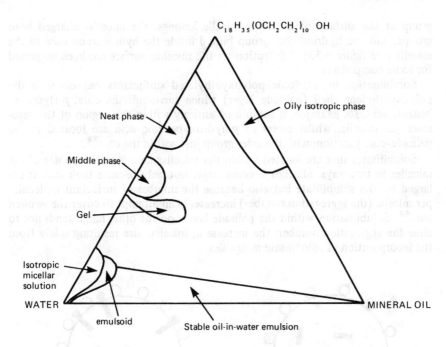

$C_{18}H_{35}(OCH_2CH_2)_{10}OH$

Oily isotropic phase

Neat phase

Middle phase

Gel

Isotropic
micellar
solution

WATER

MINERAL OIL

emulsoid

Stable oil-in-water emulsion

figure 6.32 Partial phase diagram for the polyoxyethylene (10) oleyl ether–water–mineral oil solubilised system. From R. Lachampt and R. M. Vila. *Am. Perfumer Cosmetics*, 82, 29 (1967) with permission

Solubility data are expressed as a solubility versus concentration curve or as phase diagrams. The latter are preferable since a three-component phase diagram completely describes the effect of varying all three components of the system – namely, the solubilisate, the solubiliser and the solvent. The axes of the phase diagram form an equilateral triangle (see figure 6.31), each side of which is divided into 100 parts to correspond to percentage composition.

A typical phase diagram of a solubilised system is shown in figure 6.32. In solutions of high water content the oil is solubilised in the surfactant micelles, forming an isotropic micellar solution (often referred to as the L_1 region). When the concentration of the oil is increased, stable oil-in-water emulsions may be formed, whilst an increase in the surfactant concentration results in the formation of the liquid crystalline regions, labelled middle and neat phases (section 6.4). It is important in formulation to avoid boundary regions, otherwise there is a danger of unwanted phase transitions.

6.6.2 Location of the solubilisate

The site of solubilisation within the micelle is closely related to the chemical nature of the solubilisate. It is generally accepted that non-polar solubilisates (aliphatic hydrocarbons, for example) are dissolved in the hydrocarbon core. Water-insoluble compounds containing polar groups are orientated with the polar group at the surface of the ionic micelle amongst the micellar charged head groups, and the hydrophobic group buried inside the hydrocarbon core of the micelle (see figure 6.33). Adsorption on the micellar surface has been suggested for some compounds.

Solubilisation in non-ionic polyoxyethylated surfactants can occur in the polyoxyethylene shell (palisade layer) which surrounds the core. *p*-Hydroxybenzoic acid, for example, is solubilised entirely within this region of the ceto-macrogol micelle, whilst esters of *p*-hydroxybenzoic acid are located at the palisade–core junction, with the ester group just within the core[39].

Solubilisates that are located within the micellar core increase the size of the micelles in two ways. Micelles become larger not only because their core is enlarged by the solubilisate but also because the number of surfactant molecules per micelle (the aggregation number) increases in an attempt to cover the swollen core[40]. Solubilisation within the palisade layer, on the other hand, tends not to alter the aggregation number, the increase in micellar size resulting solely from the incorporation of solubilisate molecules.

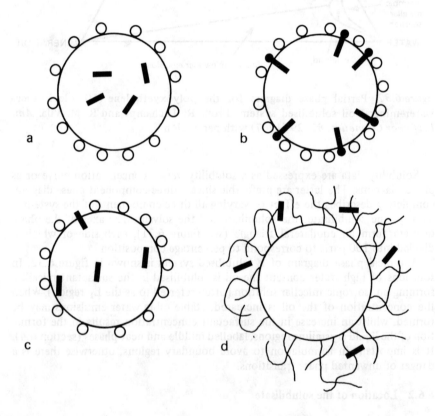

figure 6.33 Schematic representation of sites of solubilisation in ionic and non-ionic micelles. (a) Non-polar solubilisate; (b) amphipathic solubilisate; (c) slightly polar solubilisate; and (d) polar solubilisate in polyoxyethylene shell of a non-ionic micelle

6.6.3 Factors affecting solubilisation

(A) Nature of the surfactant
It is difficult to generalise about the way in which the structural characteristics of a surfactant affect its solubilising capacity because this is influenced by the solubilisation site within the micelle. In cases where the solubilisate is located within the core or deep within the micelle structure the solubilisation capacity increases with increase in alkyl chain length as might be expected. Table 6.10 clearly shows an increase of solubilising capacity of a series of polysorbates for selected barbiturates as the alkyl chain length is increased from C_{12} (Polysorbate 20) to C_{18} (Polysorbate 80). Similar effects have been noted for the solubilisation of barbiturates in polyoxyethylene surfactants with general structure

table 6.10 Solubilising capacity of polysorbates for the barbiturates at 30 °C*

Drug	Surfactant	Solubility (mg drug/g surfactant)	Solubility (mol drug/mol surfactant × 10^2)
Barbitone	Polysorbate 20	30.0	19.9
	Polysorbate 40	33.0	23.0
	Polysorbate 60	35.3	25.1
	Polysorbate 80	35.0	24.6
Diallylbarbituric acid	Polysorbate 20	24.0	14.4
	Polysorbate 40	27.0	16.4
	Polysorbate 60	28.0	17.3
	Polysorbate 80	28.0	17.4
Butethone	Polysorbate 20	100.0	57.5
	Polysorbate 40	–	–
	Polysorbate 60	–	–
	Polysorbate 80	115.0	71.1
Cyclobarbitone	Polysorbate 20	52.4	27.2
	Polysorbate 40	58.0	31.6
	Polysorbate 60	61.0	33.8
	Polysorbate 80	61.0	34.0
Phenobarbitone	Polysorbate 20	55.1	29.1
	Polysorbate 40	61.0	33.7
	Polysorbate 60	63.0	35.5
	Polysorbate 80	66.0	37.2
Amobarbitone	Polysorbate 20	32.0	17.2
	Polysorbate 40	38.0	21.7
	Polysorbate 60	–	–
	Polysorbate 80	40.0	22.9
Secobarbitone	Polysorbate 20	111.0	57.0
	Polysorbate 40	–	–
	Polysorbate 60	–	–
	Polysorbate 80	144.0	78.8

*From reference 43

table 6.11 **Micellar solubilisation parameters for steroids in n-alkyl polyoxyethylene surfactants at 25 °C***

Surfactant	Partial specific volume $(ml\,g^{-1})$	Aggre- gation number	Micelles per mole $(\times 10^{-21})$	Steroid molecules per micelle			
				Hydro- cortisone	Dexa- methasone	Testos- terone	Proges- terone
$C_{16}E_{17}$	0.9376	99	6.08	9.1	6.7	6.0	5.6
$C_{16}E_{32}$	0.9171	56	10.8	7.6	5.3	4.6	4.3
$C_{16}E_{44}$	0.8972	39	15.4	5.8	4.2	3.6	3.3
$C_{16}E_{63}$	0.8751	25	24.1	4.0	3.3	2.4	2.3

*From reference 44

$CH_3(CH_2)_m(OCH_2CH_2)_nOH$ with increase of the alkyl chain length, m[41]. There is a limit however to the improvement of solubilising capacity caused by increase of alkyl chain length in this way. Arnarson and Elworthy[42] showed that an increase of m from 16 to 22, although producing larger micelles, did not result in a corresponding increase of solubilisation.

The effect of an increase in the ethylene oxide chain length of a polyoxy-ethylated non-ionic surfactant on its solubilising capacity is again dependent on the location of the solubilisate within the micelle and is complicated by corresponding changes in the micellar size. Table 6.11 shows the solubilisation capacity of a series of polyoxyethylene non-ionic surfactants with a hydrocarbon chain length of 16 (C_{16}) and an increasing number of ethylene oxide units (E) in the polyoxyethylene chain. As seen from this table the aggregation number decreases with increase in the hydrophilic chain length so, although the number of steroid molecules solubilised per micelle also decreases, the total amount solubilised per mole of surfactant actually increases because of the increasing number of micelles.

(B) Nature of the solubilisate
Although many possible relationships between the amount solubilised and various physical properties of the solubilisate molecule (for example, molar volume, polarity, polarisability and chain length) have been explored, it has not been possible to establish simple correlations between them. In general, a decrease in solubility occurs when the alkyl chain length of a homologous series is increased. Unsaturated compounds are generally more soluble than their saturated counterparts. Branching of the hydrocarbon chain of the solubilisate has little effect, but increased solubilisation is often noted following cyclisation. These generalisations unfortunately only apply to very simple solubilisates. More specific rules can be formulated for particular series of solubilisates. Table 6.12 shows the effect of steroid structure on solubilisation by a series of surfactants. It is clear that the more hydrophilic —OH group in position 17 of the ring structure decreases the quantity of surfactant required to effect solubilisation of the hormone. The results of Ekwall et al.[45] also show that the more hydrophilic the substituent in position 17, the greater the solubilisation. Thus the extent of solubilisation of hormones in sodium lauryl sulphate follows the series progesterone < testosterone < deoxycorticosterone, the C_{17} substituents being —COCH$_3$, —OH and —COCH$_2$OH, respectively.

table 6.12 **Maximum solubilising power of surfactants for oestrone and 17β-oestradiol***

Surfactant	Concentration range (mol ℓ^{-1})	Temperature (°C)	Moles micellar surfactant per mole hormone	
			Oestrone	17β-Oestradiol
Sodium caproate	0.1–0.5	20	202	99
Sodium lauryl sulphate	0.01–0.15	40	72.5	58.1
Tetradecyltrimethylammonium bromide	0.005–0.08	20	44.6	13.3
Polysorbate 20	1–20%	20	179×10^3 g mol^{-1}	95.5×10^3 g mol^{-1}

*From L. Sjöblöm. *Acta Acad. Aboensis, math. phys.*, 21, No. 7 (1958)

Several authors have noted a relationship between the lipophilicity of the solubilisate, expressed by the partition coefficient between octanol and water, $P_{octanol}$, and its extent of solubilisation. Rank order correlations between the $P_{octanol}$ values of a series of substituted barbituric acids and the amount solubilised by polyoxyethylene stearates have been reported[43]. Similarly the partition coefficients of several steroids between ether and water have been correlated with their solubilisation by long-chain polyoxyethylene non-ionic surfactants[44]. Linear relationships between the partition coefficients of a series of substituted benzoic acids between micelles of Polysorbate 20 and water and their octanol/H_2O partition coefficient was reported by Collett and Koo[46] and Tomida and co-workers[47]. Similar relationships have been noted for the solubilisation of steroids by some polyoxyethylene non-ionic surfactants[48].

(C) Effect of temperature
In most systems the amount solubilised increases as temperature increases. The effect is particularly pronounced with some non-ionic surfactants where it is a consequence of an increase in the micellar size with temperature increase. Table 6.13 shows the effect of temperature on solubilisation by bile salts.

A complicating factor when considering the effect of temperature on the amount solubilised is the change in the aqueous solubility of the solubilisate with temperature increase. In some cases, although the amount of drug which can be taken up by a surfactant solution increases with temperature increase, this may simply reflect an increase in the amount of drug dissolved in the aqueous phase rather than an increased solubilisation by the micelles. This point is illustrated by a study of the solubilisation of benzoic acid by a series of polyoxyethylene non-ionic surfactants[49], details of which are given in table 6.14. Although the extent of solubilisation of benzoic acid increases with temperature increase, the micelle/water distribution coefficient, P_m, shows a minimum at about 27 °C. The decrease in P_m with temperature increase up to this temperature is possibly due to the increase in aqueous solubility of benzoic acid. The increase of P_m with temperature increase is due to a rapid increase of micellar size as the cloud point is approached.

table 6.13 **The effect of temperature on the maximum additive concentrations of griseofulvin, hexoestrol, and glutethimide in bile salts***

		MAC $\times 10^3$ (moles solubilisate per mole surfactant)		
Solubilisate	Surfactant	$27\,^\circ$C	$37\,^\circ$C	$45\,^\circ$C
Griseofulvin	None	4.59×10^{-4}	7.14×10^{-4}	10.2×10^{-4}
	Sodium cholate	5.36	6.18	6.80
	Sodium desoxycholate	4.68	6.18	7.54
	Sodium taurocholate	3.77	4.90	6.15
	Sodium glycocholate	3.85	5.13	5.29
Hexoestrol	None	4.66×10^{-4}	6.66×10^{-4}	9.32×10^{-4}
	Sodium cholate	187	195	197
	Sodium desoxycholate	164	167	179
	Sodium taurocholate	220	225	223
	Sodium glycocholate	221	231	251
		$27\,^\circ$C	$32\,^\circ$C	$37\,^\circ$C
Glutethimide	None	7.13×10^{-2}	8.32×10^{-2}	9.94×10^{-2}
	Sodium cholate	59.8	96.2	104
	Sodium desoxycholate	103	119	163
	Sodium taurocholate	61.2	100	108
	Sodium glycocholate	54.3	92.0	71.8

*From Bates, Gibaldi and Kanig. *J. pharm. Sci.*, 55, 191 (1966)

table 6.14 **Micelle/water distribution coefficient, P_m, for the solubilisation of benzoic acid by n-alkyl polyoxyethylene surfactants as a function of temperature***

Surfactant formula	P_m				
	291 K	298 K	304 K	310 K	318 K
$C_{16}E_{16}$	59.51	50.07	43.75	44.11	–†
$C_{16}E_{30}$	47.80	45.55	35.42	38.23	38.66
$C_{16}E_{40}$	37.07	32.72	28.76	29.90	37.06
$C_{16}E_{96}$	31.22	27.43	25.43	27.46	32.25

†No P_m value determined at this temperature for $C_{16}E_{16}$ because the cloud point tempera-temperature was exceeded
*From reference 49

6.6.4 Pharmaceutical applications of solubilisation

A wide range of insoluble drugs have been formulated using the principle of solubilisation. A few representative examples only will be discussed in this section.

Phenolic compounds such as cresol, chlorocresol, chloroxylenol and thymol are frequently solubilised with soap to form clear solutions which are widely used for disinfection. Solution of Chloroxylenol BP, for example, contains 5% v/v chloroxylenol with terpineol in an alcoholic soap solution.

Non-ionic surfactants are efficient solubilisers of iodine, and will incorporate up to 30 per cent by weight, of which three-quarters is released as available iodine on dilution. Such iodine–surfactant systems (referred to as iodophors) are more stable than iodine–iodide systems. They are preferable in instrument sterilisation since corrosion problems are reduced. Loss of iodine by sublimation from iodophor solutions is significantly less than from simple iodine solutions such as iodine solution NF. There is also evidence of an ability of the iodophor solution to penetrate hair follicles of the skin, so enhancing the activity.

The low solubility of steroids in water presents a problem in their formulation for ophthalmic use. The requirement of optical clarity precludes the use of oily solutions or suspensions and there are many examples of the use of non-ionic surfactants as a means of producing clear solutions which are stable to sterilisation. In most formulations, solubilisation has been effected using polysorbates or polyoxyethylene sorbitan esters of fatty acids.

Essential oils are extensively solubilised by surfactants, Polysorbates 60 and 80 being particularly well suited to this purpose.

The polysorbate non-ionics have also been employed in the preparation of aqueous injections of the water-insoluble vitamins A, D, E and K. Table 6.15 shows the solubility of these vitamins in 10 per cent polysorbate solutions.

table 6.15 Solubilisation of vitamins by 10 per cent polysorbate solutions*

Polysorbate	Vitamin D_2 (IU $m\ell^{-1}$)	Vitamin E (mg $m\ell^{-1}$)	Vitamin K_3 (mg $m\ell^{-1}$)	Vitamin A alcohol (IU $m\ell^{-1}$)
20	20 000	5.7	4.7	80 000
40	16 000	3.8	4.0	60 000
60	15 000	3.2	3.7	60 000
80	20 000	4.5	4.5	80 000

*From Gstirner and Tata. *Mitt. dt. pharm. Ges.*, 28, 191 (1958)

One of the problems encountered with the use of non-ionic surfactants as solubilisers is that they are prone to clouding (see section 6.3), and the presence of solubilisate can reduce the cloud point. For this reason, sucrose esters have been suggested as alternative solubilisers for the vitamins, although they do have the disadvantage of a slightly higher haemolytic activity.

It has only been possible to give a brief description of the types of drugs that have been formulated using solubilisation. Many other drugs have been formulated in this way, including the analgesics, sedatives, sulphonamides and antibiotics. The reader is referred to reference 3 for a more complete survey of this topic and for a discussion of the effects of solubilisation on drug activity and absorption characteristics.

References

1. B. A. Pethica. *Trans. Farad. Soc.*, 50, 413 (1954)
2. D. Attwood. In *Aggregation Processes in Solution* (eds E. Wyn-Jones and J. Gormally), Elsevier Scientific, Amsterdam, 1982, chapter 9

3. D. Attwood and A. T. Florence. *Surfactant Systems*, Chapman and Hall, London, 1983

4. G. Zografi and M. V. Munshi. *J. pharm. Sci.*, 59, 819 (1970)

5. P. M. Seeman and H. S. Bialy. *Biochem. Pharmacol.*, 12, 1181 (1963)

6. D. J. Crisp. *Surface Phenomena in Chemistry and Biology* (ed. J. F. Danielli, K. G. A. Pankhurst and A. C. Riddiford), Pergamon Press, Oxford, 1958, p. 23

7. J. L. Zatz, N. D. Weiner and M. Gibaldi. *J. pharm. Sci.*, 58, 1493 (1969)

8. J. L. Zatz and B. Knowles. *ibid.*, 57, 1188 (1970)

9. G. Colacicco. *J. colloid interface Sci.*, 29, 345 (1969)

10. J. D. Arnold and C. Y. Pak. *J. Am. oil chem. Soc.*, 45, 128 (1968)

11. G. Zografi, D. E. Auslander and P. L. Lytell. *J. pharm. Sci.*, 53, 573 (1964)

12. G. Zografi and D. E. Auslander. *ibid.*, 54, 1313 (1965)

13. J. C. Skou. *Acta Pharmacol. Toxicol.*, 10, 280, 317, 325 (1954)

14. J. M. Ritchie and P. Greengard. *Annu. Rev. Pharmacol.*, 6, 405 (1966)

15. R. A. Demel, F. J. L. Crombag, L. L. M. van Deenen and S. C. Kinsky. *Biochim. Biophys. Acta*, 150, 1 (1968)

16. S. El-Masry and S. A. H. Khalil. *J. Pharm. Pharmacol.*, 26, 243 (1974)

17. P. A. Bristow. *The Quality Control of Medicines* (ed. P. Deasy and R. Timoney), Elsevier Scientific, Amsterdam, 1976, chapter 21

18. E. Shotton and K. Ridgeway. *Physical Pharmaceutics*, Clarendon Press, Oxford, 1974, p. 198

19. L. Chin *et al. Toxicol. appl. Pharmacol.*, 26, 103 (1973)

20. W. J. Decker, H. J. Combs and D. G. Corby. *ibid.*, 13, 454 (1968)

21. W. J. Decker *et al. Clin. Pharm. Ther.*, 10, 710 (1969)

22. H. Yatzidis. *Proc. Eur. dialysis transplant Ass.*, 1, 88 (1964)

23. J. Kolthammer. *J. Pharm. Pharmacol.*, 27, 801 (1975)

24. J. Vale, A. J. Rees, B. Widdop and R. Goulding. *Br. med. J.*, 1, 5 (1975)

25. T. M. S. Chang. *Can. J. Physiol. Pharmacol.*, 47, 1043 (1969)

26. W. Saski. *J. pharm. Sci.*, 52, 264 (1963)

27. G. Bauer and E. Ullmann. *Arch. Pharm.*, 306, 86 (1973)

28. C. B. Macfarlane. *Kolloid-Z. Z. Polym.*, 239, 682 (1970)

29. P. Mukerjee. *J. pharm. Sci.*, 63, 972 (1974)

30. G. Gregoriadis. *Lancet*, ii, 241 (1981)

31. G. Poste, R. Kirsh and T. Koestler. In *Liposome Technology* (ed. G. Gregoriadis), vol 3, CRC Press, Cleveland, 1984, pp. 2–28

32. K. Deguchi and J. Mino. *J. colloid interface Sci.*, 65, 155 (1978)

33. C. D. Tran, P. L. Klahn, A. Romero and J. H. Fendler. *J. Am. chem. Soc.*, 100, 1622 (1978)

34. T. Kunitake and Y. Okahata. *Bull chem. Soc. Japan*, 51, 1877 (1978)

35. Y. Okahata, S. Tanamachi, M. Nagai and T. Kunitake. *J. colloid interface Sci.*, 82, 401 (1981)

36. R. M. Handjani-Vila, A. Ribier, B. Rondot and G. Vanlerberghe. *Int. J. cosmetic Sci.*, 1, 303 (1979)

37. M. N. Azmin, A. T. Florence, R. M. Handjani-Vila, J. F. B. Stuart, G. Vanlerberghe and J. S. Whittaker. *J. Pharm. Pharmacol.*, 37, 237 (1985)

38. A. J. Baillie, A. T. Florence, L. R. Hume, G. T. Muirhead and A. Rogerson. *J. Pharm. Pharmacol.*, 37, 863 (1985)

39. P. H. Elworthy and T. Corby. *J. Pharm. Pharmacol.*, 23, 45S (1971)

40. D. Attwood and S. B. Kayne. *ibid.*, 23, 77S (1971)
41. N. N. Salib, A. A. Ismail and A. S. Geneidi. *Pharm. Ind.*, 36, 108 (1974)
42. T. Arnarson and P. H. Elworthy. *J. Pharm. Pharmacol.*, 32, 381 (1980)
43. A. A. Ismail, M. W. Gouda and M. M. Motawi. *J. pharm. Sci.*, 59, 220 (1970)
44. B. W. Barry and D. I. D. El Eini. *J. Pharm. Pharmacol.*, 28, 210 (1976)
45. P. Ekwall, L. Sjöblöm and J. Olsen. *Acta chem. Scand.*, 7, 347 (1953)
46. J. H. Collett and L. Koo. *J. pharm. Sci.*, 64, 1253 (1975)
47. H. Tomida, T. Yotsuyanagi and K. Ikeda. *Chem. pharm. Bull.*, 26, 2824 (1978)
48. H. Tomida, T. Yotsuyanagi and K. Ikeda. *Chem. pharm. Bull.*, 26, 2832 (1978)
49. K. J. Humphries and C. T. Rhodes. *J. pharm. Sci.*, 57, 79 (1968)

7 Disperse Systems

This chapter is devoted to the study of emulsions and suspensions and systems in the colloidal state, that is those in which the particles range from molecular size to coarse dispersions. The word colloid derives from the Greek *kolla* (glue) and was coined from the impression that colloidal substances were amorphous or glue-like rather than crystalline forms of matter. The colloidal state was recognised by Thomas Graham in 1861 and described by Wolfgang Ostwald some fifty years later as the 'world of neglected dimensions', a reference both to the fact that colloid science has somehow remained a cinderella topic, and to the special world of systems in which the particles are extremely small, below about 1 μm in diameter. Colloids can be broadly classified as those that are *lyophobic* (solvent hating) and those that are *lyophilic* (solvent loving). The appropriate terms in aqueous media are, respectively, *hydrophobic* and *hydrophilic*. Surfactant molecules, because of their affinity for water and their tendency to associate into micelles, form hydrophilic colloidal dispersions in water. Proteins and gums also form lyophilic colloidal systems. Water-insoluble drugs or clays and oily phases will form lyophobic dispersions, the topic covered in this chapter. Lyophobic colloidal dispersions have the tendency to coalesce as they are thermodynamically unstable because of their high surface energy. Lyophilic dispersions are inherently stable.

Many natural systems — suspensions of microorganisms, blood, isolated cells (in culture, for example) — are thus lyophobic colloidal dispersions. Pharmaceutical colloids such as emulsions and suspensions and aerosols are readily identified. Colloid science is interdisciplinary for although dealing with complex systems it is nevertheless a unifying discipline as it bridges the physical and biological sciences. The concepts of the stability of colloidal systems derived for suspensions can be applied with little modification to our understanding of interactions between living cells, for example.

It is because of the subdivision of matter in colloidal systems that these have special properties. The large surface to volume ratio of the particles dispersed in a liquid medium results in a tendency for particles to associate to reduce their surface area of contact with the medium. Emulsions and aerosols are thermodynamically unstable two-phase systems which can only reach equilibrium when the globules have coalesced to form one macro-phase, and the surface area is at a minimum. Many of the pharmaceutical problems revolve around the stabilisation of colloidal systems; some biological phenomena can be understood in terms of the association of cells with other cells or with inanimate or other substrates. This, too, is a sort of colloidal instability. This chapter is an attempt to describe the various colloidal systems, to deal in outline with the current theories of colloid stability and to discuss the pharmaceutical problems encountered with colloidal dosage forms. At the close of the chapter some of the biological implications of the subject are indicated.

7.1   Classification of colloidal systems

Graham's classification of states of matter into 'crystalloids' and 'colloids' was germinal and, although no longer absolute, it drew attention to those systems in which there were present units intermediate in size between simple molecules at one end of the scale and particles visible to the naked eye at the other. Obviously the dividing line between the colloidal and non-colloidal state is sometimes difficult to draw. Table 7.1 classifies common colloidal systems in terms of the nature of the dispersing medium and the dispersed material.

table 7.1   **Classification of colloidal systems**

Dispersing medium	Dispersed matter	Name for system
Gas	Liquid	Aerosol
	Solid	Aerosol
Liquid	Gas	Foam
	Liquid	Emulsion
	Solid	Sol*/suspension
Solid	Gas	Solid foam
	Liquid	Solid foam
	Solid	Solid sol

*If liquid is water, colloidal sols are terms hydrosols

7.2   Colloid stability

In dispersions of fine particles in a liquid (or of particles in a gas) frequent encounters between the particles occur due to Brownian movement, to creaming, sedimentation or to convection. The rate of creaming depends on the difference in density between the dispersed particles and the dispersion medium, the particle size, a, and the viscosity of the dispersion medium η. According to Stokes' law the rate of sedimentation (or creaming) of a spherical particle, v, in a fluid medium is given by

$$v = \frac{2ga^2 (\rho_1 - \rho_2)}{9\eta} \qquad (7.1)$$

where ρ_1 is the density of the particle and ρ_2 is the density of the medium. Creaming of emulsions or sedimentation of suspensions can be reduced by forming small particles, by increasing the viscosity of the continuous phase or by decreasing the density difference between the two phases. However, particles will still collide, although we can reduce the likelihood of the collision. What happens when the particles do come into close contact? The encounters sometimes lead to permanent contact of solid particles or of coalescence of liquid droplets following their collision. If allowed to go unchecked the colloidal system destroys itself through growth of the disperse phase and excessive creaming or sedimentation of the large particles. Whether these collisions result in permanent contact between the particles or whether the particles rebound and remain free depends on the forces of interaction, both attractive and repulsive, between the particles.

There are five possible types of force which can be identified: (i) electrostatic forces of repulsion; (ii) van der Waals forces or electromagnetic forces of attraction; (iii) Born forces — essentially short-range and repulsive; (iv) steric forces which are dependent on the geometry and conformation of molecules at the particle interfaces; and (v) solvation forces due to changes in quantities of adsorbed solvent on approach of neighbouring particles. In the 1940s consideration of the electrostatic repulsion and van der Waals forces of attraction by Deryagin and Landau and Verwey and Overbeek produced a satisfactory quantitative approach to the stability of hydrophobic suspensions. Their theory is known as the DLVO theory of colloid stability (see references 1, 2 and 3 for details).

Van der Waals forces are always attractive between particles of the same kind. The multiplicity of interactions between pairs of atoms or molecules on neighbouring particles must be taken into account in the calculation of attractive forces. Hamaker first determined equations for these forces on the basis of the additivity of van der Waals energies between neighbouring molecules, assuming that the energies of attraction varied with the inverse sixth power of the distance between them. At greater separations of the particles, the power law changes to the inverse seventh power. The model considers two spherical particles of radius a at a distance H, R being $2a + H$ (figure 7.1).

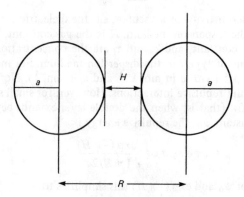

figure 7.1 Model used to consider forces of interaction between two spherical particles of radius a at a distance of separation H and centre-to-centre distance of R. $R = H + 2a$

Hamaker calculated the energy of attraction, V_A, to be

$$V_A = -\frac{A}{6}\left\{\frac{2a^2}{R^2 - 4a^2} + \frac{2a^2}{R^2} + \frac{R^2 - 4a^2}{R^2}\right\} \qquad (7.2)$$

The Hamaker constant A depends on the properties of the particles and of the medium in which they are dispersed. When H/a is small, that is when the particles are large relative to their separation, equation 7.2 reduces to

$$V_A = -\frac{Aa}{12H}$$

Charge on particles arises either from ionisation of groups on the surface or from adsorption of ions, which confer their charge onto the surface. A particle surface with a negative charge is shown in figure 7.2 along with the layer of positive ions that are attracted to the surface in the Stern layer, and the diffuse or electrical double layer which accumulates that contains both positive and negative ions.

Electrostatic forces arise from the interaction of the electrical double layers surrounding particles in suspension (see figure 7.2). This interaction leads to repulsion if the particles have surface charges and surface potentials of the same sign and magnitude. When the surface charge is produced by the adsorption of potential-determining ions the surface potential, ψ_0, is determined by the activity of these ions and remains constant during interaction with other particles, if the extent of adsorption does not change. The interaction therefore takes place at constant surface potential. In emulsion systems where the adsorbed layers can desorb, or in conditions of low availability of potential-determining ions, the interaction takes place not at constant surface potential but at constant surface charge (or at some intermediate state.) The electrostatic repulsive force decays as an exponential function of the distance and has a range of the order of the thickness of the electrical double layer, equal to the Debye-Hückel length, $1/\kappa$:

$$1/\kappa = \left(\epsilon \, \epsilon_0 RT / F^2 \sum c_i z_i \right)^{1/2} \tag{7.3}$$

where ϵ_0 is the permittivity of a vacuum, ϵ is the dielectric constant (or relative permittivity) of the dispersion medium, R is the gas constant, T is temperature, F is the Faraday constant, and c_i and z_i are the concentration and the charge number of the ions of type i in the dispersion medium. For monovalent ions in water, $c = 10^{-15} \kappa^2$ (with c in mol ℓ^{-1} and κ in cm^{-1}). No simple equations can be given for the repulsive interactions. However, for small surface potentials and low values of κ (that is, when the double layer extends beyond the particle radius) and at constant ψ_0, the repulsive energy is

$$V_R = 2\pi \, \epsilon \, \epsilon_0 \, a \, \psi_0^2 \, \frac{\exp(-\kappa H)}{1 + H/2a} \tag{7.4}$$

For small values of ψ_0 and $\exp(-\kappa H)$ this simplifies to

$$V_R = 2\pi \, \epsilon \, \epsilon_0 \, a \, \psi_0^2 \, \exp(-\kappa H) \tag{7.5}$$

The equations do not take into account the finite size of the ions, and the potential to be used is ψ_0, the potential at the Stern plane (the plane of closest approach of ions to the surface) which is difficult to obtain. The nearest experimental approximation to ψ_0 is often the zeta (ζ) potential.

In the DLVO theory the combination of the electrostatic repulsive energy V_R with the attractive potential energy V_A gives the total potential energy of interaction

$$V_{total} = V_A + V_R \tag{7.6}$$

V_{total} plotted against the distance of separation H gives a potential energy curve showing certain characteristic features, illustrated in figure 7.3. The maximum and minimum energy states are shown. At small and at large distances the van der Waals energy (proportional to H^{-x}, where x varies from 1 to 7) is greater

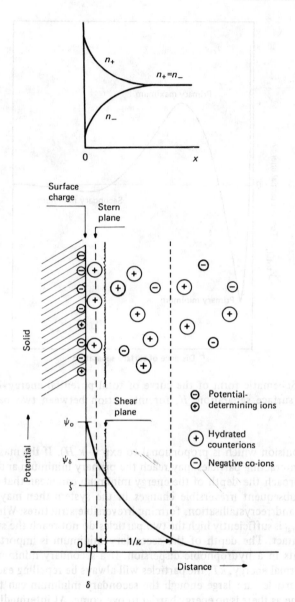

figure 7.2 Representation of the conditions at a negative surface, with a layer of adsorbed positive ions in the Stern plane. The number of negative ions increases and the number of positive ions decreases (see top diagram) as one moves away from the surface, the electrical potential becoming zero when the concentrations are equal. The surface potential ψ_0 and the potential at the Stern plane, ψ_δ, are shown. As the particle moves the effective surface is defined as the surface of shear, which is a little further out from the Stern plane, and would be dependent on surface roughness, adsorbed macromolecules, etc. It is at the surface of shear that the zeta potential, ζ, is located. The thickness of double layer is given by $1/\kappa$

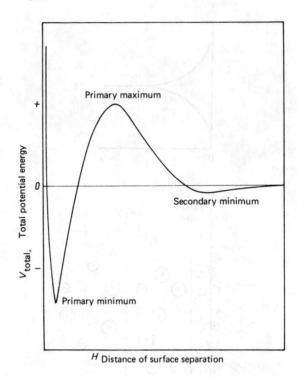

figure 7.3 Schematic form of the curve of total potential energy V_{total} against distance of surface separation, H, for interaction between two particles, with $V_{total} = V_R + V_A$

than the repulsion which is proportional to exp $(-\kappa\ H)$. If the maximum is too small, two interacting particles may reach the primary minimum and in this state of close approach the depth of the energy minimum can mean that escape is improbable. Subsequent irreversible changes in the system then may occur, such as sintering and recrystallisation, forming irreversible structures. When the maximum in V_{total} is sufficiently high the two particles do not reach the stage of being in close contact. The depth of the secondary minimum is important in determining events in a hydrophobic dispersion. If a secondary minimum is smaller than the thermal energy, kT, the particles will always be repelling each other, but when the particles are large enough the secondary minimum can trap particles for some time as there is no energy barrier to overcome. At intermediate distances the energy of repulsion may be the larger of the two. Pharmaceutical colloids are rarely simple systems. The influence of additives including simple and complex electrolytes has to be considered.

Electrolyte concentration and valence (z) are accounted for in the term $(\Sigma c_i z_i^2)$ in equation 7.3 and thus in equations 7.4 and 7.5. Figure 7.4 gives an example of the influence of electrolyte concentration on the interaction energy. In this example, $a = 10^{-5}$ cm, $A = 10^{-19}$ J, and $\psi_0 = RT/F \approx 25.6$ mV. As the electrolyte concentration is increased, κ increases due to compression of the double layer with consequent decrease in $1/\kappa$.

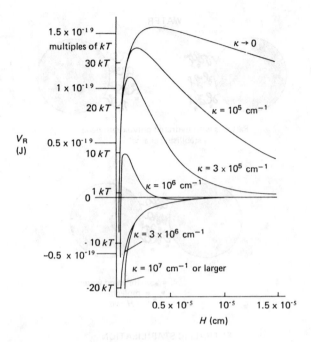

figure 7.4 Energy of interaction of two spherical particles as a function of the distance, H, between the surfaces. For monovalent ions c (in mol ℓ^{-1}) = 10^{-15} κ^2 (in cm^{-1}). In this example $a = 10^{-5}$ cm, $A = 10^{-19}$ J and $\psi_0 = RT/F =$ 26.5 mV. From J. Th. Overbeek. *J. colloid interface Sci.*, 58, 408 (1977)

The increasing usage of non-ionic macromolecules as stabilisers, which has occurred since the development of the DLVO theory, has led to the awareness of other stabilising forces. The approach of particles with adsorbed hydrated macromolecules leads, on the interaction of the polymer layers, to repulsion (figure 7.5). The origin of the repulsive forces is the positive enthalpy change which ensues. In more general terms, the approach of two particles with adsorbed stabilising chains leads to a steric interaction when the chains interact. The repulsive forces may not always be enthalpic in origin. Loss of conformational freedom leads to a negative entropy change. Each chain loses some of its conformational freedom and its contribution to the free energy of the system is increased, leading to the repulsion. This volume restriction is compounded by an 'osmotic effect' which arises as the macromolecular chains on neighbouring particles crowd into each other's space and the concentration of chains in the overlap region increases. The repulsion which arises is due to the osmotic pressure of the solvent attempting to dilute out the concentrated region: this can only be achieved by the particles moving apart. Several quantitative assessments have been made of the steric effect. One such relates the repulsion to the polymer chain length, δ, the interaction of the solvent with the chains and the number of chains per unit area of interacting surface. As this third component of the total interaction does not come into play until $H = 2\delta$, the interaction suddenly in-

WATER

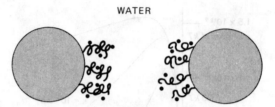

Particles with hydrated polyoxyethylene
stabilising chains

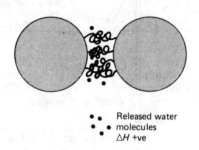

• • Released water
• • molecules
      ΔH +ve

ENTHALPIC STABILISATION

figure 7.5 Representation of enthalpic stabilisation of particles with adsorbed hydrophilic chains. The hydrated chains of the polyoxyethylene molecules $H(OCH_2 CH_2)_n OH$ protrude into the aqueous dispersing medium. On close approach of the particles to within 2δ (twice the length of the stabilising chains) hydrating water is released, resulting in a positive enthalpy change which is energetically unfavourable

creases with decreasing distance. There are many problems in applying such equations in practice, the main ones being the lack of detailed knowledge of δ, and the difficulty in taking account of desorption and conformational or solvational changes during interaction. When the steric contribution is combined with the electrostatic and van der Waals interactions, a minimum in the energy at large separations still obtains, but repulsion is generally evident at all shorter distances, provided that the adsorbed macromolecules or surfactants do not desorb or move from the points of interaction (figure 7.6).

For particles with a hydrated layer of thickness δ, the volume of the overlapping region (V_S) is as derived in figure 7.7

$$V_S = \frac{2\pi}{3} \left(\delta - \frac{H}{2}\right)^2 \left(3a + 2\delta + \frac{H}{2}\right) \qquad (7.7)$$

The difference in chemical potential in the overlap volume (μ_H) and that when the particles are at an infinite distance apart (μ_∞) gives the repulsive force which is an osmotic force, due to the increased concentration of the polymer chains in this region:

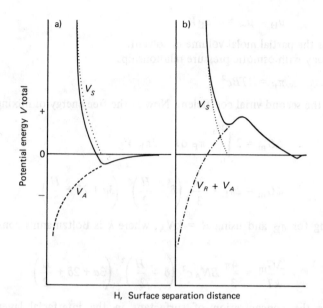

figure 7.6 The schematic form of the potential energy against distance of surface separation curves for sterically stabilised particles, V_S being the steric stabilising potential energy: (a) in the absence of electrostatic repulsion — the full curve represents $V_{total} = V_S + V_A$; (b) in the presence of electrostatic repulsion — the full curve represents $V_R + V_A + V_S$.

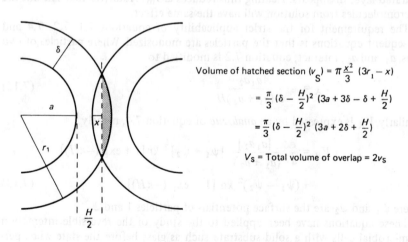

Volume of hatched section $(v_s) = \pi \dfrac{x^2}{3} (3r_1 - x)$

$= \dfrac{\pi}{3} (\delta - \dfrac{H}{2})^2 (3a + 3\delta - \delta + \dfrac{H}{2})$

$= \dfrac{\pi}{3} (\delta - \dfrac{H}{2})^2 (3a + 2\delta + \dfrac{H}{2})$

V_S = Total volume of overlap = $2v_s$

figure 7.7 The model used in the derivation of equation 7.7 by Ottewill in reference 3. Particles of radius a with adsorbed layer of thickness δ approach to a distance H between the particle surfaces. $r_1 = (a + \delta)$ and x is the distance between the surface and the line bisecting the volume of overlap

$$\mu_H - \mu_\infty = - \pi_E \bar{V}_1 \tag{7.8}$$

where $\bar{V}_1$ is the partial molal volume of solvent.

By analogy with osmotic pressure relationships

$$\pi_E = RTBc^2 \tag{7.9}$$

where B is the second virial coefficient. Now as the free energy of mixing is

$$\Delta G_m = 2 \int_0^{V_S} \pi_E \, dV = 2\pi_E V_S$$

therefore

$$\Delta G_m = 2\pi_E \cdot \frac{2\pi}{3} \left(\delta - \frac{H}{2} \right)^2 \left(3a + 2\delta + \frac{H}{2} \right) \tag{7.10}$$

Substituting for π_E and using $R = kN_A$, where k is Boltzmann's constant, we obtain

$$\frac{\Delta G_m}{kT} = \frac{4\pi}{3} BN_A c^2 \left(\delta - \frac{H}{2} \right)^2 \left(3a + 2\delta + \frac{H}{2} \right) \tag{7.11}$$

where c is the concentration of surfactant in the interfacial layer, N_A is Avogadro's constant. The equation looks more complex than it is. Apart from telling us the effect of changing δ, which we can achieve by the use of different polymers of different size, we can find out the effect of temperature and additives, as B is proportional to $1 - \theta/T$, where θ is the theta temperature, the temperature at which the polymer and solvent have no affinity for each other (section 8.2.7). Thus when $T = \theta$, $B \to$ zero, and the stabilising influence of the hydrated layer disappears. Heating thus reduces ΔG_m. Additives that salt out the macromolecules from solution will have the same effect.

The requirement for the strict applicability of equations 7.1, 7.2, 7.4 and subsequent equations is that the particles are monosized. Where particles of two sizes, a_1 and a_2, interact, equation 7.2 is modified to

$$V_A = - \frac{Aa_1 a_2}{6 (a_1 + a_2)H} \tag{7.12}$$

Similarly V_R is expressed by an *analogue* of equation 7.4, namely

$$V_R = \frac{\epsilon}{4} \frac{[a_1 a_2]}{(a_1 + a_2)} [\psi_1 + \psi_2]^2 \ln[1 + \exp(-\kappa H)]$$

$$+ (\psi_1 - \psi_2)^2 \ln [1 - \exp(-\kappa H)] \tag{7.13}$$

where ψ_1 and ψ_2 are the surface potentials of particles 1 and 2.

These equations have been applied to the study of the reversible interaction of microbial cells with a solid substrate such as glass before the state when permanent adhesion occurs due to the formation of polymeric bridges between cell and surface. Changes in the bacteria on adsorption are hazards in application of model equations; so too are the presence of cilia on the surface and protuberances which affect the closeness of approach and the nominal radius of curvature of the 'particles' (see section 7.6).

7.3 Emulsions

Emulsions are liquid dispersions generally of an oil and water. Their use as pharmaceutical dosage forms necessitates an understanding of the factors governing the stability of oil-in-water (o/w) and water-in-oil (w/o) emulsions.

7.3.1 Stability of emulsions

Adsorption of a surfactant at the oil–water interface, by lowering the interfacial tension, aids the dispersal of the oil into droplets of a small size and helps to maintain the particles in a dispersed state. Unless the interfacial tension is zero, there is a tendency for the oil droplets to coalesce to reduce the area of oil–water contact, but the presence of the surfactant monolayer at the surface of the droplet reduces the possibility of collisions leading to droplet coalescence. Charged surfactants will lead to an increase in negative or positive zeta potential and will thus help to maintain stability by increasing V_R. However, non-ionic surfactants are widely used in pharmaceutical emulsions because of their lack of toxicity and lower sensitivity to additives. The non-ionic stabilisers adsorb onto the emulsion droplets and although they generally reduce zeta potentials they maintain stability by creating a hydrated layer on the hydrophobic particle in o/w emulsions. That is, they effectively convert a hydrophobic colloidal dispersion into a hydrophilic dispersion. In a w/o emulsion the hydrocarbon chains of the adsorbed molecules protrude into the oily continuous phase. Stabilisation arises from steric repulsive forces as described above, although the emulsion is a much more complex system than a suspension. This is so because of the possibility of movement of the surfactant into either continuous or disperse phase, the possibility of micelle formation in both phases and also because of the formation under suitable conditions of liquid crystalline phases between the disperse droplets (see section 6.4).

It is usually observed that mixtures of surfactants form more stable systems than do single surfactants, even with very dilute emulsions. This may be because complex formation at the interface results in a rigid film. Certainly where complex films can be formed, such as between sodium lauryl sulphate and cetyl alcohol, the stability of emulsions prepared with such mixtures is high. Theory has not developed to an extent that it can cope with mixtures of stabiliser molecules. Complex formation in the bulk phase of emulsion systems is dealt with in section 7.3.5.

7.3.2 HLB system

In spite of many advances in the theory of stability of lyophobic colloids, resort has still to be made to an empirical approach to the choice of emulsifier devised in 1949 by Griffin in the United States. In this system one calculates the hydrophile–lipophile balance (HLB) of surfactants and matches the HLB of the surfactant mixture in the case of an o/w system to that of the oil being emulsified. The HLB number of a surfactant is calculated according to a certain empirical formula and for non-ionic surfactants the values range from 0 to 20 on an arbitrary scale

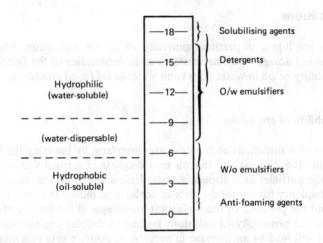

figure 7.8 HLB scale and approximate range into which solubilising agents, detergents, emulsifiers and antifoaming agents fall

(see figure 7.8). At the higher end of the scale the surfactants are hydrophilic and act as solubilising agents, detergents and o/w emulsifiers. Oil-soluble surfactants with a low HLB act as w/o emulsifiers. In the stabilisation of oil globules it is essential that there is a degree of hydrophilicity to confer an enthalpic stabilising force and a degree of hydrophobicity to secure adsorption at the interface. The balance between the two will depend on the nature of the oil and the mixture of surfactants, hence the need to apply the HLB system. The HLB of polyhydric alcohol fatty acid esters such as glyceryl monostearate may be obtained from the equation

$$HLB = 20 \left(1 - \frac{S}{A}\right) \tag{7.14}$$

where S is the saponification number of the ester and A is the acid number of the fatty acid. The HLB of Polysorbate 20 calculated using this formula is 16.7, S being 45.5 and $A = 276$.

The polysorbates (Tween) have HLB values in the range 9.6–16.7; the sorbitan esters (Span) have HLBs in the lower range of 1.8–8.6. For those materials for which it is not possible to obtain saponification numbers, for example beeswax and lanolin derivatives, HLB is calculated from

$$HLB = (E + P)/5 \tag{7.15}$$

where E is the percentage by weight of oxyethylene chains, and P is the percentage by weight of polyhydric alcohol groups (glycerol or sorbitol) in the molecule. If the hydrophile consists only of oxyethylene groups, a simpler version of the equation is

$$HLB = E/5 \tag{7.16}$$

Some HLB values of typical surfactants used in pharmacy are given in table 7.2. A more detailed list is given in tables 6.8 and 6.9.

table 7.2 **Typical HLB numbers of some surfactants**

Compound	HLB
Glyceryl monostearate	3.8
Sorbitan monooleate (Span 80)	4.3
Sorbitan monolaurate (Span 20)	8.6
Triethanolamine oleate	12.0
Polyoxyethylene sorbitan monooleate (Tween 80)	15.0
Polyoxyethylene sorbitan monolaurate (Tween 20)	16.7
Sodium oleate	18.0
Sodium lauryl sulphate*	40.0

*Although applied mainly to non-ionic surfactants it is possible to obtain numbers for ionic surfactants

The HLB system has been put on a more quantitative basis by Davies[4] who calculated group contributions to the HLB number such that the HLB was calculable from

$$HLB = \sum (\text{hydrophilic group numbers})$$
$$- \sum (\text{lipophilic group numbers}) + 7 \qquad (7.17)$$

Some group numbers are given in table 7.3.

The appropriate choice of emulsifier or emulsifier mixture can be made by preparing a series of emulsions with a range of surfactants of varying HLB. It is assumed that the HLB of a mixture of two surfactants containing fraction f of A and $(1 - f)$ of B is an algebraic mean of the two HLB numbers

$$HLB_{mixture} = f HLB_A + (1 - f) HLB_B \qquad (7.18)$$

For reasons not explained by the HLB system, but from other approaches, mixtures of surfactants give more stable emulsions than single surfactants. In the

table 7.3 **Group contributions to HLB numbers***

	Group number
Hydrophilic groups	
$COO^- Na^+$	19.1
Ester	2.4
Hydroxyl	1.9
Hydroxyl (sorbitan)	0.5
Lipophilic groups	
$-CH-$ $-CH_2-$ $-CH_3$ $=CH-$	0.475

*After reference 4

table 7.4 **Required HLB for different oils for o/w emulsion formation**

Oil	HLB
Cottonseed oil	7.5
Vaseline oil	8.5
Dodecane	9–9.5
Mineral oil	10–12
Cyclohexane	12

table 7.5 **HLB values for emulsifiers of natural origin***

Compound	HLB
Gum acacia	8.0–11.9
Gelatin	9.8
Methylcellulose	10.5
Gum tragacanth	11.9–13.2

*From A. H. C. Chun *et al.*, *Drug Cosmetic Ind.*, 82, 164 (1958)

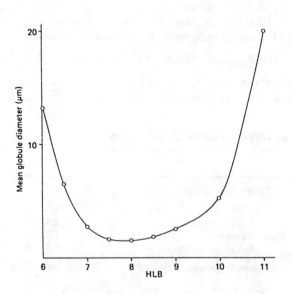

figure 7.9 Variation of mean globule size in a mineral oil-in-water emulsion as a function of the HLB of the surfactant mixtures present at a level of 2.5 per cent. Surfactants: Brij 92–Brij 96 mixtures. From P. Depraetre, M. Seiller, A. T. Florence and F. Puisieux, unpublished results

experimental set up creaming of the emulsion is observed and is taken as an index of stability. The system with the minimum creaming or separation of phases is deemed to have an optimal HLB. It is therefore possible to determine optimum HLB numbers required to produce stable emulsions of a variety of oils (table 7.4).

In spite of the optimum HLB values for forming o/w emulsions, it is possible to formulate stable systems with mixtures of surfactants well below the optimum. This is because of the formation of a viscous network in the continuous phase. The viscosity of the medium surrounding the droplets prevents their collision and this overrides the influence of the interfacial layer and barrier forces due to the presence of the adsorbed layer. Natural products such as acacia and tragacanth which have been traditionally used as emulsifiers in pharmacy have some surfactant properties but their activity is not restricted to the interface. It is possible to obtain HLB values for natural products, some of which are shown in table 7.5.

It has been found that at the optimum HLB the mean particle size of the emulsion is at a minimum (figure 7.9) and this factor would explain (see equations 7.1 and 7.2, for example) the stability of the system.

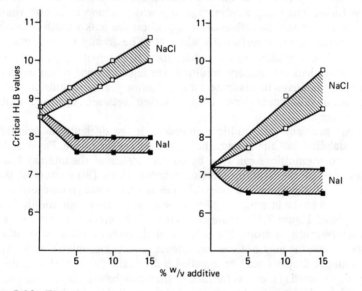

figure 7.10 The change in critical HLB values as a function of added salt concentration, where the salt is either NaCl or NaI. Results were obtained from measurements of particle size, stability, viscosity and emulsion type as a function of HLB for liquid paraffin-in-water emulsions stabilised by Brij 92–Brij 96 mixtures. Data from different experiments showed different critical values hence on each diagram hatching represents the critical regions while data points actually recorded are shown. Results on the left-hand diagram show, respectively, particle size and stability data; those on the right-hand diagram show the HLB at transition from pseudoplastic to Newtonian flow properties and emulsion type (o/w → w/o) transitions. From A. T. Florence, F. Madsen and F. Puisieux. *J. Pharm. Pharmacol.*, 27, 385 (1975)

The HLB system has several drawbacks. The calculated HLB, of course, cannot take account of the effect of temperature nor of additives. The presence in emulsion systems of agents which salt-in or salt-out non-ionic surfactant chains will respectively increase and decrease the effective (as opposed to the calculated) HLB values. Salting-out the surfactant (for example, with NaCl) will make the molecules less hydrophilic and one can thus expect a higher optimal calculated HLB value for the stabilising surfactant for o/w emulsions containing sodium chloride. Examples are shown in figure 7.10 in which NaCl and NaI are compared.

7.3.3 Multiple emulsions

Multiple emulsions are emulsions whose disperse phase contains droplets of another phase (figure 7.11). Water-in-oil-in-water (w/o/w) or o/w/o emulsions may be prepared, both forms being of interest as drug delivery systems. Water-in-oil emulsions in which a water-soluble drug is dissolved in the aqueous phase may be injected by the subcutaneous or intramuscular routes to produce a delayed-action preparation, as the drug has to diffuse through the oil to reach the tissue fluids. The main disadvantage of a w/o emulsion is its high viscosity brought about through the influence of the oil on the bulk viscosity. Emulsifying a w/o emulsion using surfactants which stabilise an oily disperse phase can produce multiple emulsion systems which have an external aqueous phase and lower viscosity than the primary emulsion. On injection, one might expect the external aqueous phase to dissipate rapidly leaving the w/o emulsion, yet biopharmaceutical differences have been observed between w/o and multiple emulsion systems.

The main problem with multiple emulsions is the difficulty in maintaining long-term stability, but the systems hold promise for the future. Physical degradation of w/o/w emulsions can arise by (i) coalescence of the internal droplets, (ii) coalescence of the large droplets surrounding them, (iii) rupture of the oil film separating the internal and external aqueous phases and (iv) osmotic flux of water to and from the internal droplets, possibly associated with micellar species in the oil phase. Figure 7.11*b* represents some of the breakdown pathways that occur in w/o/w multiple drops. The external oil drop may coalesce with other oil drops (which may or may not contain internal aqueous droplets), as in (a); the internal aqueous droplets may be expelled individually (b, c, d, e) or more than one may be expelled (f), or less frequently they may be expelled in one step (g); the internal droplets may coalesce before being expelled (h, i, j, k); or water may pass by diffusion through the oil phase gradually resulting in shrinkage of the internal droplets (l, m, n). Figure 7.11*b* is oversimplified; in practice the number of possible combinations is large. A number of factors will determine the breakdown mechanism in a particular system, but one of the main driving forces behind each step will be the reduction in the free energy of the system brought about by the reduction in the interfacial area. Mechanisms of drug release from multiple emulsion systems include diffusion of the drug molecules from the internal droplets (1), from the medium of the external droplets (2), or by mass transfer due to the coalescence of the internal droplets (3), as shown in figure 7.11*c*. Multiple emulsions are discussed in more detail in references 5 and 6.

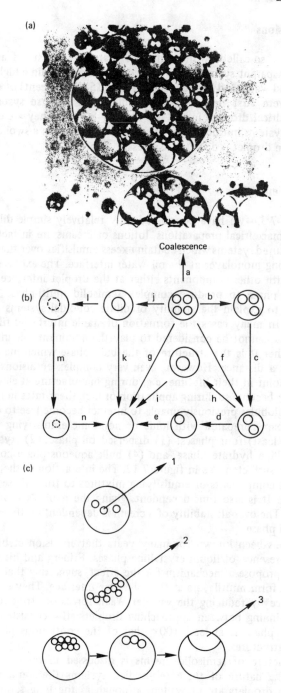

figure 7.11 (a) Photograph of multiple emulsion droplets. (b) Possible break-down pathways in w/o/w multiple emulsions (see text for explanation). From reference 5. (c) Diagrammatic representation of mechanisms of drug release, from S. S. Davis. *J. clin. Pharm.*, 1, 11 (1976)

7.3.4 Microemulsions

Microemulsions, or so-called swollen micellar systems, consists of apparently homogeneous transparent systems of low viscosity which contain a high percentage of both oil and water and high concentrations (15–25 per cent) of emulsifier mixture. They were first described by Schulman as disperse systems, with spherical or cylindrical droplets in the size range 8–80 nm. They are essentially swollen micellar systems, but obviously the distinction between a swollen micelle and small emulsion droplet is difficult to assess.

7.3.5 Structured emulsions

In sections 7.3.1–7.3.4 we have considered only relatively simple dilute emulsions. Many pharmaceutical preparations, lotions or creams are in fact complex semisolid or structured systems which contain excess emulsifier over that required to form a stabilising monolayer at the oil–water interface. The excess surfactant often interacts with other components either at the droplet interface or in the bulk continuous phase to produce complex semisolid multi-phase systems[7]. Theories derived to explain the stability of dilute colloidal systems cannot be applied directly. In many cases the formation of stable interfacial films at the oil–water interface cannot be considered to play the dominant role in maintaining stability. Rather it is the structure of the bulk phase which maintains the disperse phase at a distance. However, even very complex emulsions are often mobile at some point in their lifetime, e.g. during manufacture at elevated temperatures, or may become so during application of high shear rates in use. Under these conditions globules previously unable to interact become free to do so.

Stable o/w creams prepared with ionic or non-ionic emulsifying waxes are composed of (at least) four phases: (1) dispersed oil phase, (2) crystalline gel phase, (3) crystalline hydrate phase, and (4) bulk aqueous phase containing a dilute solution of surfactant, as in figure 7.12. The interaction of the surfactant and fatty alcohol components of emulsifying mixtures to form these structures (body) is critical. It is also time-dependent, giving the name 'self-bodying' to these emulsions. The over-all stability of a cream is dependent on the stability of the crystalline gel phase.

Although it has been known for many years that emulsion stability is increased by the presence of liquid crystalline phases, Friberg and his colleagues were the first to propose a mechanism for the effect, suggesting that the liquid crystal structures form multilayers at the oil–water interface. They thus protect against coalescence by reducing the van der Waals forces of attraction and by retarding film thinning between approaching droplets: the consistency of the liquid crystalline phase is at least 100× that of the continuous phase in the absence of these structures.

The microstructure of semisolid creams is discussed in reference 7. It is apparent from the nature of these self-bodied systems that, at equilibrium, contacts between droplets are prevented, although in the long term processes equivalent to syneresis may occur due to rearrangement of the matrix structure and the 'squeezing out' of the oil phase.

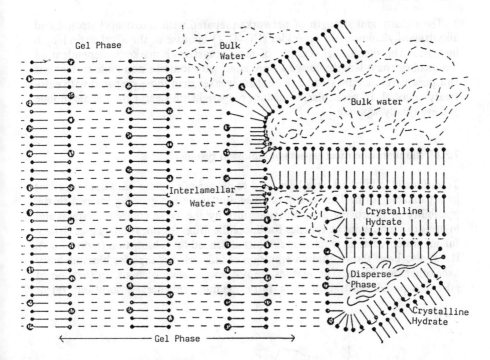

figure 7.12 Schematic diagram of a typical semisolid cream prepared with cetostearyl alcohol and ionic surfactant. Note the four phases: (1) the dispersed oil phase; (2) the crystalline gel phase containing interlamellar-fixed water; (3) phase composed of crystalline hydrates of cetostearyl alcohol; (4) bulk water phase. From reference 7 with permission

Certain guidelines have been devised[8] for the formulation of self-bodied emulsions:

The *lipophilic component* should be an amphiphile which by itself promotes the formation of water-in-oil emulsions and is capable of complexing with the hydrophilic surfactant at the oil–water interface. Its concentration should be at least sufficient to form a close-packed mixed monolayer with the hydrophilic component. To promote the formation of semisolid emulsions at room temperature, it should be near or above its saturation concentration in the oil. Excess material should diffuse readily from the warm oil phase into the warm micellar phase and there be solubilised. The melting point should be sufficiently high to precipitate solubilised material at room temperature.

The *hydrophilic component* should be a surfactant which promotes the formation of oil-in-water systems and is capable of complexing with the lipophilic component at the oil–water interface. Its concentration should be at least sufficient to form a close-packed monolayer with the lipophilic component and it should be in excess of its cmc in the aqueous phase. It should be capable of solubilising the lipophilic component when warm.

The rigidity and strength of networks prepared with cetostearyl alcohol and alkyltrimethylammonium bromides (C_{12}–C_{18}) increase as the alkyl chain length increases. The rheological stability of ternary systems is markedly dependent on the alcohol chain length; networks prepared with ionic or non-ionic surfactants and pure cetyl or pure stearyl alcohol are weaker than those prepared with cetostearyl alcohol. In particular, emulsions prepared with stearyl alcohol are mobile and eventually separate.

7.3.6 Biopharmaceutical aspects of emulsion systems

Traditionally, emulsions have been used to deliver oils (castor oil, liquid paraffin) in a palatable form. This is still one use, but there is a growing interest in the possibility of improving drug bioavailability by the use of emulsion systems and in using o/w emulsions as vehicles for lipophilic drugs for intravenous use. Griseofulvin, presented as an emulsion, exhibits enhanced oral absorption (figure 7.13). It has also been shown that an emulsion of indoxole has superior bioavailability over other oral forms. The reasons are not immediately obvious. In the case of griseofulvin, administration in a fatty medium enhances absorption; fat is

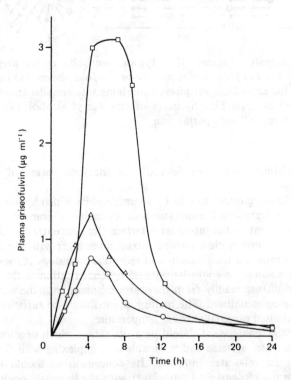

figure 7.13 Plasma levels following oral administration of griseofulvin in three different dosage forms. ○, Aqueous suspension; △, corn oil suspension; □, corn oil-in-water emulsion containing suspended griseofulvin. After P. J. Carrigan and T. R. Bates. *J. pharm. Sci.*, 62, 1476 (1973)

emulsified by the bile salts. Perhaps the administration of an already emulsified form increases the opportunity for solubilisation and hence transport across the microvilli by fat absorption pathways. The influence of the emulsifier on membrane permeability is one factor that must be considered. There have been reports of the absorption of macromolecules such as heparin and insulin administered orally in emulsion form. The reports that particles may be 'persorbed' from the gut and the increasing reports of absorption of macromolecules suggest that we may have to revise our views on the nature of absorption of many drugs from the gastro-intestinal tract.

The relationship of the HLB number of the surfactant system used in an emulsion to the extent of release of drug has been studied, mainly in relation to the release of medicament from topical and rectal semisolid emulsion formulations. This problem is related to that of preservative availability in emulsified systems, discussed below. The aim is to calculate the concentration of preservative in the non-aqueous, the aqueous and the micellar phase of the emulsion and to relate this to antimicrobial activity, as the antimicrobial agent is active in the aqueous phase.

7.3.7 Preservative availability in emulsified systems

Microbial spoilage of emulsified products is avoided by the inclusion of appropriate amounts of a preservative in the formulation. Infected emulsions have been the cause of outbreaks of pseudomonal and other bacterial skin infections. The incorporation of preservatives into pharmaceutical emulsions is not without problems as most agents partition to the oily or micellar phases of complex systems; some are inactivated by surfactants. In water, the antimicrobial activity of chlorocresol is superior to that of phenol. In arachis oil emulsions, phenol is the better agent since greater amounts of chlorocresol both interact with surfactant and partition to the oily phase.

Bean, Konning and Malcolm[9-11] have derived an equation for determining the amount of preservative (C_w) that remains in the aqueous phase (in which the preservative is active). The equation relates this value to the total amount (C) of preservative with a partition coefficient, P, added to an emulsion with an oil/water phase ratio of Φ

$$C_w = \frac{C(\Phi + 1)}{R(P\Phi + 1)} \tag{7.19}$$

where R is the preservative/emulsifier ratio or interaction ratio. If the volume of oil is V_o and the total volume of the emulsion is V_t, then the volume of the aqueous phase is $V_t - V_o$ and therefore

$$\Phi = \frac{V_o}{V_t - V_o} \tag{7.20}$$

and $$V_o = \Phi V_t / (1 + \Phi) \tag{7.21}$$

Since V_o or Φ is known from the composition of the emulsion only P and R need be determined experimentally.

The presence of surfactant micelles alters the native partition coefficient of the preservative molecule because the micellar phase offers an alternative site for

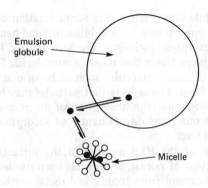

figure 7.14 Equilibrium established between emulsion globule and micelle. The preservative molecule is shown as ●

preservative molecules (see figure 7.14). The partitioning then occurs between the oil globule and the aqueous micellar phases.

For preservatives that are less soluble in the oily phase ($P < 1$), the concentration in the oily, micellar or aqueous phases increases when the proportion of oil is increased. In contrast, for those preservatives that are more soluble in oil than in water ($P > 1$), the concentration in all phases decreases when the oil phase is increased. This is the case with phenol and chlorocresol (see data given in table 7.6). In an emulsion containing 60 per cent arachis oil ($\Phi = 1.5$) and 1 per cent polysorbate, 9.6 per cent of the phenol resides in the free water phase, but only 0.3 per cent of the more lipophilic chlorocresol is free in water. As much as 93 per cent of the phenol and 99.9 per cent of chlorocresol are locked up in the oily phase or the micellar phases in emulsions containing 10 per cent Polysorbate 80. The use of equation 7.19 has been criticised because of the manner in which R has been measured and defined. For further discussion the paper by Mitchell and Kazmi[12] should be consulted. When the emulsified system

table 7.6 Percentage (W) of phenol and chlorocresol partitioned to various sites in arachis oil emulsions*

	Phenol				Chlorocresol			
Arachis oil in water Φ (phase ratio)	Oil W_o	Aqueous phase W_a	Micelles W_m	Water W_w	Oil W_o	Aqueous phase W_a	Micelles W_m	Water W_w
0.18	38	62	15.6	46.4	74.8	25.2	21.3	3.9
0.25	46.8	53.2	13.6	39.6	80.5	19.5	17	2.5
0.5	64	36	10.4	25.6	89.3	10.7	9.5	1.2
1.0	78.2	21.8	7.7	14.1	94.4	5.6	5.1	0.5
1.5	84.2	15.8	6.2	9.6	96.2	3.8	3.5	0.3

Note that $W_a = W_m + W_w$. Phenol and chlorocresol are present in 2.5 per cent concentration
Stabiliser: 1 per cent Polysorbate 80
*From reference 9

is very complex, containing not one but at least two emulsifying agents as most do, the determination of the parameters of the equation is a lengthy process, and a direct experimental approach to the determination of free aqueous concentration, such as a dialysis technique, may be preferred.

7.3.8 Mass transport in oil-in-water emulsions

Obviously, not only do preservative molecules traverse phase boundaries but drug molecules, flavouring and colouring agents do also. Interest in the extent and rate of flavour release on ingestion of a food emulsion has resulted in quantitative studies of the topic. The model used (see figure 7.15) is equally applicable to drug release in the gastro-intestinal tract, as dilution of the emulsion system occurs in both instances.

The concentration of the drug flavour, etc., in the aqueous phase immediately after dilution (C_w^o) and on re-establishing equilibrium (C_w^e) depends on the various properties of the solutes, such as partition coefficient, and the dissolution of the system.

Usually drugs dissolved in oils are absorbed mainly via the aqueous phase. Transport from one phase to the other and partitioning are therefore of importance. In the absorption of drugs in o/w emulsions when the drug partition coefficients are greater than unity, the amount of drug in the aqueous phase (rather than the concentration) is a critical factor for absorption. In the absorption of poorly oil-soluble drugs, drug absorption from emulsions is greater than from aqueous solution. In an emulsion of volume ratio Φ, the drug concentration in the aqueous phase (C_w) is related to the over-all concentration of the drug (C) by the expression of Bean and Heman–Ackah[13] :

$$C_w = C \; \frac{\Phi + 1}{P\Phi + 1} \qquad (7.22)$$

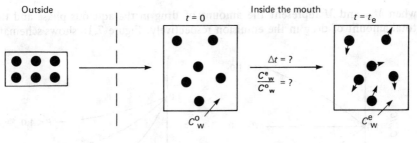

figure 7.15 Simulation of oral consumption of a simple model o/w emulsion food. Assumption: Only aqueous flavour is perceived. Questions: 'How much flavour can be transferred?' 'How fast is the rate of transfer?' From P. B. McNully and M. Kovel. *J. food Technol.*, 8, 309 (1973)

where P is the oil/water partition coefficient. This equation ignores the micellar phase. Equation 7.22 may be derived as follows.

We define P simply as C_o/C_w and the volume fractions of oil and water as ϕ_o and ϕ_w.

The total concentration $C = C_o \, \phi_o + C_w \, \phi_w$

If for convenience we write ϕ for ϕ_o, and as $\phi_w = 1 - \phi$ we can write

$$C = C_o \, \phi + C_w \, (1 - \phi)$$

Dividing by C_w

$$\frac{C}{C_w} = \frac{C_o}{C_w} \, \phi + (1 - \phi) = P \, \phi + (1 - \phi) = P\phi - \phi + 1$$

Therefore

$$C_w = \frac{C}{P\phi - \phi + 1}$$

Using $\phi = \phi_o = V_o/V_t$ and equation 7.21, we get $\phi = \Phi/(1 + \Phi)$ and so substituting this in the above equation gives

$$C_w = \frac{C(1 + \Phi)}{P\Phi + 1}$$

In an emulsion, the drug partitions between oily and aqueous phases but, unlike the preservatives which are active on their targets in the aqueous phase, drugs must be transported through an absorptive membrane and not only their concentrations but the total amount has to be taken into consideration. Bean and Heman-Ackah's equation is thus modified to give

$$M_w = M \, \frac{1}{P\Phi + 1} \tag{7.23}$$

when M_w and M represent the amount of drug in the aqueous phase and the total amount of drug in the emulsion respectively. Figure 7.16 shows schemati-

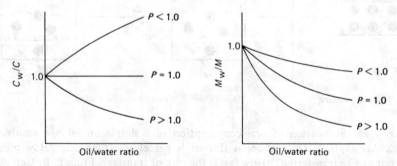

figure 7.16 Schematic representation of drug distribution as a function of oil/water phase ratio, Φ. Note that $M_w/M = C_w V_w/C V_t = C_w/C(1 + \Phi)$

cally the effect of P and Φ on C_w/C and M_w/M ratios.

Absorption of acetanilide from aqueous solution and from emulsions has been compared; drug absorption decreased with increase of Φ in isopropyl myristate emulsions containing 0.1 per cent Polysorbate 80, where P is 1.05. It appears that the amount of drug in the aqueous phase, rather than its concentration, was the dominant factor and that drugs were absorbed mainly via the aqueous phase.

7.3.9 Intravenous fat emulsions

Fat emulsions are used to supply in a small volume of isotonic liquid a large amount of energy; they supply the body with essential fatty acids and triglycerides. Fat emulsions for intravenous nutrition contain vegetable oil and a phospholipid emulsifier. There are several commercial fat emulsions available in Great Britain and Europe e.g. Intralipid, Lipiphysan, Lipofundin and Lipofundin S. They contain either cottonseed oil or soybean oil. In Intralipid, purified egg-yolk phospholipids are used as the emulsifiers. Isotonicity with blood is obtained by the addition of sorbitol, xylitol or glycerol. Intralipid has been used as the basis of an intravenous drug carrier, for example for diazepam (Diazemuls) and propofol (ICI), as an alternative to solubilisation in non-ionic micellar systems.

To avoid adverse effects on injection it is important that the particle size of the emulsions is small and remains so on storage. After storage of Intralipid for two years at $4\,^\circ\mathrm{C}$, more than 99 per cent of the particles visible by light microscopy had a diameter of less than 1 μm; that is, there was practically no change in mean diameter. Fat which finds its natural way into plasma occurs in three forms – either as lipoprotein complexes, as free fatty acids bound to albumin, or as an emulsion of particles in the size range 0.4–3.0 μm. These natural emulsion globules are called chylomicrons. Studies have shown pronounced physical similarities between natural chylomicrons and the fat particles of the Intralipid emulsion.

The addition of electrolyte or drugs to intravenous fat emulsions is contra-indicated because of the risk of destabilising the emulsion. Addition of cationic local anaesthetics reduces the electrophoretic mobility of the dispersed fat globules, and this is undoubtedly one contribution to instability[14].

One study[14] found both minimum stability and minimum zeta potential on addition to Intralipid of 3×10^{-3} mol dm^{-3} CaCl$_2$ and 2.5×10^{-1} mol dm^{-3} NaCl, which are thus recommended to be the maximum levels of these additives (see figure 7.17).

7.3.10 The rheological characteristics of emulsions

Most emulsions, unless very dilute, do not exhibit simple Newtonian flow, and display both plastic and pseudoplastic flow behaviour. The flow properties of fluid emulsions should have little influence on their biological behaviour, although the rheological characteristics of semisolid emulsions may affect their performance. The pourability and spreadability of an emulsion is directly determined by its rheological properties. The high viscosity of w/o emulsions leads to problems in injectable formulations and even in pouring these preparations; con-

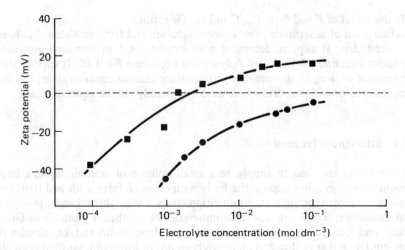

figure 7.17 Zeta potential of Intralipid 20% diluted into varying concentrations of (●) NaCl, (■) CaCl$_2$. From reference 14 with permission

version to a multiple emulsion (w/o/w), where the external oil phase is replaced by an aqueous phase, leads to a dramatic decrease in viscosity.

The influence of phase volume on the flow properties of an emulsion is shown in figure 7.18. In this diagram the relative viscosity (η_{rel}) of the system increases with increasing ϕ, and at any one phase volume increases with decreasing mean particle size, D_m. These and other factors which affect emulsion viscosity are listed in table 7.7.

Several equations for the viscosity of emulsion systems take the form

$$\eta_{rel} - 1 = \eta_{sp} = \frac{a\phi}{1 - h\phi} \tag{7.24}$$

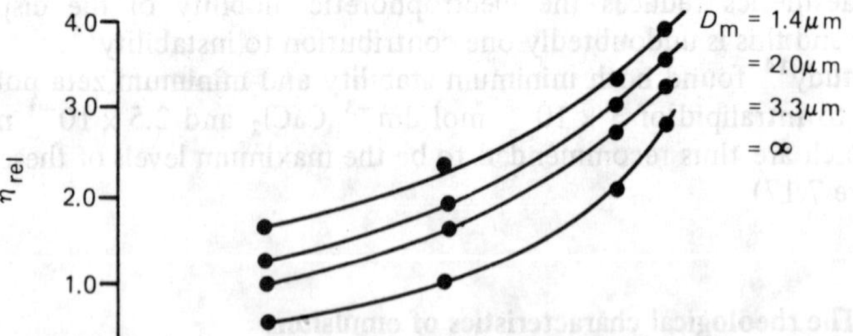

figure 7.18 The relative viscosities of w/o emulsions stabilised with sorbitan trioleate. Four emulsions have been studied with different mean diameters, D_m. From reference 15, with permission

table 7.7 **Factors that influence emulsion viscosity***

---

1. Internal (disperse) phase

    (i)   Volume fraction (ϕ)
    (ii)   Viscosity
    (iii)   Particle size and size distribution
    (iv)   Chemical nature

2. Continuous phase

    (i)   Viscosity
    (ii)   Chemical constitution and polarity

3. Emulsifier

    (i)   Chemical constitution and concentration
    (ii)   Solubility in the continuous and internal (disperse) phase
    (iii)   Physical properties of the interfacial film
    (iv)   Electroviscous effects

4. Presence of additional stabilisers, pigments, hydrocolloids, etc.

---

*From reference 15

where $a = 2.5$, and h is a measure of the fluid immobilised between the particles in concentrated emulsions and dispersions, which therefore reduces the total volume of liquid available for the particles to move around in. Immobilised liquid attached to solvated macromolecular stabilisers thus effectively increases the concentration of the particles and increases viscosity. In emulsions in which ϕ does not exceed 0.65 an equation of the form

$$\eta_{rel} = 1 + \frac{2.5}{6(1-\phi)} \tag{7.25}$$

may be used, while for emulsions in which η_{rel} becomes infinite when $\phi \to 0.74$

$$\eta_{rel} = 1 + \frac{2.5\phi}{2(1-h\phi)} \tag{7.26}$$

where h has a value of 1.28–1.35.

As most emulsions are polydisperse, the influence of particle size and, in particular, particle size distribution on viscosity is important. Figure 7.18 shows viscosity data for w/o emulsions of varying mean particle size (D_m) stabilised with sorbitan trioleate.

When an emulsion is aged its mean globule size increases. The ensuing changes in D_m and globule size distribution cause a fall in emulsion viscosity at high rates of shear. Provided no other changes have occurred in the system the viscosity at any given time should be predictable from viscosity-D_m relationships derived from fresh emulsion of the same formulation. Viscosity changes at low rates of shear are more difficult to predict because of the complication of particle aggregation which may change with time. Concentrated viscoelastic o/w emulsions stabilised by mixtures of non-ionic emulsifiers and fatty alcohols have been termed 'self-bodying' as the rheological characteristics change with time as a result of interactions among the components of the system.

An attempt has been made[16] to calculate the shear conditions for simple pharmaceutical operations such as the spreading of an ointment or cream on the skin, ointment milling, and the flow of liquid through a hypodermic needle.

7.5   Pharmaceutical suspensions

Pharmaceutical suspensions tend to be coarse rather than colloidal dispersions. The problems that arise when a solid drug is dispersed in a liquid include sedimentation, caking (leading to difficulty in resuspension), flocculation and particle growth (through dissolution and recrystallisation). Adhesion of suspension particles to container walls has also been identified as a problem. The formulation of pharmaceutical suspensions requires that caking is minimised and this can be achieved by the production of flocculated systems. A flocculate or floc is a cluster of particles held together in a loose open structure. A suspension consisting of particles in this state is termed flocculated and we discuss states of flocculation and deflocculation. Unfortunately flocculated systems clear rapidly and the preparation appears unsightly, so a partially deflocculated formulation is the ideal requirement. The viscosity of a suspension is obviously affected by flocculation.

7.5.1   Settling of suspended particles

In order to quantify the sedimentation of suspended particles, the ratio R of sedimentation layer volume (V_s) to total suspension volume (V_t)

$$R = \frac{V_s}{V_t} \approx \frac{h_\infty}{h_0} \tag{7.27}$$

may be used. A measure of sedimentation may also be obtained from the height of the sedimented layer (h_∞) in relation to the initial height of the suspension (h_0). In a completely deflocculated system the particles are not associated; particles settle under gravitational forces and the sediment layer builds up. The pressure on the individual particles can lead, in this layer, to close packing of the particles to such an extent that the secondary energy barriers are overcome and the particles become irreversibly bound together. In flocculated systems where the repulsive barriers have been reduced the particles settle as flocs and not as individual particles; the tendency is for the flocs to settle further; the supernatant clears but because of the random arrangement of the particles in the flocs the sediment is not compact. Caking therefore does not readily occur. In flocculated or concentrated suspensions, zone settling occurs (figure 7.19). In the region A–B of figure 7.19b there is hindered settling of the particle interface at a constant rate; at B–C a transitional settling occurs; from C to D consolidation of the sediment occurs.

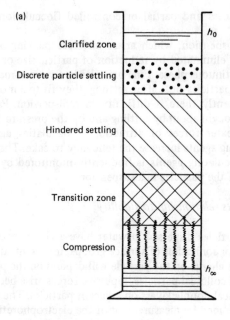

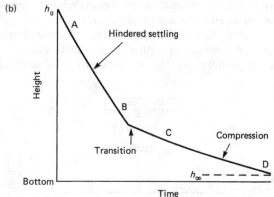

figure 7.19 Zone sedimentation in a suspension. (a) The various zones are de-
lineated, showing the clear layer at the top of the suspension and the sedimented
layer at the bottom. Immediately above this layer is a region in which the particles
are crowded and begin to be compressed to form sediment. (b) The height of the
interface between the clarified zone and the suspension as a function of time

7.5.2 Suspension stability

Suspension stability is governed by the same forces as for other disperse
systems such as emulsions. There are differences, however, as coalescence
obviously cannot occur in suspensions and the adsorption of polymers and surfactants also occurs in a different fashion. Flocculation, unlike coalescence, can

be a reversible process and partial or controlled flocculation is attempted in formulation.

Caking of the suspension, which arises on close packing of the sedimented particles, cannot be eliminated by reduction of particle size or by increasing the viscosity of the continuous phase. Fine particles in a viscous medium settle more slowly than coarse particles but, after settling, they fit to a more closely packed sediment and frequently there is difficulty in redispersion. Particles in a close packed condition brought about by settling and by the pressure of particles above thus experience greater forces of attraction. Flocculating agents can prevent caking; deflocculating agents increase the tendency to cake. The addition of flocculating and deflocculating agents is frequently monitored by measurement of the zeta potential of the particles in a suspension.

Zeta potential and its relationship to stability

Most suspension particles dispersed in water have a charge acquired by specific adsorption of ions or ionisation of surface groups, if present. If the charge arises from ionisation, the charge on the particle will depend on the pH of the environment. As with other colloidal particles, repulsive forces arise because of the interaction of the electrical double layers on adjacent particles. The magnitude of the charge can be determined by measurement of the electrophoretic mobility of the particles in an applied electrical field. A microelectrophoresis apparatus in which this mobility may be measured is shown schematically in figure 7.20.

The velocity of migration of the particles (μ_E) under unit applied potential can be determined microscopically with a timing device and eye-piece graticule. For non-conducting particles, the Henry equation is used to obtain ζ from μ_E.

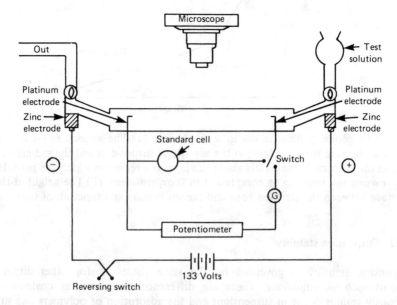

figure 7.20 Schematic drawing of a microelectrophoresis apparatus

This equation can be written in the form

$$\mu_E = \frac{\zeta\epsilon}{4\pi\eta} \, f(\kappa a) \tag{7.28}$$

where $f(\kappa a)$ varies between 1, for small κa, and 1.5, for large κa. ϵ is the dielectric constant of the continuous phase and η is its viscosity. In systems with low values of κa the equation can be written in the form

$$\mu = \frac{\zeta\epsilon}{4\pi\eta}$$

The zeta potential (ζ) is not the surface potential (ψ_0) discussed earlier but is related to it. Factors which alter ζ will affect ψ_0 and therefore ζ can be used as a reliable guide to the magnitude of electric repulsive forces between particles. Changes in ζ on the addition of flocculating agents, surfactants and other additives can then be used to predict the stability of the system. Martin[17] has described the changes in a barium subnitrate suspension system on addition of dibasic potassium phosphate as flocculating agent (figure 7.21). Bismuth subnitrate has a positive zeta potential; addition of phosphate reduces the charge and the zeta potential falls to a point where maximum flocculation is observed. In this zone there is no caking. Further addition of phosphate leads to a negative zeta potential and a propensity towards caking. Flocculation can therefore be controlled by the use of ionic species with a charge opposite to the charge of the particles dispersed in the medium.

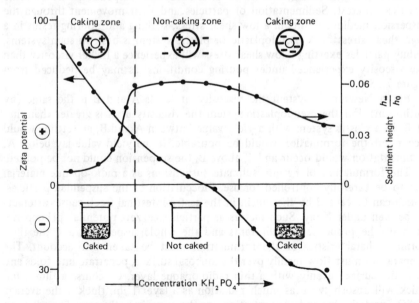

figure 7.21 Caking diagram showing controlled flocculation of a bismuth subnitrate suspension employing dibasic potassium phosphate as the flocculating agent. From reference 17

The rapid clearance of the supernatant in a flocculated system is undesirable in a pharmaceutical suspension. The use of thickeners such as tragacanth, sodium carboxymethylcellulose or bentonite hinders the movement of the particles by production of a viscous medium, so that sedimentation is slowed down. The incompatibility of these anionic agents with cationic flocculating agents has to be considered. Martin has suggested a technique to overcome the problem − the conversion of the particle surfaces into positive surfaces so that they require anions and not cations to flocculate them. Negatively charged or neutral particles can be converted into positively charged particles by addition of a surface-active amine. Such a suspension can then be treated with phosphate ions to induce flocculation.

It is perhaps not surprising that with some complex systems the interpretation of behaviour differs and is open to debate. Consider the system shown in figure 7.22. One starts with a clumped suspension of sulphamerazine, a flocculated system which produces non-caking sediments. Addition of sodium dioctylsulphosuccinate confers a greater negative charge on the suspension particles and deflocculation results. Addition of aluminium chloride as a flocculating agent reduces the negative charge in a controlled way to produce the loose clusters illustrated in the diagram. These are the observable results of the procedures. The different interpretations (of Haines and Martin, on the one hand, and of Wilson and Ecanow, on the other) are diagrammatically realised. The difference lies in the manner in which the aluminium ions adsorb onto the sulphamerazine particles; in Haines' view they adsorb directly; in the other view they interact with the surfactant ions on the surface.

The rheological characteristics of pharmaceutical suspensions are of some practical interest. Sedimentation of particles and their movement through the dispersion medium results in a low shear stress; shaking and pouring result in a high shear stress[18]. A pseudoplastic or plastic system is best. In such systems, falling particles exerting a low shear stress will experience a higher viscosity than the viscosity experienced under pouring conditions, as may be deduced from figure 7.23.

For a Newtonian system the viscosity at levels A and B is the same (by definition). For the pseudoplastic system the viscosity at A is greater than that at B. In a plastic system with a yield value between A and B, no settling would occur but the formulation would be pourable. If the yield value lay below A, sedimentation would occur and, if above B, the suspension could not be poured.

The formulation of barium sulphate suspensions as a radio-opaque material has to be carefully controlled for use. Flocculation of the suspended particles, which can be caused by the mucin in the gastro-intestinal tract, causes artefacts to be seen under X-ray. Such factors as particle size, zeta potential, pH-dependence of the properties of adjuvants and the whole suspension, and the film-forming characteristics of the formulation must be taken into account. The preparation must flow readily over the mucosal surface, penetrate into folds and coat the surface evenly with a thin radio-opaque layer. Of course a film 2 μm thick will absorb twice as much radiation as a layer 1 μm thick if the average $BaSO_4$ particle size and concentration are the same in both cases. The adhesion of the wet film of barium sulphate suspensions to surfaces and their thickness has been assessed *in vitro* by a simple method in which a clean microscope slide is dipped into the suspension and allowed to drain for 30 s. The gross appearance

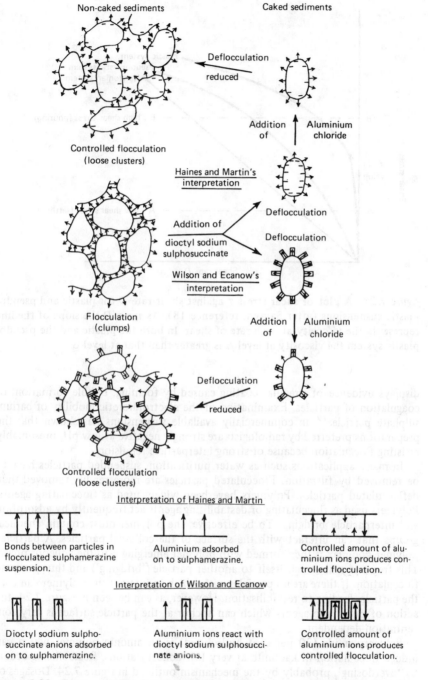

figure 7.22 Diagrammatic drawing of flocculation and controlled flocculation in a sulphamerazine suspension. The effect of the addition of dioctyl sodium sulphosuccinate and aluminium chloride is shown and two interpretations of the results are outlined. From R. Woodford, *Pharmacy Digest*, 29, 17 (1966)

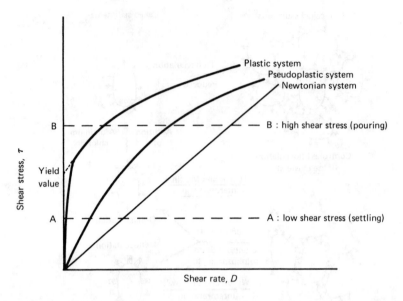

figure 7.23   A plot of shear stress τ against shear rate D for plastic and pseudo-plastic suspensions (after Samyn, reference 18). As $\eta = \tau/D$ the slope of the line represents the viscosity at each rate of shear. In both the plastic and the pseudo-plastic system the viscosity at level A is greater than that at level B

displays evidence of irregular coating caused by foaming, bubble formation, or coagulation of particles. Examination of the electrophoretic mobility of barium sulphate particles[19] in commercially available suspensions has shown that the preparations preferred by radiologists are strongly negative at low pH, presumably resisting flocculation because of strong interparticle repulsion.

In many applications such as water purification, suspended particles have to be removed by filtration. Flocculated particles are more readily removed than deflocculated particles. Polymers have been widely used as flocculating agents. Polymers used as flocculating or destabilising agents act frequently by adsorption and interparticle bridging. To be effective, the polymer must contain chemical groups that can interact with the surface of the colloidal particles. A particle-polymer complex is then formed with polymer emerging into the aqueous phase. This free end will attach itself to another particle ('bridging') and thus promote flocculation. If there are no particles with which to interact the polymer can coat the particle, leading to restabilisation. However, as can be seen in figure 7.24, the action of polymeric agents which can anchor at the particle surface is very con-centration-dependent.

Polyacrylamide (30 per cent hydrolysed) is an anionic polymer which can induce flocculation in kaolinite at very low concentrations. Restabilising occurs by 'overdosing', probably by the mechanism outlined in figure 7.24. Dosages of polymer which are sufficiently large to saturate the colloidal surfaces produce a stable colloidal system, since no sites are available for the formation of inter-

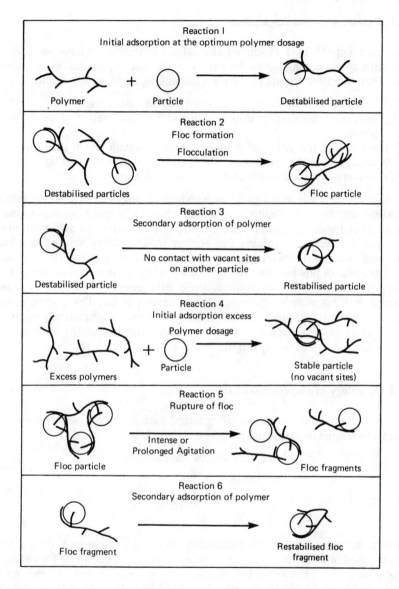

figure 7.24 Schematic representation of the bridging model for the destabilisation of colloids by polymers. The concentration dependence of the process is illustrated. From W. J. Weber, *Physicochemical Processes for Water Quality Control*, Wiley, New York, 1972

particle bridges. Under certain conditions physical agitation of the system can lead to breaking of polymer–suspension bonds and to a change in the state of the system.

7.5.3 Extemporaneous suspensions

Farley and Lund[20] have examined alternatives to traditional materials such as tragacanth as suspending agents for the extemporaneous preparation of suspensions. They listed the criteria of the ideal suspending agent, which should:

(A) be readily and uniformly incorporated in the formulation
(B) be readily dissolved or dispersed in water without resort to special techniques
(C) ensure the formation of a loosely packed system which does not cake
(D) not influence the dissolution rate or absorption rate of the drug
(E) be inert, non-toxic and free from incompatibilities.

Among the alternatives investigated were sodium carboxymethylcellulose, microcrystalline cellulose, aluminium magnesium silicate (Veegum), sodium alginate (Manucol) and sodium starch (Primojel). Preparations examined over a period of weeks for sedimentation volume and ease of redispersibility showed that pregelatinised starches, Primojel and Veegum were promising alternatives to compound tragacanth powder. Figure 7.25 shows some of the results obtained with five of the macromolecules studied, these data being compared with those obtained with tragacanth or compound tragacanth powder. The marked differences in sedimentation volume of the paediatric succinylsulphathiazole mixtures do not seem to be reflected in the redispersion numbers of the systems. No attempt has been made to explain the results on a molecular basis.

7.5.4 Rheology of suspensions of solid particles

Einstein considered a suspension of spherical particles which were far enough apart to be treated independently where ϕ is defined by

$$\phi = \frac{\text{volume occupied by the particles}}{\text{total volume of suspension}} \tag{7.29}$$

The suspension could be assigned an effective viscosity, η_*, is given by

$$\eta_* = \eta_0 \left(1 + 2.5\phi\right) \tag{7.30}$$

where η_0 is the viscosity of the suspending fluid. As we have seen, the assumptions involved in the derivation of the Einstein equation do not hold for colloidal systems subject to Brownian forces, electrical interactions and van der Waals forces. Brownian forces result from 'the random jostling of particles by the molecules of the suspending fluid due to thermal agitation and fluctuation on a very short time scale'.

A charged particle in suspension with its inner immobile Stern layer and outer diffuse Gouy (or Debye–Hückel) layer presents a different problem than a smooth and small non-polar sphere. Such particles when moving experience electroviscous effects which have two sources: (i) the resistance of the ion cloud to deformation; (ii) the repulsion between particles in close contact. When particles interact, for example, to form pairs in the system the new particle will have a different shape from the original and will have different flow properties. The co-

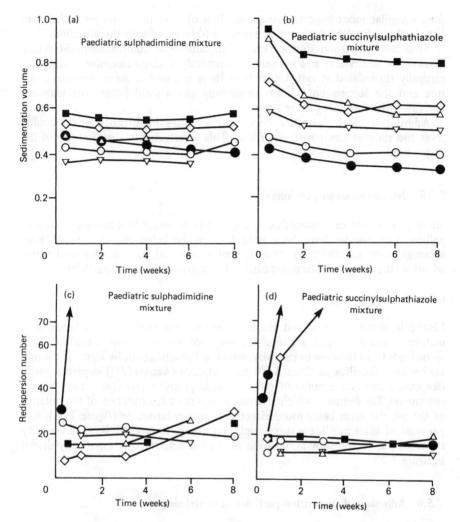

figure 7.25 (a) (b): Sedimentation volumes of BPC mixtures made with alternative suspending agents. (c) (d): Effect of storage on redispersibility of BPC mixtures made with alternative suspending agents. From Farley and Lund[20]
■, BPC mixture made with tragacanth or compound tragacanth powder;
○, Primojel (1 per cent); ◇ Veegum (2 per cent); △, Clearjel/sorbitol (1 per cent
of each; ▽, Snowflakes 12016 (1 per cent); ●, Alginate YZ (1 per cent)

efficient 2.5 in Einstein's equation 7.30 applies only to spheres; asymmetric
particles will produce coefficients greater than 2.5.

Other problems in deriving *a priori* equations result from the polydisperse
nature of pharmaceutical suspensions. The particle size distribution will determine η. Experimentally a polydisperse suspension of spheres has a lower viscosity

than a similar monodisperse suspension. It is obvious that a simple undifferentiated volume fraction ϕ cannot be expected to be of value in this situation.

Structure formation during flow is an additional complication. Structure breakdown occurs also and is evident particularly in clay suspensions, which are generally flocculated at rest. Under flow there is a loss of the continuous structure and the suspension exhibits thixotropy and a yield point. The viscosity decreases with increasing shear stress.

Addition of electrolytes to a suspension decreases the thickness of the double layer and reduces electroviscous effects. This is reflected in the viscosity of the suspension.

7.5.5   Non-aqueous suspensions

Many pharmaceutical aerosols consist of solids dispersed in a non-aqueous propellant. Few studies have been published on the behaviour of such systems, although their sensitivity to water is well established. Low amounts of water adsorb at the particle surface and can lead to aggregation of the particles.

Oleogels

Lipophilic ointment bases and non-aqueous suspensions may be thickened with materials such as Aerosil, a coagulated silica sol. Incorporation of the silica into an oil leads to an increase in viscosity, which is brought about by hydrogen bonding between the silica particles. 8–10 per cent silica (Aerosil 200) imparts a paste-like consistency on a range of oils such as isopropyl myristate, peanut oil and silicone oil. The degree to which viscosity is increased is a function of the polarity of the oil, the silica being more effective in non-polar media (figure 7.26). Suspensions of silica in oils are thixotropic: on storage for several days the viscosity increases due to the slow aggregation of the silica particles schematically shown in figure 7.27.

7.5.6   Adhesion of suspension particles to containers

When the wall of a container is wetted repeatedly an adhering layer of suspension particles may build up, which subsequently dries to a hard and thick layer. In figure 7.28 three types of wetting are shown. Where the suspension is in constant contact with the container wall, immersional wetting occurs, in which particles are pressed up to the wall and may or may not adhere. Above the liquid line spreading of the suspension during shaking or pouring may also lead to adhesion of the particles contained in the spreading liquid. Adhesional wetting occurs when a liquid drop remains suspended, like a drop of water on a clothes line. Obviously the surface tension of the suspension plays a part in the spreading and wetting processes. Adhesion increases with increase in suspension concentration, and with the number of contacts the suspension makes with the surfaces in question.

Additives, especially surfactants, will modify the adhesion of suspension particles. They will act in two ways: (i) by decreasing the surface tension; and

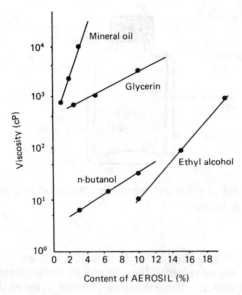

figure 7.26 Influence of the 'polarity' of the medium on the increase in viscosity attainable with Aerosil 200, a coagulated silica sol. From *Degussa's Technical Bulletin* No. 4

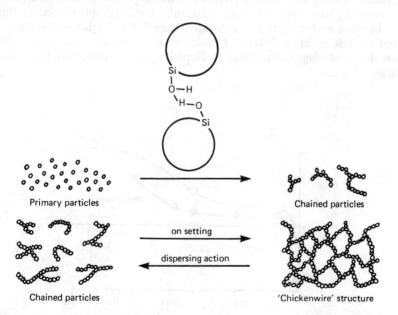

figure 7.27 Schematic representation of the interaction between two Aerosil particles (top), the formation of a chain structure (centre), and the development of a 'chickenwire' structure as well as the thixotropy (bottom). Mean diameter of Aerosil is 10 nm. From *Degussa's Technical Bulletin* No. 4 on Aerosil

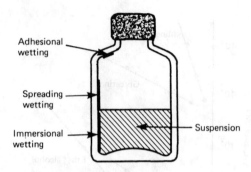

figure 7.28 Three types of wetting giving rise to adhesion of suspension particles. After Uno and Tanaka, reference 21

(ii) by adsorption modifying the forces of interaction between particle and container. The example below refers to the addition of benzethonium chloride to chloramphenicol suspensions. Benzethonium chloride converts both the glass surface and the particles into positively charged entities (see figure 7.29). Adhesion in the presence of this cationic surfactant is concentration-dependent; the process is akin to flocculation. At low concentrations the surfactant adsorbs by its cationic head to the negative glass and to the suspension particle. The glass is thus made hydrophobic. At higher concentrations hydrophobic interactions occur between coated particle and surface (figure 7.30). Further increase in concentration results in multilayer formation of surfactant, adsorbed hydrophobically rendering the surfaces hydrophilic. In this condition they repel, reducing adhesion.

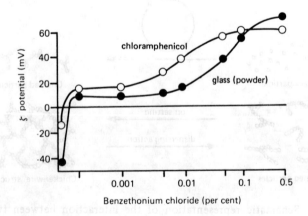

figure 7.29 Zeta potential of glass (powder) and chloramphenicol in aqueous benzethonium chloride solutions. From reference 21

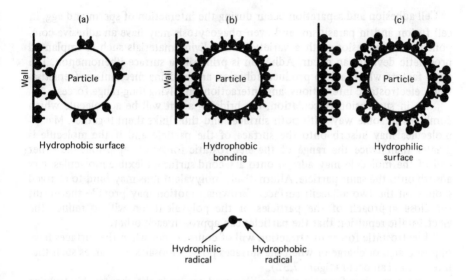

figure 7.30 Adsorption of benzethonium chloride on particle and wall surfaces, and hydrophobic bonding formation. After Uno and Tanaka, reference 21

7.6 Application of colloid theory to other systems

There are several colloidal systems, other than man-made emulsions and suspensions, which are of interest — systems such as blood and other cell suspensions, whose behaviour can now be better understood by application of DLVO theory. The adhesion of cells to surfaces, aggregation of platelets, the spontaneous sorting out of mixed cell aggregates and other such phenomena, depend to a large extent on interaction between the surfaces of the objects in question, although the surfaces are frequently more labile than those encountered in model colloids.

7.6.1 Cell–cell interactions

In considering interactions between free-floating cells it is appropriate to treat the cells as spheres, but when adhesion between cells is to be considered the interactions between planar surfaces more adequately describe the situation (figure 7.31).

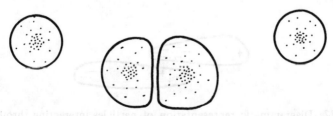

figure 7.31 Interacting cells with planar interface formed between the cells

Cell adhesion and separation occur during the interaction of sperm and egg, in cell fusion and in parasitism, and even phagocytosis may have an adhesive component. Cell adhesion with a variety of non-living materials such as implanted prosthetic devices can occur. Adhesion is primarily a surface phenomenon. The main factors which act to produce adhesion are bridging mechanisms (sensitisation), electrostatic interactions, and interactions involving long-range forces.

As in suspension flocculation, the bridging agent will be a molecule which combines in some way with both surfaces and thus links them together. Macromolecules may adsorb onto the surface of the particle and if the molecule is greater than twice the range of the electrostatic forces of repulsion, the other end of the molecule may adsorb onto a second surface. Flexible molecules may adsorb onto the same particle. Alternatively, polyvalent ions may bind to charged groups on the two adjacent surfaces. Brownian motion may provide the means for close approach of the particles or the polyvalent ion will so reduce the electrostatic repulsion that the particles can approach each other.

Electrostatic forces of attraction will of course occur when the surfaces have opposite sign or charge or when the surfaces possess mosaics of charges such that interaction can occur (figure 7.32).

The long-range forces are those discussed earlier in the chapter. In studying echinoderm egg attachment to glass, an inverse relationship was found between zeta potential and adhesiveness, when zeta potential was controlled by the addition of monovalent cations. Low pH, high ionic strength and the presence of covalent cations all favour cell–cell adhesion.

Artificial bridging adhesions between cells are known. All immunological procedures which involve the agglutination of cells by an antibody (or antigen when the cells are coated with antibody) seem to be bridging reactions. Polycations may be used to flocculate cells; erythrocytes have been flocculated with poly-lysine; polyvinyl amine hydrochloride and protamine sulphate clump cells. Lectins (phytohaemagglutinins) stimulate cell adhesion by combining with certain sugar groupings on the plasmalemma glycolipids or glycoproteins.

It is not possible to discuss all cell–cell reactions rigorously. Often adhesion results in the secretion of complex chemicals which further induce interaction which cannot be treated by any physicochemical model.

Drug treatment, if it results in the adsorption of drugs on cell surfaces, may result in changes in the behaviour of cells, but investigation of this effect *in vitro* is fraught with difficulties unless the exact metabolic products of the drugs are known. Hydrocortisone treatment has been found to increase the adhesiveness of cells to glass; presumably the drug reduces the surface charge density and increases the hydrophobic nature of the surfaces. The subject has been reviewed by Curtis[22].

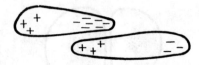

figure 7.32 Diagrammatic representation of particles interacting through overlap of regions of opposite charge

7.6.2 Sorption of microbial cells on surfaces

The contamination of pharmaceutical suspensions for oral use has recently caused concern. The behaviour of organisms in such suspensions is therefore of some interest before interpretation of experimental results can be undertaken. Under some conditions, bacteria may be strongly adsorbed and therefore more resistant to the effects of preservatives; in other cases, the bacteria may be free in suspension.

Two distinct phases of bacterial sorption onto glass have been observed[23]; the first reversible phase may be interpreted in terms of DLVO theory. Reversible sorption of a non-mobile strain (*Achromobacter*) decreased to zero as the electrolyte concentration of the media was increased, as would be expected. The second irreversible phase is probably the result of polymeric bridging between bacterial cell and the surface in contact with it. It is obviously not easy to apply colloid theory directly but the influence of factors such as ψ_0, pH and additives can be predicted and experimentally confirmed.

The flexibility of surfaces and the formation in the cells of pseudopodia with small terminal radii of curvature will obviously complicate the application of theory. In the agglutination of erythrocytes and the adsorption of erythrocytes

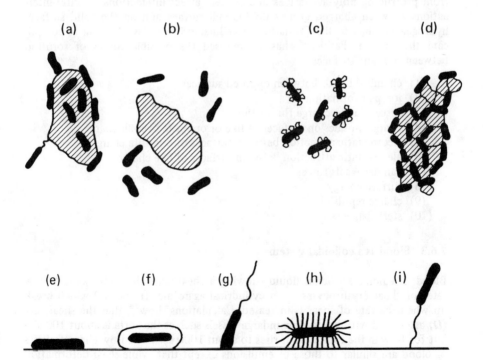

figure 7.33 Various types of bacterial sorption on solid surfaces. From D. Zvyagintsev. *Interaction between Microorganisms and Solid Surfaces*, Moscow University Press, Moscow, 1973

to virus-infected cells, projections of small radius of curvature have been observed. Such highly curved regions could well account for local penetration of the energy barrier and strong adhesion at the primary minimum. The various modes of cell sorption are shown diagrammatically in figure 7.33.

Sorption of microbial cells is selective and there is no obvious relation between gram-staining characteristics and attachment. In figure 7.33, bacterial cells are shown adsorbing onto larger solid particles (*a*) and in suspension (*b*); in (*c*) small particles are shown adsorbing onto the bacterial cell. The bacterial cells are adsorbed onto flocculated particles in (*d*), onto solid surfaces in (*e*); (*f*)–(*i*) show the more complicated behaviour of bacterial forms with coat, cilia and flagella. The adsorption affects growth partly by masking the cell surface and partly by altering the release of metabolites from the cell.

Addition of HCl and NaOH to drug suspensions alters the adhesion of *E. coli* to pyrophyllite and kaolinite (figure 7.34). The negative charge of the cell surfaces will decrease with decrease in pH; the isoelectric point of many bacteria lies between pH 2 and 3. At pH values lower than the isoelectric point the cell surface carries a positive charge. The flat surface of the clay also carries a negative charge, which also diminishes with decrease of pH; the positive charge localised on the edge of the clay platelet will be observed only in acidic solution.

In addition to the van der Waals and electrical forces, steric forces resulting from protruding polysaccharides and protein affect interactions; specific interactions between charged groups on the cell surface and on the solid surface, hydrogen bonding or the formation of cellular bridges may all occur to complicate the picture. Pethica[24] has summarised the possible forces of sorption between cells and surfaces:

(1) chemical bonds between opposed surfaces
(2) ion-pair formation
(3) forces due to charge fluctuation
(4) charge mosaics on surfaces of like or opposite over-all charge
(5) electrostatic attraction between surfaces of opposite charge
(6) electrostatic attraction between surfaces of like charge
(7) van der Waals forces
(8) surface energy
(9) charge repulsion
(10) steric barriers

7.6.3 Blood as a colloidal system

Blood is a non-Newtonian liquid showing a shear-dependent viscosity. At low rates of shear erythrocytes form cylindrical aggregates (rouleaux) which break up when the rate of shear is increased. Calculations show[25] that the shear rate (*D*) associated with blood flow in large vessels such as the aorta is about 100 s^{-1} but for flow in the capillaries it rises to about 1000 s^{-1}. The flow characteristics of blood are similar to those of emulsions except that, while shear deformation of oil globules can occur with a consequent change in surface tension, no change in membrane tension occurs on cell deformation. Figure 7.35 shows the viscosity of blood at low shear rates, measured in a Brookfield LVT micro cone–plate viscometer.

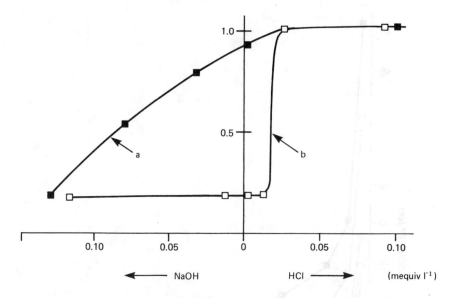

figure 7.34 Effect of HCl and NaOH on the formation or stability of *E. coli*–clay complexes. The ordinate signifies the number of cells adhered to one clay particle. Addition of sodium hydroxide decreases adhesion. The results of curve a refer to *E. coli*–pyrophyllite and of curve b to *E. coli*–kaolinite. From T. Hattori. *J. gen. appl. Microbiol.*, 16, 351 (1970)

Figure 7.35 also shows the influence of a surfactant on blood flow: low concentrations of sodium oleate decrease the viscosity and concentrations higher than 60 mg per 100 mℓ increase the viscosity. One would anticipate changes in viscosity on addition of an anionic surfactant; interpretation is complicated by the fact that a reversible morphological effect takes place. Surfactants such as sodium oleate are also able to disaggregate erythrocytes, irrespective of rate of shear, and consequently would be expected to reduce the viscosity. At higher concentrations the increase in viscosity may be due to an electroviscous effect or the altered shape of the particle.

Velocity gradients in blood vessels are reduced in cases of retarded peripheral circulation, especially in shock. Under these conditions erythrocytes may aggregate and the discovery of agents that are capable of reducing this structural viscosity are thus of clinical interest. Dextrans and polyvinylpyrrolidones diminish attraction between individual cells in blood and alter flow properties.

White cells are phagocytic and adhesive whereas red cells are normally dispersed under circumstances where white cells become sticky. The differential adhesive properties of blood cells might be attributed to differences in surface ionic groups. Although erythrocytes and leucocytes may have the same charge density at pH 7.2 their effective pK_a values might be different and they therefore have a different affinity for materials such as Ca^{2+}. On the other hand, the adhesive properties may be due to local protuberances having a small radius of

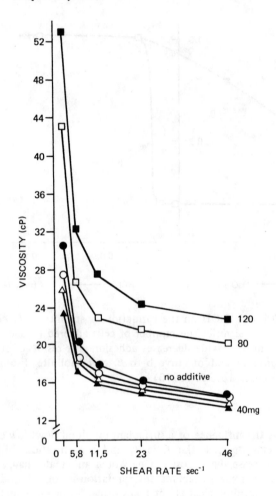

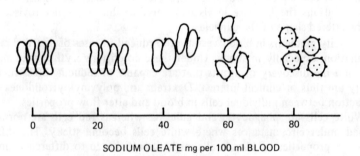

figure 7.35 The influence of varying concentrations of sodium oleate on the viscosity of blood at different rates of shear. Each value represents an average of 8 subjects. From A. M. Ehrly. *Biorheology*, 5, 209 (1968)

curvature (say of about 0.1 μm), spacing out surface charges as shown diagrammatically below

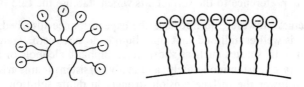

By reducing surface charge density it would allow two such tips to approach to the necessary distance. Such a close approach might then result in successful bridging by Ca^{2+}, provided the surface groups were COO^- (as in leucocytes).

Aggregation of platelets involves a contact and an adhesive phase, shown diagrammatically in figure 7.36. There is good evidence that most thrombi that form within the arterial tree after endothelial injury consist initially of a mass of associated particles on the surface of the vessel. The shearing effects of blood may dislodge platelets.

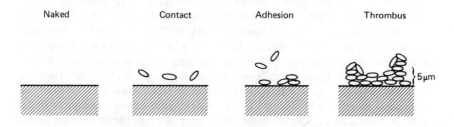

figure 7.36 Sequence in the formation of a thrombus at a surface involving three stages: contact, adhesion and thrombus growth

Interest in platelet interactions with simpler surfaces has been stimulated by the increasing use of plastic prosthetic devices which come into contact with blood. While we do not have a clear idea of the physical and chemical properties of surfaces which are responsible for the attraction of platelets, a relationship between adhesion and the critical surface tension of uncharged hydrophobic surfaces has been demonstrated. The number of platelets adsorbed increases as the critical surface tension increases. There is other evidence that platelets adhere readily to a high-energy hydrophilic surface and less readily to low-energy hydrophobic surfaces. Thus platelet adhesion to glass is reduced following coating with dimethylsiloxane.

7.7 Foams

The breaking and prevention of liquid foams is less well understood than the stabilisation of foams. It is recognised, however, that small quantities of specific agents can reduce foam stability markedly. There are two types of such agent:[26]

(1) Foam breakers, which are thought to act as small droplets forming in the foam lamellae.
(2) Foam preventatives, which are thought to adsorb at the air–water interface in preference to the surfactants which stabilise the thin films.

These molecules, however, do not have the capacity, once adsorbed, to stabilise the foam. It is well established that pure liquids do not foam. Transient foams are obtained with solutes such as short-chain aliphatic alcohols or acids which lower the surface tension moderately; and really persistent foams arise only with solutes which lower the surface tension strongly in dilute solution — the highly surface active materials such as detergents and proteins. The physical chemistry of the surface layers of the solutions is what determines the stability of the system. Foam is a disperse system with a high surface area, consequently foams tend to collapse spontaneously. Ordinarily, three-dimensional foams of surfactant solutes persist for a matter of hours in closed vessels. Gas slowly diffuses from the small bubbles to the large ones (since the pressure and hence thermodynamic activity of the gas within the bubbles is inversely proportional to bubble radius). Diffusion of gas leads to a rearrangement of the foam structures and this is often sufficient to rupture the thin lamellae in a well-drained film.

The most important action of an antifoam agent is to eliminate surface elasticity — the property that is responsible for the durability of foams. To do this it must displace foam stabiliser. It must therefore have a low interfacial tension in the pure state to allow it to spread when applied to the foam and it must be present in sufficient quantity to maintain a high surface concentration. Many foams can be made to collapse by applying drops of ether, octyl alcohol, etc. Ether has a low surface tension. Addition of ether to an aqueous foam will locally produce regions which have a low tension; these regions are rapidly pulled out by surrounding regions of higher tension. The foam breaks as the ethereal region cannot stretch. Long-chain alcohols also break foams because the surface is swamped by rapidly diffusing molecules so that changes in surface tension are rapidly reversed (that is, elasticity disappears).

Generally more effective and more versatile than any soluble antifoams are the silicone fluids which have surface tensions as low as 20 mN m^{-1}. Quantities of the order of 1–60 ppm prevent foaming in fermentation vats, sewage tanks, and dyebaths. Polyfluorinated hydrocarbons will lower surface tensions to the order of 10 mN m^{-1}.

7.7.1 Foams: clinical considerations

X-ray studies have clearly shown the presence of foam in the upper gastrointestinal tract in man.

Since the report in 1949 that bloat in cattle should be treated by the use of antifoaming agents[27], the clinical use of silicone antifoaming agents has increased. These derive their value from their ability to change the surface tension of the mucus-covered gas bubbles in the gut and thus to cause the bubbles to coalesce. A range of polydimethylsiloxanes are available commercially. Dimethicones 20, 200, 350, 500 and 1000 may be procured, the numbers referring to the viscosity of the oil in centistokes. In simple *in vitro* tests polydimethyl-

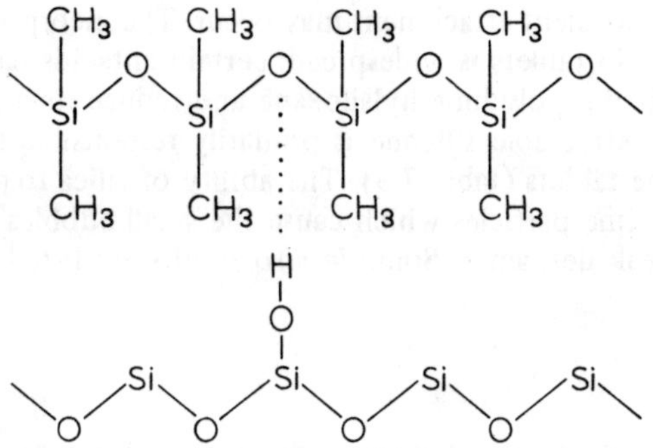

Polydimethyl siloxane

Surface of silica

siloxane, of molecular weight used in pharmaceutical formulations (Dimethicone 1000), has poor antifoaming properties. The addition of a small percentage (2-8 per cent) of hydrophobic silica increases the activity of the polydimethylsiloxane, the finely divided silica being suspended in the silicone fluid. The product is a simple physical mixture of silica and polydimethylsiloxane. At

table 7.8 *In vitro* froth test: time taken for antifoam agent to remove experimental foam*

Frothing system	Antacid preparation**	Froth reduction Time(s)	Extent (per cent)
Cetomacrogol–0.1 N HCl	Asilone tablet	17	100
	Asilone (ether-extracted)	60	80
Cetomacrogol–saturated NaHCO$_3$	Asilone tablet	20	100
	Asilone (ether-extracted)	–	10–0

*From reference 29
**Ether extraction removes dimethicone

table 7.9 *In vivo* effects of polydimethylsiloxane (PDMS)*

Treatment by mouth	Volume (ml)	Dose (mg/rat) PDMS	Silica	Mean percentage reduction† in foam height (± s.e.)
PDMS	0.25	250		19 ± 4
	0.50	500		29 ± 6
	1.00	1000		56 ± 6
	2.00	2000		84 ± 3
PDMS containing 6% w/v silica	0.005	4.7	0.2	45 ± 6
	0.01	9.4	0.6	58 ± 5
	0.02	18.9	1.1	61 ± 5
	0.04	37.7	2.0	87 ± 2

*From reference 30
†10 rats in each group. Foam induced by saponin

elevated temperatures (~300 °C) covalent attachment may occur. The incorporation of these materials into antacid tablets is widespread; certain antacids have
been found, however, to adsorb the polydimethylsiloxane and reduce its antifoaming potential[28]. Unbound extractable silicone is primarily responsible for
the antifoaming properties of the tablets (table 7.8). The ability of silica to de-
foam has been attributed to the fine particles which cause the small bubbles to
coalesce. Alone the silica is a weak defoamer. Some *in vivo* results are listed in
table 7.9.

References

1. D. J. Shaw. *Introduction to Colloid and Surface Chemistry*, 3rd Edition, Butterworths, London, 1980
2. R. H. Ottewill. In *Colloid Science*, vol. 1, The Chemical Society, London, 1973, p. 173 *et seq.*
3. R. H. Ottewill. In *Non-ionic Surfactants* (ed. M. J. Schick), Marcel Dekker, New York, 1967
4. J. T. Davies and E. K. Rideal. *Interfacial Phenomena*, 2nd Edition, Academic Press, London, 1963
5. A. T. Florence and D. Whitehill. *J. colloid interface Sci.*, 79, 243 (1981)
6. A. T. Florence and D. Whitehill. In *Macro- and Micro-Emulsions* (ed. D. O. Shah), ACS Symp. Ser. No. 272, American Chemical Society, Washington, DC, 1983, pp. 359–380
7. G. M. Eccleston. *Pharm. Int.*, 7, 63 (1986)
8. B. W. Barry. *Rheol. Acta*, 10, 96 (1971)
9. G. K. Konning. *Can. J. pharm. Sci.*, 9, 103 (1974)
10. G. K. Konning. *ibid.*, 9, 107 (1974)
11. H. S. Bean, G. K. Konning and S. M. Malcolm. *J. Pharm. Pharmacol.*, 21, Suppl., 173 (1969)
12. A. Mitchell and S. J. A. Kazmi. *Can. J. pharm. Sci.*, 10, 67 (1975)
13. H. S. Bean and S. M. Heman-Ackah. *J. Pharm. Pharmacol.*, 16, 58T (1964)
14. T. L. Whateley, G. Steele, J. Urwin and G. A. Smail. *J. clin. hosp. Pharm.*, 9, 113 (1984)
15. P. Sherman. *J. Pharm. Pharmacol.*, 16, 1 (1964)
16. N. L. Henderson, P. M. Mee and H. B. Kostenbauder. *J. pharm. Sci.*, 50, 788 (1961)
17. A. N. Martin. *J. pharm. Sci.*, 50, 513 (1961)
18. J. C. Samyn. *J. pharm. Sci.*, 50, 517 (1961)
19. W. Anderson, J. E. Harthill, B. James and D. Montgomery. *J. Pharm. Pharmacol.*, 30, 76P (1972)
20. C. A. Farley and W. Lund. *Pharm. J.*, 216, 562 (1976)
21. H. Uno and S. Tanaka. *Kolloid Z. Z. Polym.*, 250, 238 (1972)
22. A. S. G. Curtis. *Prog. Biophys. mol. Biol.*, 27, 317 (1973)
23. K. C. Marshall *et al. J. gen. Microbiol.*, 68, 337 (1971)
24. B. Pethica. *Exp. cell Res.*, 8, 123 (1961)
25. P. Sherman. *Chem. Br.*, 321 (1975)
26. K. Roberts, G. Axberg and R. Osterlund. In *Foams* (ed. R. J. Ackers), Academic Press, London, 1976

27. A. H. Quinn. *J. Am. vet. med. Ass.*, 114, 313 (1949)
28. M. J. Rezak. *J. pharm. Sci.*, 55, 538 (1966)
29. J. E. Carless, J. B. Stenlake and W. D. Williams. *J. Pharm. Pharmacol.*, 25, 849 (1973)
30. R. D. N. Birtley, J. S. Burton, *et al. ibid.*, 25, 859 (1973)

Further reading

On surfactants and surfactant systems

D. Attwood and A. T. Florence. *Surfactant Systems: Their Chemistry, Pharmacy and Biology*, Chapman and Hall, London, 1983

On adhesion in biological systems

M. A. Longer and J. R. Robinson. *Pharm. Int.*, 7, 114 (1986)

On foams in pharmacy

N. Pilpel. *Endeavour* (NS), 9, 87 (1985)

8 Polymers and Macromolecules

Polymers are used widely in pharmaceutical systems as adjuvants, suspending and emulsifying agents, flocculating agents, adhesives, packaging and coating materials, and increasingly as the basis of drug delivery systems.

This selective account of the pharmaceutics of polymeric molecules will deal with recent developments in the applications of polymers and with some advances in our understanding of their uses in a variety of pharmaceutical fields.

The synthesis of custom polymers offers exciting possibilities, especially as it holds out the hope of obtaining new polymers for drug delivery devices, so essential for the efficient use of today's potent and toxic drugs.

Polymers are substances of high molecular weight made up of repeating monomer units. They owe their unique properties to their size and often to their asymmetry. The chemical reactivity of polymers depends on the chemistry of their monomer units, but their properties depend to a large extent on the way the monomers are put together; it is this fact that leads to the versatility of synthetic polymers. The polymer molecules may be linear or branched, and separate linear or branched chains may be joined by cross-links. Extensive cross-linking leads to a three-dimensional and often insoluble polymer network. More than one monomer type is involved in the synthesis of *copolymers*; various arrangements of the monomers A and B (figure 8.1) can be produced with consequent effects on the physical properties of the resulting polymer.

Polymers that have fairly symmetrical chains and strong interchain forces can be drawn into fibres. *Plastics* are polymers with lower degrees of crystallinity which can be moulded. Further down the rigidity scale are *rubbers* and *elastomers*, whose properties are well known.

It is apparent that polymer molecules will have a much wider range of physical properties than small chemical entities. Even when considering one chemical type (for example, polyethylene) the properties of the product may be altered by increasing or decreasing the molecular weight (figure 8.2). There is a possible

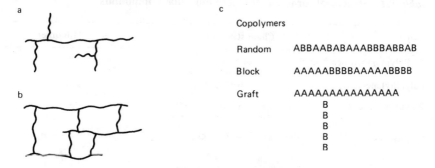

figure 8.1 Diagrammatic representation of some alternative shapes and construction of polymer molecules. Branched and cross-linked polymer chains are shown in (a) and (b) and various copolymers of monomers A and B are shown in (c)

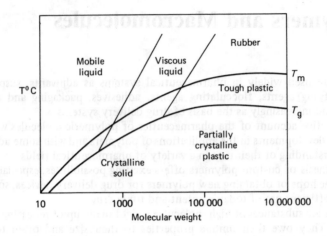

figure 8.2 Approximate relations between molecular weight, T_g (glass transition temperature), T_m (melting point), and polymer properties. From reference 1

degree of control over properties which there is not with simple organic materials and, generally, it is because of this that synthetic polymers have an advantage over the natural polymers. Natural materials can be modified chemically and this approach can lead to useful new products, as with the celluloses.

We can conveniently divide the polymers used in pharmacy into those that are water-soluble and those that are water-insoluble. This division, while not rigid, is useful in that it separates the two main areas of use which form the basis of this account.

The structural formulae of some common macromolecules are given in table 8.1. The table is divided into two parts showing compounds with a $-C-C-$ backbone and those with a heterochain backbone.

table 8.1 **Structural formulae of macromolecular compounds**

Name	Chain structure	Monomer
1. Polymers with carbon chain backbone		
Polyethylene	~ $CH_2-CH_2-CH_2-CH_2-CH_2$ ~	$CH_2{=}CH_2$
Polypropylene	~ $CH_2-\underset{CH_3}{CH} - CH_2-\underset{CH_3}{CH} - CH_2-\underset{CH_3}{CH}-CH_2$ ~	$CH_2{=}\underset{CH_3}{CH}$
Polystyrene	~ $CH_2-CH - CH_2-CH - CH_2-CH-CH_2$ ~	$CH{=}CH_2$
Poly(vinyl chloride)	~ $CH_2-\underset{Cl}{CH}-CH_2-\underset{Cl}{CH}-CH_2-\underset{Cl}{CH}-CH_2$ ~	$CH_2{=}\underset{Cl}{CH}$

Name	Chain structure	Monomer
Polytetrafluoroethylene	$-\overset{\overset{F}{\mid}}{\underset{\underset{F}{\mid}}{C}}-\overset{\overset{F}{\mid}}{\underset{\underset{F}{\mid}}{C}}-\overset{\overset{F}{\mid}}{\underset{\underset{F}{\mid}}{C}}-\overset{\overset{F}{\mid}}{\underset{\underset{F}{\mid}}{C}}-\overset{\overset{F}{\mid}}{\underset{\underset{F}{\mid}}{C}}-\overset{\overset{F}{\mid}}{\underset{\underset{F}{\mid}}{C}}-$	$\overset{F}{\underset{F}{}}C=C\overset{F}{\underset{F}{}}$
Polyacrylonitrile	$\sim CH_2-\underset{\underset{CN}{\mid}}{CH}-CH_2-\underset{\underset{CN}{\mid}}{CH}-CH_2-\underset{\underset{CN}{\mid}}{CH}-CH_2\sim$	$CH_2=\underset{\underset{CN}{\mid}}{CH}$
Poly(vinyl alcohol)	$\sim CH_2-\underset{\underset{HO}{\mid}}{CH}-CH_2-\underset{\underset{HO}{\mid}}{CH}-CH_2-\underset{\underset{HO}{\mid}}{CH}-CH_2\sim$ $\big\uparrow[OH^-]$	$CH_2=\underset{\underset{OH}{\mid}}{CH}$
Poly(vinyl acetate)	$\sim CH_2-\underset{\underset{\underset{\underset{CH_3}{\mid}}{C=O}}{\underset{\mid}{O}}}{CH}-CH_2-\underset{\underset{\underset{\underset{CH_3}{\mid}}{C=O}}{\underset{\mid}{O}}}{CH}-CH_2-\underset{\underset{\underset{\underset{CH_3}{\mid}}{C=O}}{\underset{\mid}{O}}}{CH}-CH_2\sim$	$CH_2=\underset{\underset{\underset{\underset{CH_3}{\mid}}{C=O}}{\underset{\mid}{O}}}{CH}$
Polyacrylamide	$\sim CH_2-\underset{\underset{CONH_2}{\mid}}{CH}-CH_2-\underset{\underset{CONH_2}{\mid}}{CH}-CH_2-\underset{\underset{CONH_2}{\mid}}{CH}-CH_2\sim$	$\underset{\underset{O=C-NH_2}{\mid}}{CH_2=CH}$
Poly(methyl methacrylate)	$\sim CH_2-\overset{\overset{CH_3}{\mid}}{\underset{\underset{COOCH_3}{\mid}}{C}}-CH_2-\overset{\overset{CH_3}{\mid}}{\underset{\underset{COOCH_3}{\mid}}{C}}-CH_2-\overset{\overset{CH_3}{\mid}}{\underset{\underset{COOCH_3}{\mid}}{C}}-CH_2$	$CH_2=\overset{\overset{CH_3}{\mid}}{\underset{\underset{COOCH_3}{\mid}}{C}}$
Polyvinylpyrrolidone	$\sim CH_2-\underset{\mid}{CH}-CH_2-\underset{\mid}{CH}-CH_2-\underset{\mid}{CH}-CH_2\sim$	$CH_2=\underset{\mid}{CH}$

2. Polymers with heterochain backbone

Name	Chain structure	Monomer
Polyethylene oxide	$\sim O-CH_2-CH_2-O-CH_2-CH_2-O-CH_2-CH_2-O\sim$	$CH_2\overset{O}{\diagup\diagdown}CH_2$
Polypropylene oxide	$\sim O-CH_2-\underset{\underset{CH_3}{\mid}}{CH}-O-CH_2-\underset{\underset{CH_3}{\mid}}{CH}-O-CH_2-\underset{\underset{CH_3}{\mid}}{CH}-O$	$CH_2-\underset{O}{\diagup\diagdown}CH-CH_3$
Cellulose (Poly-glucoside, β- 1,4)		Glucose
Amylose (Poly-glucoside, α- 1,4) (component of starch)		Glucose
Pectinic acid (Polygalacturonoside, α- 1,4) (jelly-forming component of fruits)		Galacturonic acid
Polyethylene glycol terephthalate		
Polydimethylsiloxane	$\sim O-\overset{\overset{CH_3}{\mid}}{\underset{\underset{CH_3}{\mid}}{Si}}-O-\overset{\overset{CH_3}{\mid}}{\underset{\underset{CH_3}{\mid}}{Si}}-O-\overset{\overset{CH_3}{\mid}}{\underset{\underset{CH_3}{\mid}}{Si}}$	

Polydispersity

Nearly all synthetic polymers and naturally occurring macromolecules possess a range of molecular weights. The exceptions to this are proteins and polypeptides. The molecular weight determined is thus an average molecular weight, the value of which depends on the method of measurement. Average molecular weights may be determined by chemical analysis, osmotic pressure or light-scattering measurements. When determined by chemical analysis or osmotic pressure, a number average molecular weight, M_n, is found, which in a mixture containing $n_1, n_2, n_3 \ldots$ moles of polymer with molecular weights $M_1, M_2, M_3 \ldots$, respectively, is defined by

$$M_n = \frac{n_1 M_1 + n_2 M_2 + n_3 M_3 +}{n_1 + n_2 + n_3 +} = \frac{\Sigma n_i M_i}{\Sigma n_i} \qquad (8.1)$$

The individual weights M_1, M_2, etc. cannot be determined separately — the equation merely explains the meaning of the value M_n. In light scattering techniques larger molecules produce greater scattering, thus the weight rather than the number of the molecules is important, giving a weight average molecular weight, M_w

$$M_w = \frac{m_1 M_1 + m_2 M_2 + m_3 M_3 +}{m_1 + m_2 + m_3 +} = \frac{\Sigma n_i M_i^2}{\Sigma n_i M_i} \qquad (8.2)$$

Here $m_1, m_2, m_3 \ldots$ are the masses of each species. m_i is obtained by multiplying the molecular weight of each species by the number of molecules of that weight; that is $m_i = n_i M_i$. Thus the molecular weight appears as the square in the numerator of equation 8.2 so that the weight average molecular weight is biased towards larger molecules. Another consequence is that $M_w > M_n$; the average molecular weight of a polymer measured by light-scattering must be greater than that obtained by osmotic pressure measurements if the polymer is polydisperse (that is, contains a range of molecular weights). The ratio M_w/M_n expresses the degree of polydispersity. Table 8.2 shows actual values for the number and weight average molecular weights for dextrans — microbial polysaccharides used as plasma expanders.

table 8.2 **Number and weight average molecular weights for dextran fractions***

Fraction	M_n	M_w	M_w/M_n
A	41 000	47 000	1.14
B	38 000	50 000	1.31
C	64 000	76 000	1.18
D	95 000	170 000	1.79
E	240 000	540 000	2.25

*From M Wales *et al. J. Polymer Sci.*, 10, 229 (1953)

The solubility of polymeric substances in water is determined by the same considerations that apply to smaller molecules. Those polymers which are sufficiently polar will be able to interact with the water to provide sufficient energy to remove individual polymer chains from the solid state.

Water-soluble polymers have an ability to increase the viscosity of solvents at low concentrations, to swell or change shape in solution, and to adsorb at surfaces. These are significant features of their behaviour, which we will deal with briefly. Insoluble polymers or polymers with a low rate of solution are used more to form thin films as film-coating materials, membranes for dialysis or filtration, surgical dressings, or to form matrices for enveloping drugs to control their release properties; or simply as packaging materials.

8.1 Water-soluble polymers

The rate of solution of the water-soluble polymer depends on molecular weight: the larger the molecule, the stronger are the forces holding the chains together. More energy has to be expended to force the chains apart in the liquid.

The velocity of penetration (S) of a solvent into the bulk polymer obeys the relationship

$$S = kM^{-A} \tag{8.3}$$

where M is the polymer molecular weight, k and A being constants. The dissolution process, however, is more complicated than with ordinary crystalline materials. It is frequently observed that swollen layers and gel layers form next to the polymer (see figure 8.3). If a drug is embedded in the polymer, the drug has to diffuse through these gel layers and finally through the diffusion layer.

It is the combination of slow solution rate and the formation of viscous surface layers that makes hydrophilic polymers useful in controlling the release rate

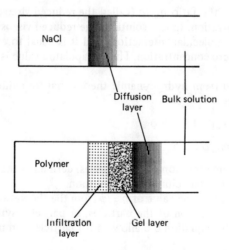

figure 8.3 Penetration of solvent into polymer and soluble crystalline material compared

of soluble drugs, which are perhaps irritant to the stomach or which cause nausea on rapid release (see section 8.5). Choice of appropriate polymer molecular weight controls the rate of solution. Molecular weight also controls the viscosity of the resulting solution. A balance between rate of polymer solution and viscosity of the solution layer must be achieved: if the polymer solution rate is too slow then soluble drug is leached out with little retardation.

The bulk viscosity of polymer solutions is an important parameter when polymers are being used as suspending agents to maintain solid particles in suspension by prevention of settling (see chapter 7) and when used to modify the properties of liquid medicines for oral and topical use.

8.2 General properties of polymer solutions

8.2.1 Viscosity

The presence in solution of large macromolecular solutes may have an appreciable effect on the viscosity of the solution. From a study of the concentration dependence of the viscosity it is possible to gain information on the shape or hydration of these polymers in solution and also their average molecular weight. The assumption is made in this section that the solution exhibits Newtonian flow characteristics.

The viscosity of solutions of macromolecules is conveniently expressed by the ratio of the viscosity of the solution η to the viscosity of the pure solvent η_0 at the same temperature. This ratio is referred to as the relative viscosity η_{rel}.

$$\eta_{rel} = \eta/\eta_0 \qquad (8.4)$$

A second useful expression is the specific viscosity η_{sp} of the solution defined by

$$\eta_{sp} = \eta_{rel} - 1 \qquad (8.5)$$

For ideal solutions, the ratio η_{sp}/c (called the reduced viscosity) is independent of solution concentration. In real solutions the reduced viscosity varies with concentration due to molecular interactions and it is usual to extrapolate plots of η_{sp}/c versus c to zero concentration. The extrapolated value is called the intrinsic viscosity $[\eta]$.

Einstein showed from hydrodynamic theory that for a dilute system of rigid, spherical particles

$$\eta_{rel} = 1 + 2.5 \, \phi \qquad (8.6)$$

$$\eta_{sp}/\phi = 2.5 \qquad (8.7)$$

where ϕ is the volume fraction of the particles, defined as the volume of the particles divided by the total volume of the solution.

Departure of the limiting value of η_{sp}/ϕ from the theoretical value of 2.5 may result from either hydration of the particles, or particle asymmetry, or both. A more general form of equation 8.6 allowing for particle asymmetry is

$$\eta_{rel} = 1 + v \, \phi \qquad (8.8)$$

where v is a shape factor related to the axial ratio of an ellipsoid by equations derived by Mehl *et al.* In the case of non-hydrated spheres v reduces to 2.5. The volume fraction is usually replaced by the weight concentration, c. For a macro-molecule of hydrodynamic volume v_h, and molecular weight, M

$$\phi = N_A \, c \, v_h / M \qquad (8.9)$$

and equation 8.8 becomes

$$\eta_{rel} - 1 = v \, \phi = v N_A \, v_h \, c/M \qquad (8.10)$$

Consider a hydrated macromolecule containing δ_1 grams of solvent per gram of dry macromolecular material. The specific volume v_0 (volume per gram) of the entrapped water may be considerably different from that of pure solvent, $v_1^\ominus$. If v_2 is the average specific volume of the macromolecular material then the total hydrodynamic volume of the particle, v_h, is

$$v_h = \frac{M}{N_A} (v_2 + \delta_1 \, v_1) \qquad (8.11)$$

The total volume V of a solution containing g_1 grams of solvent and g_2 grams of dry macromolecular solute is

$$V = \text{volume of solute}$$

$$+ \text{ volume of water of hydration}$$

$$+ \text{ volume of free solvent}$$

$$= g_2 \, v_2 + g_2 \, \delta_1 \, v_1 + (g_1 - g_2 \, \delta_1) v_1^\ominus \qquad (8.12)$$

Therefore $\qquad \bar{v}_2 = \left(\dfrac{\partial V}{\partial g_2}\right)_{P,T,g_1} = v_2 + \delta_1 \, v_1 - \delta_1 \, v_1^\ominus \qquad (8.13)$

Substituting for v_2 in equation 8.11

$$v_h = \frac{M}{N_A} (\bar{v}_2 + \delta_1 \, v_1^\ominus) \qquad (8.14)$$

Substituting for v_h in equation 8.10

$$\boxed{[\eta] = \lim_{c \to 0} \left(\frac{\eta_{rel} - 1}{c}\right) = v \, (\bar{v}_2 + \delta_1 \, v_1^\ominus)} \qquad (8.15)$$

If the particle can be assumed to be unhydrated, or if the degree of hydration can be estimated with certainty from other experimental techniques, equation 8.15 may be used to determine the asymmetry of the particle. Alternatively, if the macromolecule may be assumed to be symmetrical or its asymmetry is known from other techniques, then this equation may be used to estimate the extent of hydration of the macromolecule.

As the shape of molecules is to a large extent the determinant of flow properties, change in shape of the molecules due to changes in polymer–solvent interactions and the binding of small molecules with the polymer may lead to significant changes in solution viscosity. The nature of the solvent is thus of prime import-ance in this regard. In so-called 'good' solvents linear macromolecules will be expanded as the polar groups will be solvated. In a 'poor' solvent the intra-molecular attraction between the segments is greater than the segment–solvent

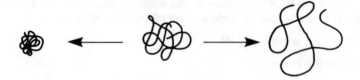

In 'poor' solvent   In 'good' solvent

figure 8.4 Change of effective volume of a polymer in 'poor' and 'good' solvents. In a good solvent, in which polymer–solvent interactions are favoured, each polymer molecule will tend to exclude all others from the volume it occupies — the excluded volume. The excluded volume becomes smaller as the solvent becomes poorer. Chains will be contracted in poor solvents as polymer–polymer contacts become more favoured

affinity and the molecule will tend to coil up (figure 8.4). The viscosity of ionised macromolecules is complicated by charge interactions which vary with concentration and additive concentration. Flexible charged macromolecules will vary in shape with the degree of ionisation. At maximum ionisation they are stretched out due to mutual charge repulsion and the viscosity increases. On addition of small counterions the effective charge is reduced and the molecules contract; the viscosity falls as a result. Some of the effects are illustrated later in this chapter in discussion of individual macromolecules, for example, with gum arabic.

The viscosity of the globular proteins which are more or less rigid is only slightly affected by change in ionic strength. The intrinsic viscosity of serum albumin varies only between 3.6 and 4.1 mℓ g^{-1} when the pH is varied between 4.3 and 10.5 and the ionic strength between zero and 0.50.

In cases where control of molecular weight is important, for example in the use of dextran fractions as plasma expanders, a viscosity method is specified, for example, in the BP monograph. Staudinger proposed that the reduced viscosity of solutions of linear high polymers is proportional to the molecular weight of the polymer or its degree of polymerisation, p.

$$\eta_{sp}/c = K_m p \tag{8.16}$$

This empirical law has been modified to

$$\lim_{c \to 0} (\eta_{sp}/c) = [\eta] = KM^a \tag{8.17}$$

where a is a constant in the range 0–2, but for most high polymers having a value between 0.6 and 0.8, $[\eta]$ is the intrinsic viscosity as defined previously, and M is the molecular weight of the polymer. For a given polymer–solvent system K and a are constant. Values of these constants may be determined from measurements on a series of fractions of known molecular weight and hence the molecular weight of an unknown fraction can be determined by measurement of the intrinsic viscosity. The viscosity average molecular weight is essentially a weight average since the larger macromolecules influence viscosity more than the smaller ones. The intrinsic viscosity of Dextran 40 BP is stated to be not less than 16 mℓ g^{-1}

and not more than 20 mℓ g^{-1} at 37 °C, while that of Dextran 110 is not less than 27 mℓ g^{-1} and not more than 32 mℓ g^{-1}.

8.2.2 Gelling tendency

Concentrated polymer solutions frequently exhibit a very high viscosity because of the interaction of polymer chains in a three-dimensional fashion in the bulk solvent. These viscous cross-linked systems are termed gels. A gel is a polymer–solvent system containing a three-dimensional network of quite stable bonds almost unaffected by thermal motion. If such a polymer network is surrounded by the solvent – the system can be arrived at by swelling of solid polymer or by reduction in the solubility of the polymer in the solution – the system is a gel regardless of whether the network is formed by chemical or physical bonds. When gels are formed from solutions, each system is characterised by a critical concentration of gelation below which a gel is not formed. This concentration is determined by the hydrophile–lipophile balance of the polymer and the degree of regularity of the structure, by polymer–solvent interaction, molecular weight and the flexibility of the chain. The more flexible the molecule the higher the critical concentration. The characteristic feature of a gel system is the considerable increase in viscosity above the gel point, the appearance of a rubber-like elasticity, and at high polymer concentration a yield point stress. Under small stress the gel should retain its shape and loss of fluidity; at higher stress considerable deformation can occur.

Gels can be divided into two groups, depending on the nature of the bonds between the chains of the network. Gels of type I are irreversible systems with a three-dimensional network formed by covalent bonds between the macromolecules. They include swollen networks which have been formed by polymerisation of a monomer in the presence of a cross-linking agent.

Gels of type II are heat-reversible, being held together by intermolecular bonds such as hydrogen bonds. Sometimes bridging by additive molecules can take place in these type II systems. Poly(vinyl alcohol) solutions gel on cooling at a temperature known as the gel point[2]. The gel point can therefore be influenced by the presence of additives which can induce gel formation by acting as bridge molecules, as, for example, with borax and poly(vinyl alcohol). The gel point of polymers can also be increased or decreased by the addition of solvents which alter the polymer's affinity for the solvent (table 8.3).

Solutions of poly(vinyl alcohols) in water are viscous mucilages which resemble those formed by methylcellulose; the viscosity of the mucilage is greatly increased by incorporating sodium perborate or silicate. Because of their gelling properties poly(vinyl alcohols) are used as jellies for application of drugs to the

table 8.3 **Gel points of 10 per cent poly(vinyl alcohol)***

Solvent	Gel point (°C)
Water	14
Glycerol	64
Ethylene glycol	102

*From reference 2

skin. On application the gel dries rapidly leaving a plastic film with the drug in intimate contact with the skin. Plastic film (Canadian Pharmacopoeia) is prepared from poly(vinyl alcohol) and other additives and is intended as a vehicle for acriflavine, benzocaine, ichthammol and other topical drugs. Gelation can occur either with a fall (as with poly(vinyl alcohol)) or a rise in the temperature depending on the type of temperature variation of solubility. While gels of type II are commonest in pharmacy, with the interest in polymers as drug delivery adjuvants some type I materials are being used.

Cross-linked polymeric systems

If water-soluble polymer chains are covalently cross-linked, gels will be formed when the drug material interacts with water. The polymer swells in water. This expansion on contact with water has been put to use as follows:

Expanding implants have been made of cross-linked hydrophilic polymers which imbibe body fluids and swell to a pre-determined volume. These materials, such as the polyglycol methacrylates, are insoluble and chemically stable because of their three-dimensional structure (see I). Implanted in the dehydrated state these polymers swell to fill a body cavity or to give form to surrounding tissues. The gels may be used as vehicles for antibiotics permitting protracted release of drug in the immediate environment of the implant. Antibiotic-loaded gels like this have been used in infections of the middle ear and other sites not readily reached by other methods of administration. Surgical suture material coated with antibiotic-containing hydrophilic gels acquires a chemotherapeutic role as the development of spread of infection along the suture fibre is prevented.

Hydrophilic contact lenses (such as Soflens) are made from polyglycol methacrylates. These have also been utilised as drug carriers. Conventional eye medication has been modified over the years through the addition to formulations of a

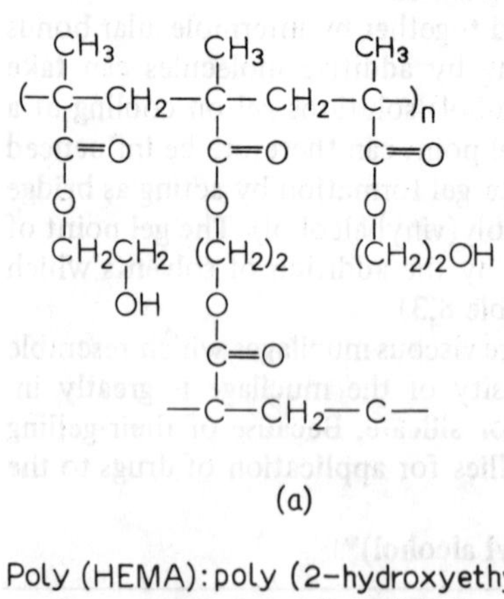

Poly (HEMA):poly (2–hydroxyethyl methacrylate) cross-linked with ethylene glycol dimethacrylate (EGDMA)

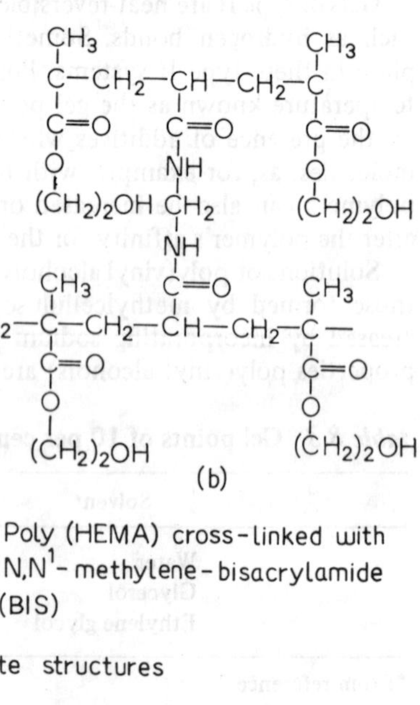

(a)

Poly (HEMA) cross-linked with N,N[1]– methylene – bisacrylamide (BIS)

(b)

(I) Polyglycol methacrylate structures

variety of viscosity-enhancing agents, polymers such as hydroxypropylmethyl cellulose, poly(vinyl alcohol) and silicones. These all prolong contact of drug with the cornea by increasing the viscosity of the medium and retarding the drainage of the tear fluid from the eye. Some of these systems are discussed in chapter 9.

8.2.3 Heterogels

As it is possible to produce macromolecular chains with segments that have different solubilities in a given solvent (copolymers), one would expect that concentrated solutions of such copolymers would behave in a manner different from that of a simple polymer. In block copolymers of the type AAABBBAAA, in which A is water-soluble and B is water-insoluble, the insoluble parts will tend to aggregate. If for instance a polystyrene–polyoxyethylene copolymer, comprising 41 per cent of polystyrene and 59 per cent of polyoxyethylene, is dissolved at 80 °C in butylphthalate (a good solvent for polystyrene), a gel with a microscopic layer structure is formed at room temperature. In nitromethane the form is somewhat different (figure 8.5a) as the nitromethane preferentially dissolves the polyoxyethylene chains.

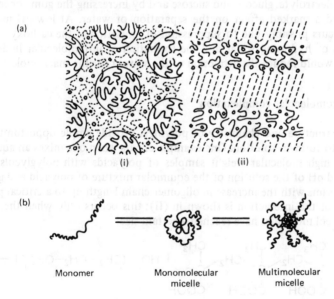

| | Monomer | Monomolecular
micelle | Multimolecular
micelle |

figure 8.5 (a) Structure of a copolymer of type A–B made from polystyrene and polyoxyethylene (i) in nitromethane (cylindrical structure) and (ii) in butyl phthalate (layer structure). Nitromethane dissolves the polyoxyethylene part preferentially, but butyl phthalate the polystyrene part. (——) polystyrene, (- - - -) polyoxyethylene, (○) solvent. From F. Sadron. *Angew. chem.*, 2, 248 (1963) (b) Representation of monomolecular and multimolecular micelle formation by copolymers of the ABA type

Polyoxyethylene–polyoxypropylene–polyoxyethylene block copolymers, known commercially as Pluronic surfactants, are used as emulsifiers. They are thought to form monomolecular micelles (see figure 8.5*b*) in aqueous solution at low concentrations, but at higher concentrations multimolecular micellar aggregates are believed to form, for the same reasons that normal micelles form, and heterogels of the type shown in figure 8.5*a* also arise. The hydrophobic central block associates with other like blocks, leaving the hydrophilic polyoxyethylene chains outside and protecting the inner core.

8.2.4 Syneresis

This is the term used for the separation of liquid from a swollen gel. Syneresis is thus a form of instability in aqueous and non-aqueous gels. Separation of a solvent phase is thought to occur because of the elastic contraction of the polymeric molecules; in the swelling process during gel formation the macromolecules involved become stretched and the elastic forces increase as swelling proceeds. At equilibrium the restoring force of the macromolecules is balanced by the swelling forces, determined by the osmotic pressure. If the osmotic pressure decreases, for example, on cooling, water may be squeezed out of the gel. The syneresis of an acidic gel from *Plantago albicans* seed gum[3] was decreased by the addition of electrolyte, glucose and sucrose and by increasing the gum concentration; pH had a marked effect on the separation of water. At low pH marked syneresis occurs possibly due to suppression of ionisation of the carboxylic acid groups, loss of hydrating water and the formation of intramolecular hydrogen bonds. This would reduce the attraction of the solvent for the macromolecule.

8.2.5 Macromolecular complexation

The varied structure and chemistry of polymers provide ample opportunity for complexes to form in solution. One example occurs when one mixes an aqueous solution of high molecular weight samples of polyacids with polyglycols. The viscosity and pH of the solution of the equimolar mixture of polyacid and glycol remains the same with the increase in oligomer chain length up to a critical point. The nature of the interaction is shown in (II); this occurs only when the polyethylene glycol molecules have reached a certain size.

(II)

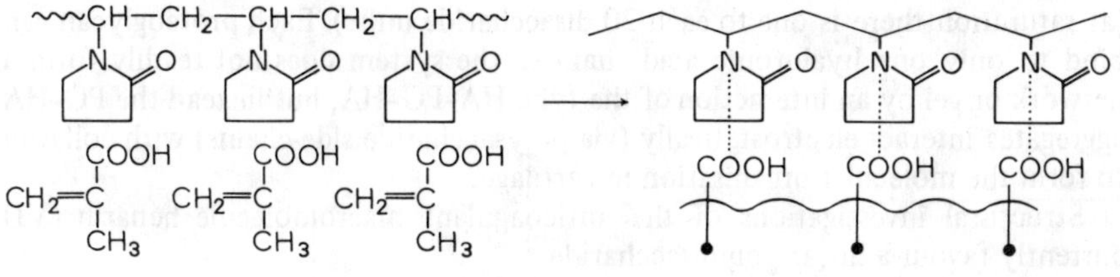

(III)

Complexes between polyvinylpyrrolidone and polyacrylic acids are also possible (III). Such intermacromolecular reactions are highly selective and strongly dependent on molecular size and conformation. An important feature of 'poly-reactions' is the all-or-nothing principle. On mixing, some of the macromolecule might be involved in the complex while the rest will be free. The reason for this compositional heterogeneity of the products could be the conformational transitions of macromolecules in the course of polycomplex formation.

Interactions between macromolecules can occur in formulations, for example, when preparations are mixed. However, they can be put to good advantage in the synthesis of novel compounds. Polyethyleneimine and polyacrylic acid form a polyelectrolyte complex with salt-like bonds as shown in (IV). If the complex is heated as a film, interchain amide bonds are formed between the groups which formed electrostatic links. The non-ionised −COOH and −NH groups in the chain are the points of structural defects in the film.

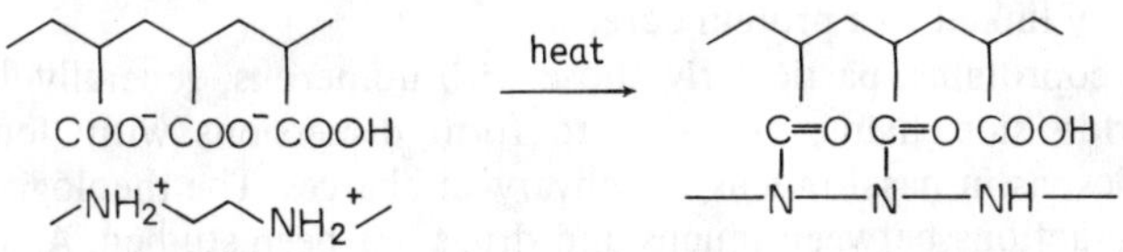

(IV)

Biological macromolecules undergo complex reactions which are often vital to their activity. Recent studies have established a specific interaction between hyaluronic acid (V) and the proteoglycans in the intracellular matrix in cartilage. An understanding of these macromolecular interactions is sometimes of value in elucidating the effects of drugs or formulations *in vivo*. The essential feature of the proposed proteoglycan–hyaluronic acid (PG-HA) complex is that many proteoglycans are able to bind along the entire length of the hyaluronic acid chain

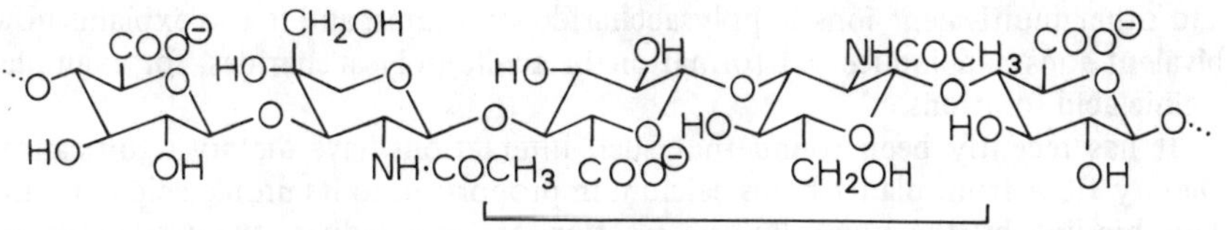

(V)

Repeating sequence of hyaluronic acid,
a high molecular weight glycosaminoglycan

(at saturation there is one to each 20 disaccharide units). Each proteoglycan can bind to only one hyaluronic acid chain so the system does not readily form a network or gel by an interaction of the type HA–PG–HA, but instead the PG–HA aggregates interact electrostatically (via polysaccharide side-chains) with collagen to form the molecular organisation in cartilage.

Structural investigations of the anticoagulant macromolecule heparin (VI) currently favour a linear polydisaccharide.

Repeating sequence of heparin

(VI)

The two types of chondroitin (chondroitin-4-sulphate and -6-sulphate) differ from hyaluronic acid by replacement of the β-acetylglucosamine with sulphated *N*-acetylgalactosamine. As with hyaluronic acid, helices of varying degrees of compactness are found. The chondroitin and dermatan sulphate chains are of lower molecular weight (about 500 000) than those of hyaluronate and occur *in vivo* covalently linked to a protein core.

Some glycoproteins, particularly those with numerous, generally distributed, oligosaccharide side-chains, are able to form dispersions with 'long stringy' characteristics, as in nasal mucus or salivary discharges. The rheological implications of interactions between mucus and drugs has been studied. As yet there is no coherent view as to what the ideal mucolytic agent, for example, should achieve. Reduction in 'viscosity' is too simple a concept to use for such complex systems, and parameters such as 'consistency' and dynamic viscosity have to be studied.

8.2.6 Binding of ions to macromolecules

Calcium is coordinated between classes of certain uronic acid containing polysaccharides (VII)[4]. Such complexation can explain the tight bonding of calcium and other multivalent ions in polysaccharide structures, and it also explains how bivalent ions can induce gel formation in acidic polysaccharides, for example, alginic acid solutions.

It has recently been found that such interactions have dietary significance. Dietary fibre from plants binds calcium in proportion to its uronic acid content. This binding by the non-cellulosic fraction of fibre reduces the availability of calcium for small-intestinal absorption although colonic digestion of uronic acids liberates the calcium[5]. The pH dependence of the binding strongly suggests the involvement of carboxylic acid groups (table 8.4).

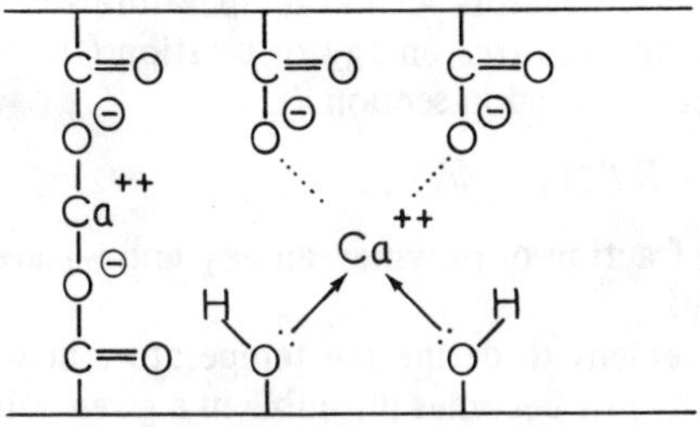

Calcium complexed in polysaccharides.

(VII)

table 8.4 **pH dependence of calcium binding by carrot fibre***

	pH							
	2.0	2.5	3.7	4.2	5.7	6.9	7.6	8.3
Percentage calcium bound								
Mean	4.5	15.0	35.6	38.4	39.1	46.8	45.5	46.3
± s.e.m.	±0.3	±0.5	±0.9	±0.5	±2.9	±2.0	±1.8	±0.8

*From reference 5

In some communities where daily fibre intakes may vary between 50 and 150 g with perhaps 30 to 110 mmol uronic acid, the binding capacity of fibre may exceed the total intake of calcium which may be less than 20 mmol (800 mg) per day.

8.2.7 Interaction of polymers with solvents including water

As a consequence of their size, polymers interact with solvents in a more complex fashion than do smaller crystalline solutes. A given polymer may have no saturation solubility; it usually either dissolves completely or is only swollen by a given liquid. If the polymer is cross-linked, solution cannot occur and the polymer will only swell by imbibition of liquid. Swelling decreases as the degree of cross-linking increases. Swelling is a function of the solubility parameter of the liquid phase.

Highly polar polymers like polyamide, poly(vinyl chloride) and some cellulose derivatives require polar liquids as solvents, in which dipole interactions or hydrogen bonding between polymer and solvent molecules occur. Solvation does not, however, necessarily lead to solution because the liquid, if it is to act as a solvent, must dissolve the solvated polymer. This process may be very slow because of the high viscosity of the solvated polymer.

In has been seen in chapter 7 that the use of macromolecules as dispersion stabilisers depends in part on the osmotic forces arising from the interaction of solvated polymer chains as neighbouring particles approach (see figure 7.7). It is thus important to know how factors such as temperature and additive affect this interaction. Flory has given the free energy of dilution (the opposite process to the concentration effect discussed in section 7.2) as

$$\Delta G_1 = RT (k_1 - \psi_1) \phi_2 \tag{8.18}$$

where ϕ_2 is the volume fraction of polymer, and k_1 and ψ_1 are heat and entropy parameters, respectively.

It is sometimes convenient to define the temperature at which a polymer of infinite molecular weight just becomes insoluble in a given solvent; this temperature is the Flory temperature or theta temperature, θ, which also may be defined by

$$\theta = \frac{k_1 T}{\psi_1} \tag{8.19}$$

so that substituting in equation 8.18 we obtain the relationship between ΔG_1 and temperature

$$\Delta G_1 = -RT \psi_1 \left(1 - \frac{\theta}{T}\right) \phi_2 \tag{8.20}$$

ΔG_1 is therefore zero at the θ temperature when deviations from ideality vanish, that is, there are no polymer–polymer or polymer–solvent interactions. When $T = \theta$ there can thus be no stabilisation as molecules will interpenetrate without net interaction and will exert no forces on each other.

Not only do most linear polysaccharides tend to form spirals in solution but in their tendency to associate they may form double helices, like carrageenan. Under certain conditions of concentration and temperature the double helices may associate, forming gels. Possibilities exist for complex gel formation as with carrageenan, or of xanthan gum with locust bean gum. The locust bean gum molecule can associate over part of its length with the helix of xanthan, for example, while the other part of the molecule associates with another xanthan molecule thereby acting as a bridging agent. Locust bean gum differs from a similar galactomannan, guaran, in being able to increase dramatically the viscosity of xanthan and carrageenan solutions. In the former the side-chains are grouped together leaving long parts of the main chain exposed and therefore free to associate with other molecules. Guaran, on the other hand, has its side-chains evenly spread on the mannose backbone.

The firmness or strength of gels produced by such interactions will depend on the degree of interaction of the complex with water and the properties of the bridging units.

The ability of carbohydrates and other macromolecules to imbibe large quantities of water is put to use both medicinally and industrially; for example, in paper and sanitary towels, nappies and surgical dressings. Medically, use is made of the swelling properties in the treatment of constipation and in appetite suppression. Three properties are said to be of importance in the *in vitro* evaluation of bulk laxatives[6]: (a) the volume of water absorbed in the various media; (b) the viscosity and texture of the gel formed; and (c) the ability of the gel to retain water. The swelling properties of a sterculia-based preparation (Normacol) in various aqueous media and a comparison of Normacol with two other agents are

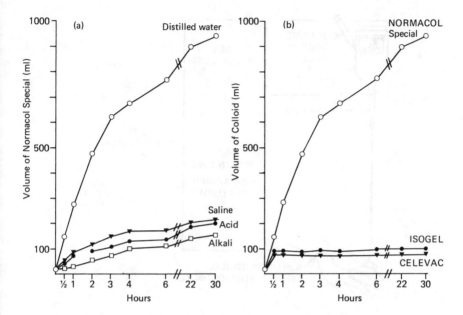

figure 8.6 (a) The volume attained by 5 g of Normacol Special in various solutions over 30 hours. (b) The volumes attained by 5 g of Normacol Special, Isogel and Celevac in distilled water. From reference 7

shown in figure 8.6. It is desirable that colloidal bulk laxatives swell in the lower part of the small intestine and in the large intestine to cause reflex peristalsis rather than in the stomach or duodenum; that is, they should swell in neutral rather than acidic or alkaline conditions. In artificial intestinal juice, psyllium seed gum increased in volume 5-14 times, locust bean gum 5-10 times and methylcellulose 16-30 times in 24 hours[8]. *In vivo* evaluation of methylcellulose and carboxymethylcellulose suggests that they have two advantages over the natural gums. Methylcellulose is more efficient as a bulk laxative because of its greater water-retentive capacity whereas carboxymethylcellulose gives uniform distribution through the intestinal contents.

8.2.8 Adsorption of macromolecules

The ability of some macromolecules to adsorb at interfaces is made use of in suspension and emulsion stabilisation (see chapter 7). Gelatin, acacia, poly(vinyl alcohol) and proteins adsorb at interfaces. Sometimes such adsorption is unwanted as in the case of insulin adsorption onto glass infusion bottles, poly(vinyl chloride) infusion containers and tubing used in giving sets. Adsorption of insulin to glass bottles and plastic i.v. tubing at slow rates of infusion is well documented[9]. Adsorption ranged from 5 to 3.1 per cent when 20 and 40 units respectively were added to 500 mℓ of isotonic sodium chloride solution, while plastic i.v. tubing adsorbed 30 per cent of 20 units and 26 per cent of 40 units added to the

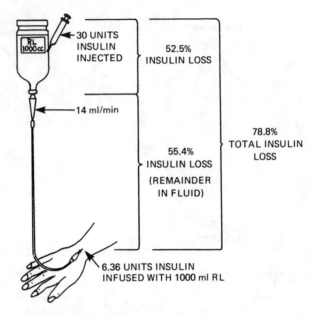

figure 8.7 Amounts of insulin lost by adsorption to glass bottles, and plastic intravenous tubing, following injection of 30 units of insulin. The patient receives only 6.36 units. From reference 9

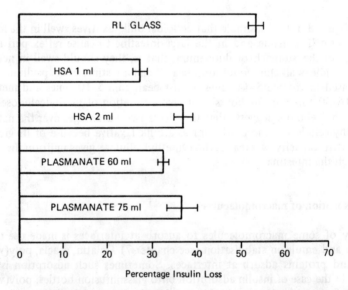

figure 8.8 Prevention of insulin loss via adsorption by the addition of human serum albumin (HSA) or Plasmanate to 1000 mℓ of Ringer's lactate (RL) solution in a glass-bottle. Insulin (30 units) was injected and measured at 5 minutes. Values represent means ± S.E. All HSA and Plasmanate values were significantly different from Ringer lactate solution–glass bottle control ($p < 0.001$). From reference 9

same infusion bottles (figure 8.7). Adsorption occurs rapidly, within 15 seconds. Addition of albumin to prevent adsorption is now common practice. Presumably the albumin itself adsorbs at the glass or plastic surface and presents a more polar surface to the solution, thus reducing but not always preventing adsorption of the insulin (figure 8.8). The binding is considered to be a non-specific phenomenon which may occur on other inert materials such as polyethylene and glass[10].

The adsorption of macromolecules at interfaces may be the reason why molecules such as those of hyaluronic acid can act as biological lubricants in joint fluids. In healthy joints only 0.5 mℓ of synovial fluid is required to provide almost perfect lubrication; in diseased joints there are faults sometimes in this system and some research has been aimed at producing synthetic substitutes for synovial fluid. Polymer solutions provide one approach as their rheological characteristics more closely approach that of the natural fluid, which is non-Newtonian.

8.3 Details of some water-soluble polymers used in pharmacy and medicine

In this section the properties of some specific polymers used in pharmacy and medicine will be discussed. This cannot be an exhaustive treatment of the subject, so choice of the macromolecules for this section has been based partly on the degree of use, but partly on the generally interesting features they display. The choice of a macromolecule for a particular pharmaceutical use is often difficult because of the diversity of properties exhibited by the materials available. Figure 8.9 illustrates how the field can be narrowed to some extent by grouping the natural and synthetic materials of interest to the formulator. This is, however, a very general guide, as the properties of individual macromolecules will often vary with pH, temperature, molecular weight and ionic strength. The most readily altered variable is, of course, the concentration of the macromolecule, whose effect on viscosity is illustrated for a range of compounds in figure 8.10. The most viscous material shown here is Carbopol 934. This is discussed first.

8.3.1 Carboxypolymethylene (Carbomer, Carbopol)

This is used as a suspending agent in pharmaceutical preparations and as a binding agent in tablets, and it is used in the formulation of prolonged-acting tablets. It is a high molecular weight polymer of acrylic acid, cross-linked with allyl sucrose and containing a high proportion of carboxyl groups. Its aqueous solutions are acidic; when neutralised the solutions become very viscous with a maximum viscosity between pH 6 and 11. Electrolytes reduce the viscosity of the system and thus high concentrations of the polymer have to be employed in vehicles where ionisable drugs are present.

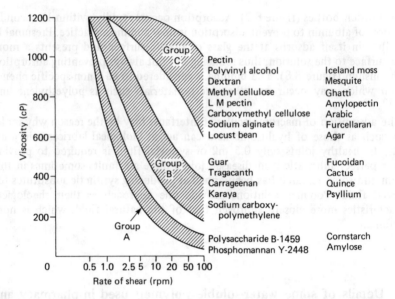

figure 8.9 Effect of shear rate on the viscosity of gum solutions grouped according to their rheological behaviour. After A. S. Szezesniak and E. H. Farkas, *J. Food Sci.*, 27, 381 (1962)

8.3.2 Cellulose derivatives[11]

Cellulose itself is virtually insoluble in water, but aqueous solubility can be conferred by partial methylation or carboxymethylation.

Ethylcellulose is an ethyl ether of cellulose containing 44–51 per cent of ethoxyl groups. It is insoluble in water but soluble in chloroform and in alcohol. It is possible to form water-soluble grades with a lower degree of substitution.

Commercial *methylcellulose* samples are prepared by heterogeneous reaction which is usually controlled to allow substitution of, on average, about one-half of the hydroxyl groups. This leads to a product in which the methylated groups are not evenly distributed throughout the chains; rather, there are regions of high density of substitution (as in structure VIII) which are hydrophobic in nature, and regions of low density of substitution which are hydrophilic in nature.

Methylcellulose is thus a methyl ether of cellulose containing about 29 per cent of methoxyl groups; it is slowly soluble in water. A 2 per cent solution of methylcellulose 4500 has a gel point of about 50 °C. High concentrations of electrolytes salt out the macromolecules and increase their viscosity; eventually precipitation may occur. Low-viscosity grades are used as emulsifiers for liquid paraffin and other mineral oils. High-viscosity grades are used as thickening agents for medicated jellies and as dispersing and thickening agents in suspensions.

Since methylcelluloses are poorly soluble in cold water, preliminary use of hot water assures wetting of all portions of the particle prior to solution in cold water. The water-soluble methylcelluloses possess the property of thermal gela-

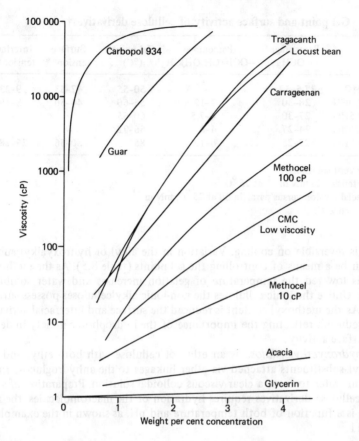

figure 8.10 Viscosity of solutions of a number of pharmaceutical polymers and gums plotted as a function of concentration. From A. N. Martin *et al.*, *Adv. Pharm. Sci.*, 1, 1 (1964)

(VIII)

tion; that is, they gel *on heating* while the natural gums gel *on cooling*. Methyl-cellulose exists in solution as long thread-like molecules hydrated by water molecules. On heating, the water of solvation tends to be lost and the 'lubricating' action of the hydration layer is also lost and the molecules lock together in a gel.

table 8.5 **Gel point and surface activity of cellulose derivatives†**

	Percentage −OCH₃	Percentage −OCH₂CH(OH)CH₃	Gel point* (°C)	Surface tension**	Interfacial tension***
Methocel MC	27.5–32	–	50–55	47–53	19–23
Methocel 60HG	28–30	7–12	55–60	44–50	18–19
Methocel 65HG	27–30	4–7.5	60–65	–	–
Methocel 70HG	24–27	4–8	66–72	–	–
Methocel 90HG	22–25	6–12	85	50–56	26–28

*For 2 per cent solutions
**Surface tension in mN m^{-1} at 25 °C
***Interfacial tension versus paraffin oil at 25 °C, mN m^{-1}
†From reference 12

Gelation is reversible on cooling. Variation in the alkyl or hydroxyalkyl substitution can be a means of controlling the gel points (table 8.5). As the methoxyl content is lowered the temperature of gelation increases and water solubility decreases. Unlike the ionic celluloses the non-ionic alkylcelluloses possess surface activity. As the methoxyl content is reduced the surface and interfacial activities are also reduced, reflecting the importance of the hydrophobic moiety in determining surface activity.

Ethylhydroxyethylcellulose is an ether of cellulose with both ethyl and hydroxyethyl substituents attached via ether linkages to the anhydroglucose rings. It swells in water to form a clear viscous colloidal solution. Preparation of solutions of cellulose derivatives requires hydration of the macromolecules, the rate of which is a function of both temperature and pH, as shown in the example in figure 8.11.

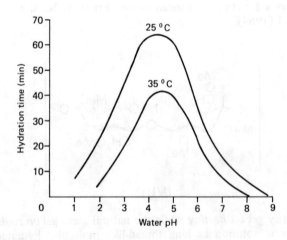

figure 8.11 Effect of pH and temperature on the hydration time of fast-dissolving grades of hydroxyethylcellulose. From reference 11

Ethylmethylcellulose contains ethyl and methyl groups, a 4 per cent solution having approximately the same viscosity as acacia mucilage. Hydroxyethylcellulose is soluble in hot and cold water but does not gel. It has been used in ophthalmic solutions. More widely used for the latter, however, is hydroxypropylmethyl-cellulose (hypromellose) which is a mixed ether of cellulose containing 27-30 per cent of $-OCH_3$ groups and 4-7.5 per cent of $-OC_3H_6OH$ groups. It forms a viscous colloidal solution. There are various pharmaceutical grades. For example, hypromellose 20 is a 2 per cent solution which has a viscosity between 15 and 25 cS at 20 °C; the viscosity of a 2 per cent hypromellose 15000 solution lies between 12000 and 18000 cS. Hypromellose prolongs the action of medicated eye drops and is employed as an artificial tear fluid.

Sodium carboxymethylcellulose is soluble in water at all temperatures. Because of the carboxylate group its mucilages are more sensitive to change in pH than are those of methylcellulose. The viscosity of a sodium carboxymethylcellulose mucilage is decreased markedly below pH 5 or above pH 10. Addition of heavy metal ions (Al^{3+}, Zn^{2+}, Fe^{2+}) causes changes in solution properties.

8.3.3   Natural gums and mucilages

Gum arabic is widely used in pharmacy as an emulsifier. It is a polyelectrolyte whose solutions are highly viscous due to the branched structure of the macro-molecular chains; its adhesive properties are also believed to be due, or in some way related, to this branched structure. Molecular weights of between 200 000 and 250 000 (M_n) have been determined by osmotic pressure, of between 250 000 and 300 000 by sedimentation and diffusion, and of one million by light scattering, which also points to the shape of the molecules as short stiff spirals with numerous side-chains. Arabic acid prepared from commercial gum arabic by precipitation is a moderately strong acid whose aqueous solutions have a pH of 2.2-2.7. It has a higher viscosity than its salts but emulsions prepared with arabic acid cream are not as stable as those made with its salts.

Whereas most gums are very viscous in aqueous solution, gum arabic is un-usual in that being extremely soluble it can form solutions over a wide range of concentrations up to about 37 per cent at 25 °C. The effect of pH and salt con-centration on the viscosity of gum arabic solutions is shown in figure 8.12. The marked variation in viscosity means that the gum arabic molecules must be flex-ible with the ionic acid carboxyl groups distributed along the chain. At low pH the carboxyl groups are unionised. On increase of pH the carboxyl groups become progressively ionised and the folded chains expand due to repulsion between the charged groups, causing an increase in viscosity. On addition of NaOH to the sys-tem the viscosity falls again as the concentration of counterion (Na^+) increases and effectively shields the acidic groups. The molecule then folds on itself. Simi-lar falls in viscosity are exhibited on addition of sodium chloride. The effect of salt addition to the gum at fixed pH reflects the decrease in effective charge on the molecules of gum with resultant contraction and reduction in viscosity.

The gum arabic molecule is, in addition, surface-active, a 4 per cent solution at 30 °C having a surface tension of 63.2 mN m^{-1}. Addition of electrolytes makes the molecule more surface-active either by causing a change in conforma-tion of the molecule at the interface, allowing closer packing, or by increasing

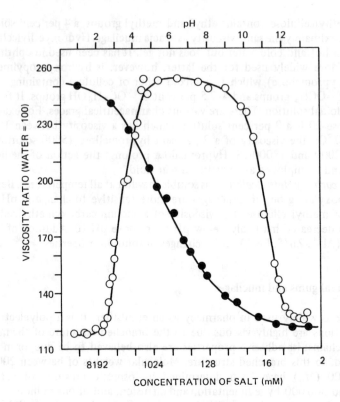

figure 8.12 The viscosity of gum arabic solutions as a function of pH (○) and salt concentration (●). Redrawn from Thomas and Murray in reference 11

the hydrophobicity of the molecule. It is an effective emulsifier, the stabilisation of the emulsion being dependent mainly on the coherence and elasticity of the interfacial film, which is by no means monomolecular.

Gum arabic is incompatible with several phenolic compounds (phenol, thymol, cresols, eugenol) and under suitable conditions forms coacervates (see section 8.5.2) with gelatin and other positively charged polyelectrolytes.

Gum tragacanth partially dissolves in water; the soluble portion is called tragacanthin and this can be purified by precipitation from water by acetone or alcohol. Tragacanthin is a highly viscous polyelectrolyte with a molecular weight of 800 000 by sedimentation. It is one of the most widely used natural emulsifiers and thickeners. As its molecules have an elongated shape its solutions have a high viscosity which, as with gum arabic, is dependent on pH. The maximum viscosity occurs at pH 8 initially but due to ageing effects it is found that the maximum stable viscosity is near pH 5.

It is an effective suspending agent for pharmaceuticals and is used in conjunction with acacia as an emulsifier, the tragacanth imparting a high structural viscosity while the gum arabic adsorbs at the oil-water interface. It is also used in spermicidal jellies acting by immobilising spermatozoa and as a viscous barrier.

Alginates — although the solutions of alginate are very viscous and set on addition of acid or calcium salts, they are less readily gelled than pectin and are used chiefly as stabilisers and thickening agents. Propylene glycol alginate does not precipitate in acid and as it is non-toxic is widely used as a stabiliser for food-stuffs. The molecules are highly asymmetric, with molecular weights in the range 47 000–370 000. Sodium alginate has the structure given in IX.

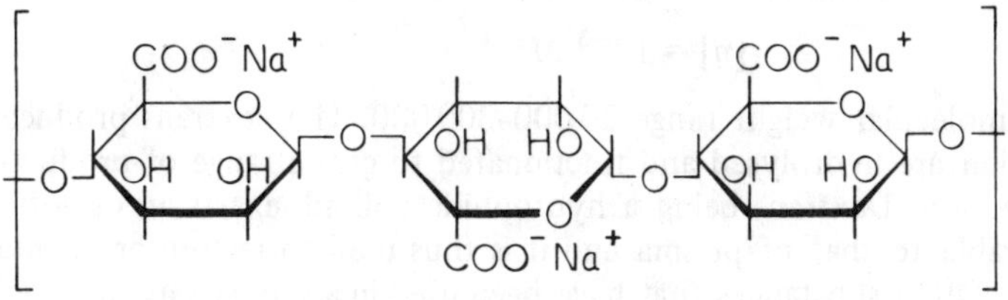

Sodium alginate

(IX)

Pectin is a purified carbohydrate product from extracts of the rind of citrus fruits and consists of partially methoxylated polygalacturonic acid[13] (X). It has remarkable gelling qualities but is also used therapeutically, often with kaolin, in the treatment of diarrhoea. It has been established that the longer the pectin

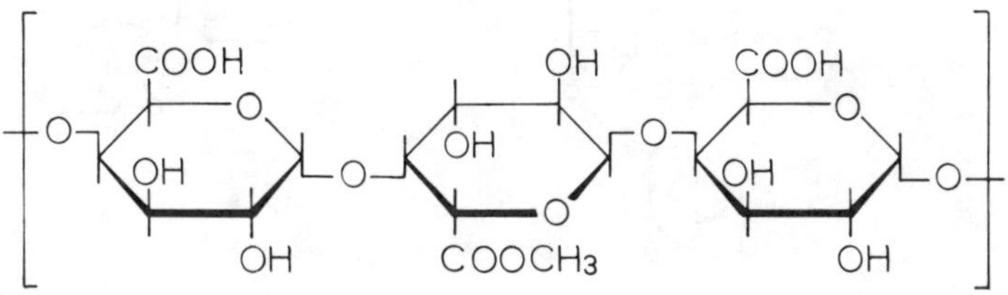

Partially methylated chain of polygalacturonic acid of pectin. Average molecular weight of pectins 20 000 – 400 000

(X)

chains, the greater its capacity for gel formation. The presence of inorganic cations and the degree of esterification of the carboxyl groups are important factors; in the case of calcium pectate gel it can be assumed that the calcium or indeed other polyvalent cations can interlink the chains by binding through COO^- ... Ca^{2+} ... ^-OOC interactions. Thus a high degree of esterification will disfavour gelation in this case. However, in the absence of inorganic cations, a high degree of esterification aids gelation, suggesting that hydrophobic interactions cause the chains to associate. The properties of the formed gels also depend on the degree of esterification; the rigidity of the 40–60 per cent ester pectin gels is higher than that of 70–80 per cent ester jellies. This suggests that rigidity is due to hydrogen bonding between the hydroxyl groups and the free carboxyls.

8.3.4 Dextran

Certain fractions of partially hydrolysed dextran are used as plasma substitutes or 'expanders'. Certain strains of *Leuconostoc meserentoides* are cultivated to synthesise dextran which is anhydroglucose, linked through α-1,6 glucosidic linkages. The chains are branched, and on the average, one branch occurs for every 10–12 glucose residues (XI). According to Wales *et al.*[14]

$$[\eta] = 10^{-3} M^{1/2} \tag{8.21}$$

in the molecular weight range 20 000–200 000. The dextrans produced by fermentation are hydrolysed and fractionated to give a range of products suitable for injection. Dextran, being a hydrophilic colloid, exerts an osmotic pressure comparable to that of plasma and it is thus used to restore or maintain blood volume. Other substances that have been used in a similar way include hydroxyethyl starch, polyvinylpyrrolidone and gelatin.

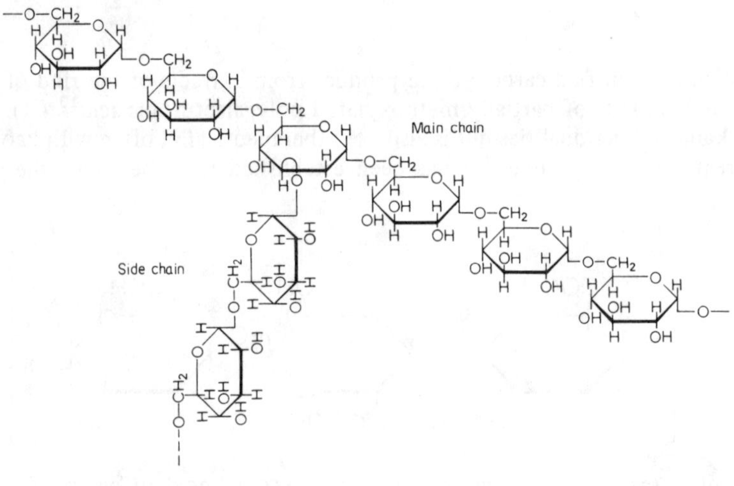

Structural formula of dextran

(XI)

Dextran injections are sterile solutions of dextran with weight average molecular weights of about 40 000–110 000. Dextrans with a molecular weight of about 50 000 or less are excreted in the urine within 48 hours of injection. Dextran molecules with higher molecular weights disappear more slowly from the blood stream and are temporarily stored in the reticulo-endothelial system.

Dextran '70' (mol wt ~ 70 000) and Dextran '110' (mol wt ~ 110 000) are used to maintain blood volume, and Dextran 40 is used primarily to prevent intravascular aggregation of blood cells and for assisting capillary blood flow. This latter effect is the result of dextran adsorption and stabilisation of the erythrocyte suspensions. If the higher molecular weight fractions exceed about 1 per cent concentration in the blood, rouleaux tend to form. The sensitivity of blood to the concentration and molecular weight of dextran is clearly seen in

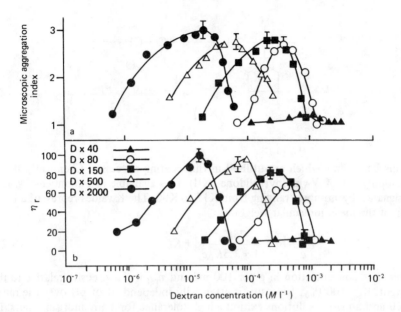

figure 8.13 Variation of indices of red-cell aggregation with the concentration of five dextran fractions of different molecular weight. (a) Microscopic aggregation index; and (b) relative viscosity at a shear rate of 0.1 s^{-1}. Note that the maximum of each curve (that is, maximum aggregation of the red cells) corresponds to a well-defined concentration of a particular dextran fraction. Maxima in both indices of aggregation of red cells occur at about the same concentration of dextran fraction. Reproduced from S. Chien. *Bibl. Anat. Basel*, 11, 244 (1973)

figure 8.13, where aggregation and relative viscosity of red-cell suspensions are shown in the presence of varying amounts of five different dextrans. Molecular weight control is thus important and may be exercised by measurement of intrinsic viscosity, $[\eta]$. Admixture of dextran with ascorbic acid, chlortetracycline hydrochloride, phytomenadione or promethazine hydrochloride leads to a loss of clarity.

Iron–dextran complexes are soluble, non-ionic and suitable for injection for the treatment of anaemia; the complex is stable on storage in the pH range 4–11. More recently aminoethyl dextran–methotrexate complexes have been prepared, the object being to influence uptake of the drug selectively into tumour cells. Attachment of the drug to the macromolecule allows selective uptake into malignant cells, as such cells are more active than normal cells in pinocytosis, the mechanism by which macromolecules are taken into many cells.

8.3.5 Polyvinylpyrrolidone[15]

Polyvinylpyrrolidone (PVP) is used as a suspending and dispersing agent, a tablet binding and granulating agent, and as a vehicle for drugs such as penicillin, cortisone, procaine and insulin to delay their absorption and prolong their action. It

(XII)

forms hard films which are utilised in film-coating processes. Chemically it is a homopolymer of *N*-vinylpyrrolidone (XII). It is available in a number of grades designated by numbers ranging from K15 to K90. The K values represent a function of the mean molecular weight as

$$\frac{\log \eta_{\mathrm{rel}}}{c} = \frac{75K_0^{\,2}}{1 + 1.5K_0 c} + K_0 \tag{8.22}$$

where c = concentration in g per 100 mℓ and η_{rel} is the viscosity relative to the solvent. K = $1000K_0$. Viscosity is essentially independent of pH over the range 0–10 and aqueous solutions exhibit a high tolerance for many inorganic salts. Its wide solubility in organic solvents is unusual. The viscosity of a range of aqueous solutions of PVP is shown in figure 8.14.

PVP forms molecular adducts with many substances. Insoluble complexes are formed when aqueous solutions of PVP are added to tannic acid, polyacrylic acid

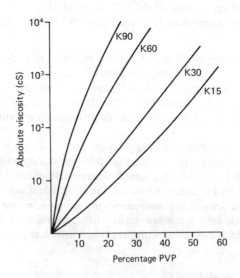

figure 8.14   Viscosity of polyvinylpyrrolidone as a function of molecular weight (K15 [mol wt 44 000] to K90 [mol wt 700 000]) and concentration of the polymer in water. From reference 15

and methyl vinyl ether-maleic anhydride copolymer. Soluble complexes (iodophors) are formed with iodine whose solubility is increased from 0.034 per cent in water at 25 °C to 0.58 per cent by 1 per cent PVP. The resulting iodophor retains the germicidal properties of iodine. It is thought that the iodine is held in a PVP helix in solution. The influence of two samples of PVP on the solubility of testosterone is shown in figure 8.15. The PVP correspondingly increases the rate of solution of the steroid from solid dispersions.

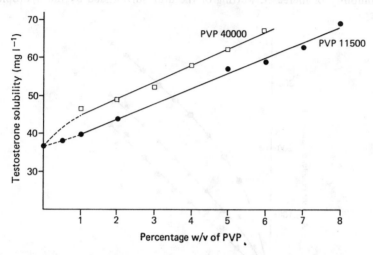

figure 8.15 The influence of PVP 11 500 and PVP 40 000 on the aqueous solubility of testosterone at 37 °C. From A. Hoelgaard and N. Muller. *Arch. pharm. chem.*, 3, 34 (1975)

8.3.6 Polyoxyethylene glycols (Macrogols)

The *macrogols* (polyoxyethylene glycols, PEGs, XIII) are liquid over the molecular weight range 200–700; the liquid members and semisolid members of the series are hygroscopic. Macrogol 200 has a hygroscopicity 70 per cent of that of glycerol but this decreases with molecular weight; Macrogol 1540 has a hygroscopicity of 30 per cent compared with glycerol. They are used as solvents for drugs such as hydrocortisone. The macrogols are incompatible with phenols and can reduce the antimicrobial activity of other preservatives. Higher molecular weight PEGs are more effective on a molecular basis as complexing agents. Up to four phenol molecules bind to each PEG molecule; the complex formed is of the donor–acceptor type. The semisolid and waxy members of the series may be

$$HO(CH_2CH_2O)_nH$$

Macrogol, polyoxyethylene glycol

(XIII)

used as suppository bases; in such cases their potential to interact with medicaments must be borne in mind.

Use of polyoxyethylene glycols, and other hydrophilic polymers, in high concentrations in formulations can influence the behaviour of drugs even when the drug is present as a physical mixture with the polymer. For example, combination of polyoxyethylene glycol 4000 with sulphathiazole increases the solution rate of sulphonamides (figure 8.16). Probable mechanisms include an increase in drug solubility or increased wetting of the drug surrounded by the hydrophilic polymer.

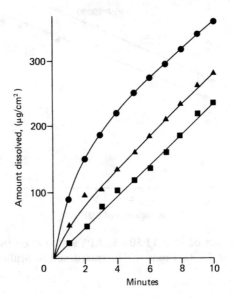

figure 8.16 Dissolution rates of sulphathiazole (form I)–polyethylene glycol 4000 physical mixtures. Key (polyethylene glycol 4000): ●, 10 per cent; ▲, 5 per cent; and ■, 2 per cent. From S. Niazi. *J. pharm. Sci.*, 65, 302 (1976)

8.3.7 Silicones

The *silicones* are examples of hydrophobic liquid polymers although in high molecular weight they exist as waxes and resins. Silicones are polymers with a structure containing alternate atoms of Si and O; the dimethicones are fluid polymers with the general formula $CH_3[Si(CH_3)_2O]_nSi(CH_3)_3$ in which each unit has two methyl groups and an oxygen atom attached to the silicon atom in the chain. The viscosity range of 0.65 cS to 3×10^6 cS exists. The PC describes five dimethicones (20, 200, 350, 500 and 1000) the numbers representing the average viscosity in cS at 25 °C. Their rheological properties have been put to use in ophthalmology and in rheumatoid arthritis. Dimethicone 200 has been used as a lubricant for artificial eyes and to replace the degenerative vitreous fluid in cases of retinal detachment. It can act as a simple lubricant in joints. More common uses are as barrier substances, silicone lotions and creams acting as water-repellent applications protecting the skin against water-soluble irritants. Methyl phenylsilicone is used as a lubricant for hypodermic syringes. Glassware which has been treated with a thin film of silicone is rendered hydrophobic; solutions and aqueous suspensions thus drain completely from such vessels.

Activated dimethicone (activated polymethylsiloxane) is a mixture of liquid dimethicones containing finely divided silica to enhance the defoaming properties of the silicone. The mechanism by which dispersion of colloidal silica in silicone antifoams improves their action is not well understood.

8.3.8 Polymers used as wound dressings

Several polymers are now used in the preparation of synthetic wound dressings. Synthaderm is a 'synthetic skin' of a modified polyurethane foam, hydrophilic on one side and hydrophobic on the other. The hydrophilic side is placed in contact with the wound. The system has been described as an 'environmental dressing'[16] as it (a) maintains a high humidity at the dressing interface, (b) removes excess exudate, (c) allows gaseous exchange and (d) provides insulation and is impermeable to bacteria, the outer surface remaining dry unlike many saturable dressings. Lyofoam is a similar product[17]. Laminates of polypeptides and elastomers have been proposed as burn wound coverings to provide a film with the appropriate strength and physical properties[18].

Cross-linked dextran gels are used in chromatography. The reaction of dextran with an α,ω-diepoxy-, an α,ω-halo- or an α-halo-ω-epoxy-compund and NaOH yields an insoluble hydrophilic gel that can be partially depolymerised to the required molecular weight. The gel is produced in a bead shape; the degree of cross-linking determines the water uptake and pore size and therefore the molecular exclusion limit. Choice of cross-linking agents leads to gels which swell in water, ethanol and chloroform, extending the separation procedures to non-aqueous solvents. *Dextranomer* (Debrisan) is a cross-linked dextran with pores large enough to allow substances with a molecular weight of less than 1000 to enter the beads[19]. Each gram of beads abstracts approximately 4 mℓ of fluid. Applied to the surface of secreting wounds, dextranomer removes by suction various exudates that tend to impede tissue repair, while leaving behind high molecular weight materials such as plasma proteins and fibrinogen.

8.3.9 Polymer crystallinity

Polymers form perfect crystals with difficulty simply because of the low probability of arranging the chains in regular fashion, especially at high molecular weights. Advantage can be taken of defects in crystals in the preparation of microcrystals. Microcrystalline cellulose (Avicel) is prepared by disruption of larger crystals. It is used as a tablet excipient and as a binder-disintegrant. Dispersed in water it forms colloidal gels, and it can be used to form heat-stable o/w emulsions. Spheroidised forms of microcrystalline cellulose with accurately controlled diameters can be prepared and drugs can be incorporated during preparation. Microcrystalline polyamides and gelatin are still experimental materials but the latter should find application in the preparation of sustained intramuscular injections. The concept of crystallinity is potentially important when considering polymer membranes, as discussed below.

8.4   Water-insoluble polymers and polymer membranes

8.4.1   Permeability of polymers

Hydrophobic polymers also play an important role in pharmacy. When these materials are used as membranes, containers, or tubing material their surfaces may come into contact with solutions. The surfaces of insoluble polymers are not as inert as might be thought. The interaction of drugs and preservatives with plastics depends on the structure of the polymer and on the affinity of the compound for the plastic. The latter is determined by the degree of crystallinity of the polymer, as permeability is a function of the fraction of amorphous polymer. The crystalline regions of the solid polymer present an impenetrable barrier to the movement of most molecules. Diffusing molecules thus have to circumnavigate the crystalline islands which act as obstructions. The greater the volume fraction of crystalline material (ϕ_c) the slower the movement of molecules.

Diffusion in non-porous solid polymer is of course a more difficult process than in a fluid because of the necessity for the movement of polymer chains to allow passage of the drug molecule, and it is therefore slower. The equation which governs the process is Fick's first law (see section 3.5, equation 3.82)

$$J = -D \frac{dc}{dx} \qquad (8.23)$$

where J is the flux, D is the diffusion coefficient of the drug in the membrane, and dc/dx is the concentration gradient across the membrane. If the membrane is of thickness l, and Δc represents the difference in solution concentration of drug at the two faces of membrane

$$J = \frac{DK\Delta c}{l} \qquad (8.24)$$

where K is the distribution coefficient of the permeant towards the polymer. Therefore alteration of polymer/membrane thickness, coupled with appropriate choice of polymer can give rise to the desired flux. Within a given polymer permeability is a function of the degree of crystallinity, itself a function of polymer molecular weight. If P is the permeability of drug in a partially crystalline polymer (see figure 8.17), the volume fraction of the crystalline regions being ϕ_c, and P_a is the permeability in an amorphous sample, then

$$P/P_a = (1 - \phi_c)^2 \qquad (8.25)$$

Permeation of drug molecules through the solid polymer, which may be acting as a drug depot, is a function of the solubility of the drug in the polymer as

$$P = DK \qquad (8.26)$$

Addition of inorganic fillers in which the drug is insoluble alters the over-all solubility of the drug in the polymer and hence alters the permeation characteristics. The equation for over-all solubility (S) is given by

$$S = S_f \phi_f + S_p \phi_p \qquad (8.27)$$

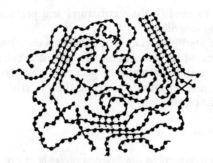

figure 8.17 The fine structure or supermolecular structure illustrating the laterally ordered (crystalline) and disordered (amorphous) placement of linear polymer chains in a film matrix

where f refers to filler and p to polymer. Thus, when $S_f \to 0$ as happens when inorganic fillers such as zinc oxide are employed, an obstruction type equation may be written

$$S/S_p = (1 - \phi_p) \qquad (8.28)$$

The natural permeabilities of polymers vary over a wide range and this widens the choice, provided one can select a polymer which is compatible with the tissues with which it comes in contact. Having chosen a polymer which gives a flux of drug sufficient to provide adequate circulating levels of drug, use of fillers and plasticisers can give fine control of permeability (table 8.6).

The method of preparation also influences the properties of the film. Cast films of varying properties can be prepared by variation *inter alia* of the solvent power of the casting solution containing the polymer, although the complex processes involved in film formation are not yet fully understood. However, it is clear that the conformation of the polymer chains in concentrated solution just prior to solvent evaporation will determine the density of the film, the number and size of pores and voids. Drug flux through dense (non-porous) polymer membranes is by diffusion; flux through porous membranes will be by diffusion and by transport in solvent through pores in the film. With porous films, control can be exercised on porosity, and hence over-all permeability, by the use of swelling agents. Dense membranes can be subjected to certain post-formation treat-

table 8.6 **Factors that influence diffusivity in polymers**

Factor	Net effect on D
Increased polymer molecular weight	↓
Increased degree of cross-linking	↓
Diluents and plasticisers	↑
Fillers	↓
Increased crystallinity of polymers	↓
Increased drug size	↓

ments which serve to modify their structural and performance characteristics, for example, thermal annealing.

The solubility of sterilising gases in polymers is important in determining the retention of residues which may, as in the case of ethylene oxide residues, be toxic. The quality control problems of polymers and plastics are considerable. Both the chemical and physical nature of the material has to be taken into account, as well as its purity.

Permeability of polymers to gases

The permeability of polymers to the gaseous phase is of importance when the use of polymers as packaging materials is considered. Figure 8.18 shows oxygen penetration through a wide range of plastic materials ranging from Teflon to dimethylsilicone rubber, which shows the highest permeability. Ether, nitrous oxide, halothane and cyclopropane diffuse through silicone rubbers, and general anaesthesia in dogs has been achieved by passing the vapours of these substances through a coil of silicone rubber tubing, each end of which is placed in an artery or vein. Aspects of the permeability of polymeric films of interest pharmaceutically include the process of gas diffusion, water sorption and permeation and dialysis processes. With few exceptions Munden et al.[20] found an inverse relationship between water-vapour transmission and oxygen permeability. Water-vapour permeability has been shown to be dependent on the polarity of the polymer. More polar films tend to be more ordered and less porous, hence less oxygen-permeable. The less polar films are more porous, permitting the permeation of oxygen but not necessarily of the larger water molecules. Being more lipophilic the less polar films have less affinity for water. Because of the importance of

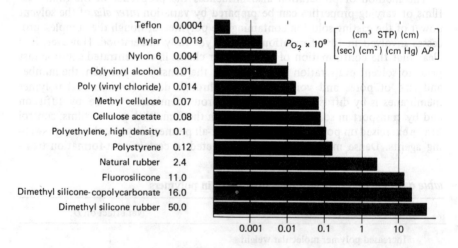

figure 8.18   Permeation of oxygen through several polymer films. This illustrates the diversity of the properties of available polymers and shows the relatively high permeability of polydimethylsiloxane, a common component of prolonged-acting drug devices. From R. Kesting. *Synthetic Polymeric Membranes*, McGraw-Hill, New York, 1971

water as a solvent and permeant species, much work has been directed towards the synthesis of polymeric membranes with controlled hydrophilic/hydrophobic balance. The hydrophilicity of cellulose acetates is directly proportional to their —OH content and inversely proportional to the hydrophobic acetyl content (table 8.7). Alternative approaches to alteration of characteristics of water permeability include the use of block copolymers where one can alter the ratio of hydrophilic polymer to increase transport rates of polar materials.

table 8.7   **Effect of hydroxyl and acetyl content on water permeability and sorption of moisture by cellulose acetates***

Hydroxyl content (%)	Acetyl content (%)	Moisture sorption (%) at 95 per cent relative humidity at 25 °C	Water permeability $D_1 C_1$ (10^{-7} g cm s^{-1})
7.2	34	19	12
5.9	36	17	7
4.6	38	14.7	4
3.3	40	12.6	2.5
2.0	42	10.5	1.5

*From C. Reid and E. Breton. *J. appl. Polymer Sci.*, 1, 133 (1959)

The affinity of drugs for plastics

The affinity of a drug for a plastic will vary with the structure of the drug. Chlorpromazine, for example, has a very high affinity for some materials used as tubing (table 8.8). Silicone is very permeable and this can be put to good use in other

table 8.8   **Concentration of chlorpromazine in buffer solution after shaking with various polymers for 1 hour at 22 °C (original concentration 100 μmol ℓ$^{-1}$)***

Material	Chlorpromazine concentration (μmol ℓ$^{-1}$)
Silicone tubing	1 ± 0.4
Latex tubing	16 ± 4.7
PVC tubing	14 ± 1.0
Polyethylene tubing	60 ± 2.0
Polyethylene test-tube	70 ± 3.8
Polyethylene stopper	77 ± 5.3
Plexiglass chippings	74 ± 4.3
Teflon chippings	81 ± 4.4
Polystyrene test-tube	89 ± 1.2

*From G. Krieglstein *et al. Arzneim. Forsch.*, 22, 1538 (1972)

areas. Polyethylene adsorbs steroids from solutions passing through polyethylene tubing. In analytical techniques such interactions can be important. Glyceryl trinitrate, which has a high affinity for lipophilic plastics, migrates from tablets in contact with plastic liners in packages until the active content of many tablets can be reduced to zero. This peculiar migratory behaviour is due to the volatility of the drug; normally the drug molecules would only be able to be significantly affected by such transfer when in the solution state. This topic is dealt with also in section 10.8.

8.4.2   Ion-exchange resins

Synthetic organic polymers comprising a hydrocarbon cross-linked network to which ionisable groups are attached have the ability to exchange ions attracted to their ionised groups with ions of the same charge present in solution (figure 8.19). These substances, usually prepared in the form of beads, are ion-exchange resins and are insoluble in water, the aqueous phase diffusing into the porous resin beads. Because ions must diffuse into and out of the resin for exchange to occur, ions larger than a given size may be excluded from reaction by altering the nature of the cross-links in the polymer. The resins may be either cation exchangers in which the resin ionisable group is acidic, for example, sulphonic, carboxylic (XIV) or phenolic groups, or anion exchangers in which the ionisable group is basic — amine or quaternary ammonium groups. The equations describing the equilibria involved are

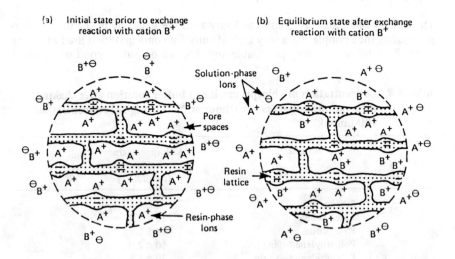

figure 8.19   Schematic diagram of a cation-exchange resin framework with fixed exchange sites prior to and following an exchange reaction. (a) Initial state prior to exchange reaction with cation B^+. (b) Equilibrium state after exchange reaction with cation B^+. From W. J. Weber. *Physicochemical Processes for Water Quality Control*, Wiley, New York, 1972

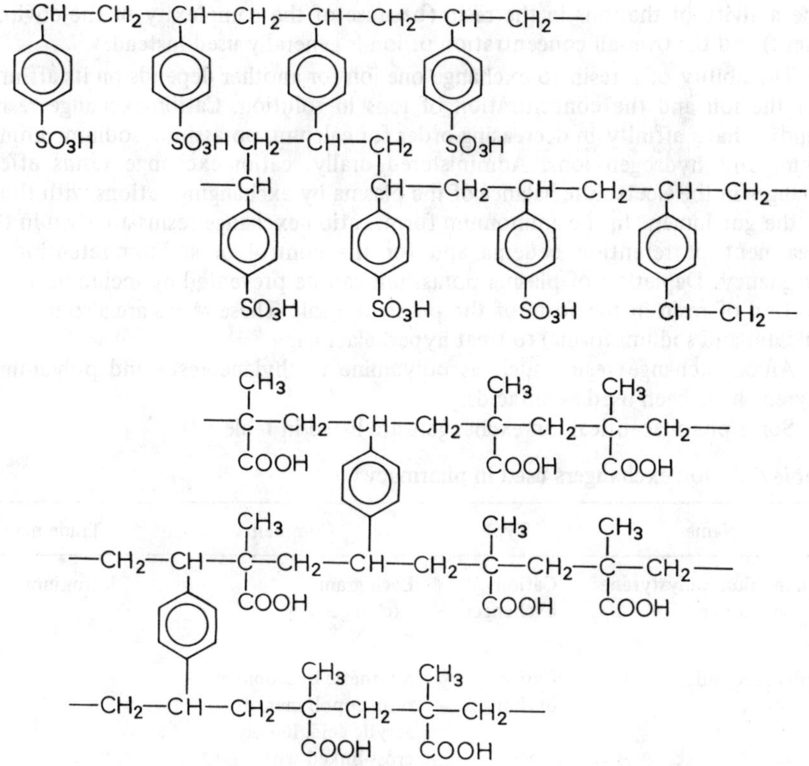

Sulphonic acid and carboxylic acid ion-exchange resins

(XIV)

Cation-exchange
resin:

$$POL–(SO_3^-)\, A^+ + B^+ \rightleftharpoons POL–(SO_3^-)\, B^+ + A^+$$

Anion-exchange
resin:

$$POL–N(CH_3)_3^+\, X^- + Y^- \rightleftharpoons POL–N(CH_3)_3^+\, Y^- + X^-$$

The equilibrium constant for the cation exchange resin is

$$K_{cation} = \frac{[POL–B^+]\,[A^+]}{[POL–A^+]\,[B^+]} \tag{8.29}$$

However, application of equation 8.29 is impossible because of the inaccessibility of the terms $[POL–B^+]$ and $[POL–A^+]$. Some estimation of a resin's affinity for ions can be made using a standard ion such as lithium for cation-exchange resins. A selectivity coefficient, k, may be defined as

$$k = \frac{[B^+]_{resin}\,[A^+]_{solution}}{[A^+]_{resin}\,[B^+]_{solution}} \tag{8.30}$$

Even here there is a problem arising from the difficulty in the determination of the activity of the ions in the resin (because of the complexity of the environment) and the over-all concentration of ion is generally used instead.

The ability of a resin to exchange one ion for another depends on its affinity for the ion and the concentration of ions in solution. Cation-exchange resins tend to have affinity in decreasing order for calcium, potassium, sodium, ammonium and hydrogen ions. Administered orally, cation-exchange resins effect changes in the electrolyte balance of the plasma by exchanging cations with those in the gut lumen. In the ammonium form, cation-exchange resins are used in the treatment of retention oedema and for the control of sodium retention in pregnancy. Depletion of plasma potassium can be prevented by including a proportion of resin in the form of the potassium salt. These resins are also used (as calcium and sodium forms) to treat hyperkalaemia.

Anion-exchange resins such as polyamine methylene resin and polyaminostyrene have been used as antacids.

Some pharmaceutical ion exchangers are listed in table 8.9.

table 8.9 **Ion exchangers used in pharmacy***

Name	Type	Comments	Trade name
Ammonium polystyrene sulphonate	Cation exchanger	Each gram exchanges 2.5 mEq Na^+	Katonium
Polycarbophil	Cation exchanger	Synthetic hydrophilic resin copolymer of acrylic acid, loosely cross-linked with divinyl glycol. Marked water-binding capacity	–
Calcium polystyrene sulphonate	Cation exchanger	Each gram exchanges about 1.3 mEq K^+	Calcium Resonium
Cholestyramine	Anion exchanger	The chloride of a strongly basic anion-exchange resin containing quaternary ammonium groups attached to styrene–divinyl benzene copolymer	
Polyamine–methylene resin	Anion exchanger	Effects a temporary binding of HCl + pepsin in the stomach, later released in the intestine	Resinat (USA) Exorbin (USA)
Sodium polystyrene sulphonate	Cation exchanger	Each gram exchanges 2.8–3.5 mEq K^+	Resonium A
Polyaminostyrene	Weak anion exchanger		

*Details from Martindale, *The Extra Pharmacopoeia*, 26th Edition

Apart from these medical uses, ion-exchange resins are used in the removal of ionised impurities from water and in the prolongation of drug action. Purified water may be obtained by passing water through two columns containing a strong cation exchanger and a strong anion exchanger, respectively, or a column containing mixed resins. Anionic impurities in the water are replaced by OH^- from the anion exchanger and cations by H^+ from the cation-exchange resin. Dissolved salts are thus removed and replaced by H_2O molecules, but of course non-ionic impurities and colloidal material are not removed. Regeneration of the resins is accomplished using NaOH and HCl for the anion and cation exchangers, respectively.

In the prolongation of drug action, ion-exchange resins are employed to form chemical complexes with drug substances, especially basic drugs such as ephedrine, pholcodine (XV) and phenyltoloxamine (XVI) (an isomer of diphenhydra-

pholcodine

(XV)

phenyltoloxamine

(XVI)

mine). Pholtex is a sustained-action liquid utilising a sulphonic acid ion-exchange resin with pholcodine and phenyltoloxamine as resin complexes. The rate of release of basic drugs from cation-exchange resins depends on the diameter of the resin beads, on the degree of cross-linking within the resin and on the pK_a of the ionisable resin group. The resin–drug complex may be tabletted and administered orally; resin complexes have been used to mask the taste of bitter drugs and to reduce the nausea produced by some irritant drugs. Some types, such as polacrilin potassium (Amberlite IRP 88, sulphonated polystyrene resin) are used as tablet disintegrants because of the degree of swelling the drug resins undergo on interaction with water.

8.5 Some applications of polymeric systems in drug delivery

Control of the rate of release of a drug when administered by oral or parenteral routes is aided by the use of polymers which function as a barrier to drug movement.

Film coating

Polymer solutions allowed to evaporate produce polymeric films which can act as protective layers for tablets or granules containing sensitive drug substances

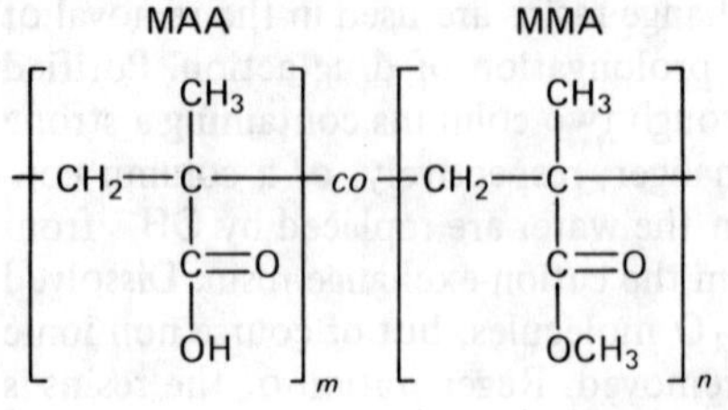

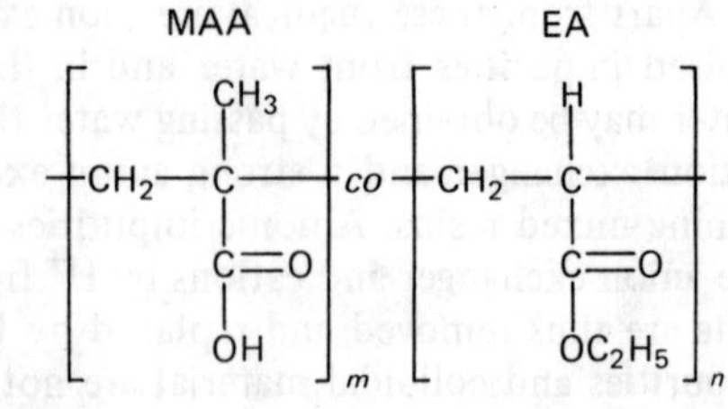

MAA   MMA   MAA   EA

$$\left[\begin{matrix} CH_2-\underset{\underset{\underset{OH}{|}}{\underset{C=O}{|}}}{\overset{\overset{CH_3}{|}}{C}} \end{matrix}\right]_m co \left[\begin{matrix} CH_2-\underset{\underset{\underset{OCH_3}{|}}{\underset{C=O}{|}}}{\overset{\overset{CH_3}{|}}{C}} \end{matrix}\right]_n$$

(XVII)

$$\left[\begin{matrix} CH_2-\underset{\underset{\underset{OH}{|}}{\underset{C=O}{|}}}{\overset{\overset{CH_3}{|}}{C}} \end{matrix}\right]_m co \left[\begin{matrix} CH_2-\underset{\underset{\underset{OC_2H_5}{|}}{\underset{C=O}{|}}}{\overset{\overset{H}{|}}{C}} \end{matrix}\right]_n$$

(XVIII)

Chemical structure of various preparations of
Eudragit L range (anionic)

XVII   Poly(methacrylic acid, methyl methacrylate) 1:1
      (Eudragit L 12.5, L 100)

XVIII   Poly(ethyl acrylate, methacrylic acid) 1:1
      (Eudragit L 30 D, L 100-55)

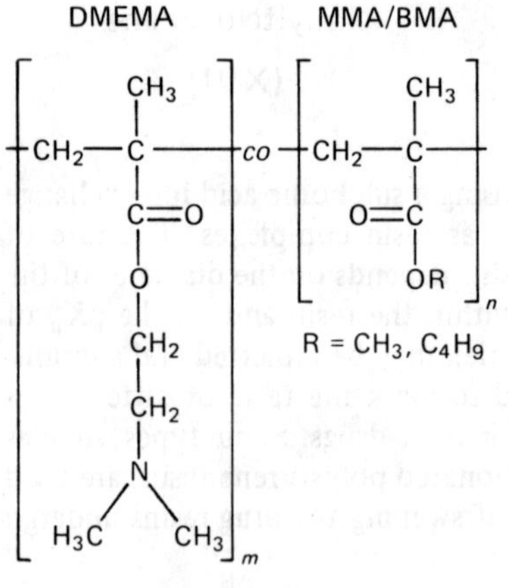

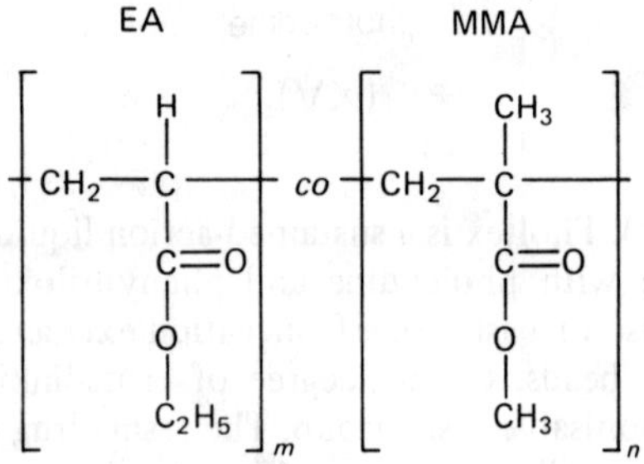

DMEMA   MMA/BMA   EA   MMA

R = CH$_3$, C$_4$H$_9$

(XIX)   (XX)

Chemical structure of various preparations of
Eudragit E range (cationic)

XIX Poly(butyl methacrylate, 2-dimethylaminoethyl methacrylate,
    methyl methacrylate) 1:2:1
    (Eudragit E 100, E 12.5)

XX Poly(ethyl acrylate, methyl methacrylate) 2:1
    (Eudragit E 30 D)

or as a rate-controlling barrier to drug release. Materials that have been used as film formers include shellac, zein, cellulose acetate phthalate, glyceryl stearates, paraffins, and a range of anionic and cationic polymers, such as the Eudragit polymers (XVII to XX). Shellac has traditionally been used as an enteric coating material, as it has a pH-dependent dissolution mechanism. Newer materials used for the same purpose include cellulose acetate phthalate.

Different film coats can cause quite different rates of solution when applied to tablet surfaces. In figure 8.20 hydroxypropylmethylcellulose, an acrylic derivative and zein are compared for their effect on sodium chloride dissolution from discs. Two plasticisers have been used: glycerin and diethyl phthalate. Times for 50 per cent dissolution range from a few minutes to 450 minutes, indicating the scope of the technique for retard formulations, and the possibility of unwittingly extending dissolution times by the injudicious choice of coating material. Film coats have been divided into two types: those that dissolve rapidly and those that behave as a dialysis membrane allowing slow diffusion of solute or some delayed diffusion by acting as gel layer.

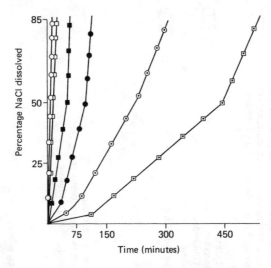

figure 8.20 Dissolution of sodium chloride from tablets coated with (o–o) hydroxypropylmethylcellulose, a vinyl polymer (•–•) and zein (⊙–⊙) with glycerin as an additive, and the same polymers with diethyl phthalate as additive (□–□, ■–■ and ▣–▣, respectively). From O. Laguna *et al. Ann. pharm. franc.*, 33, 235 (1975)

Matrices

The use of a barrier film coating is only one of several procedures that can be adopted to control release of drugs. Ritschel[21] has listed the various methods as shown in table 8.10. If hydrophobic water-insoluble polymers are used the mechanism of release is the passage of drug through pores in the plastic, or by

table 8.10 **Depot forms employing polymeric films and matrices†**

Type	Materials*	Diagrammatic representation	Mechanisms
1 Barrier coating	Beeswax, glyceryl monostearate, ethylcellulose, nylon (Ultramid IC), acrylic resins (Eudragit retard)	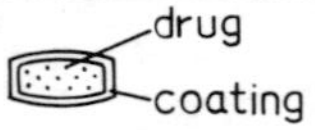 drug / coating	Diffusion
2 Fat embedment	Glycerol palmitostearate (Precirol), beeswax, glycowax, castorwax, aluminium monostearate, carnauba wax, glyceryl monostearate, stearyl alcohol	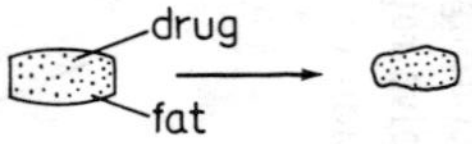 drug / fat	Erosion, hydrolysis of fat, dissolution
3 Plastic matrix	Polyethylene, Poly(vinyl acetate), Polymethacrylate, Poly(vinyl chloride), Ethylcellulose	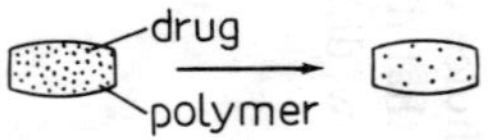 drug / polymer	Leaching, diffusion
4 Repeat action	Cellulose acetylphthalate	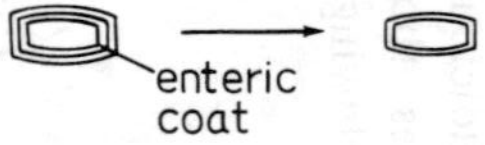 enteric coat	Dissolution of enteric coat
5 Ion exchange	Amberlite, Dowex		Dissociation of drug–resin complex
6 Hydrophilic matrix	Carboxymethylcellulose, Sodium carboxymethylcellulose, Hydroxypropylmethylcellulose	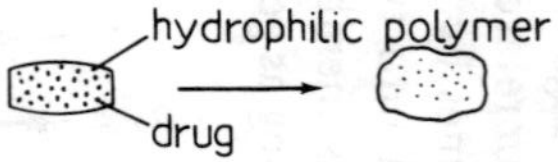 hydrophilic polymer / drug	Gelation, diffusion
7 Epoxy resin beads	Epoxy resins	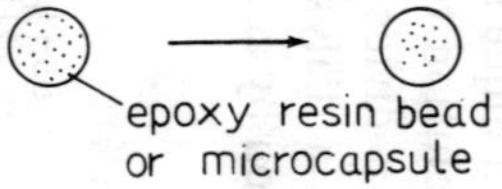 epoxy resin bead or microcapsule	Dissolution of resin or swelling, diffusion
8 Microcapsules	Polyamides, gelatin		
9 Soft gelatin depot capsules	Shellac–PEG, Poly(vinyl acetate)–PEG		Diffusion

*Materials are not all polymeric. The waxes are included for completeness; these depend on conferring a hydrophobic layer on the drug, tablet, or granule to prevent access of solvent

† After Ritschel, reference 21

leaching or slow diffusion of drug through the polymer wall, as discussed earlier in this chapter. When water-soluble polymers are employed, for example, as hydrophilic matrices, the entry of water into the polymer is followed by swelling and gelation and drug must diffuse through the viscous gel, a process obviously slower than diffusion through plain solvent.

8.5.1 Release of drugs from matrices

Equations describing the rate of drug release from hydrophobic and hydrophilic matrices are useful in determining which factors may be altered to change the measured release rate of drug. Higuchi[22] proposed the following equation for the amount of drug, Q, released per unit area of tablet surface in time t, from an insoluble matrix

$$Q = \left\{ \frac{D\epsilon}{\tau} \left(2A - \epsilon C_s \right) C_s t \right\}^{1/2} \tag{8.31}$$

D being the diffusion coefficient of the drug in the release medium, C_s the solubility of drug in the medium, ϵ the porosity of the matrix, τ the tortuosity of the matrix, and A the total amount of drug in the matrix per unit volume.

If the same matrix is saturated with a solution of the drug (as in medicated soft contact lenses, see chapter 9) the appropriate equation becomes, if C_o is the concentration of drug solution

$$Q = 2C_o \, \epsilon \left\{ \frac{Dt}{\tau\pi} \right\}^{1/2} \tag{8.32}$$

that is, for a given drug in a given matrix $Q \propto t^{1/2}$. The more porous the matrix the more rapid the release. The more tortuous the pores the longer the path for diffusing molecules, thus the lower is Q. More soluble drugs diffuse more quickly from the matrix.

The extension of these equations to hydrophilic matrices is difficult because the conditions in a hydrophilic matrix change with time as water penetrates into it. If the polymer does not dissolve but simply swells and if the drug has not completely dissolved in the incoming solvent, diffusion of drug commences from a saturated solution through the gel layer.

$$Q = \frac{D\epsilon}{\tau} \left\{ \left(\frac{2W_o}{V} - \epsilon C_s \right) t \, C_s \right\}^{1/2} \tag{8.33}$$

Equation 8.33 is similar to equation 8.31 except that the effective volume V of the hydrated matrix is used as this is not a fixed quantity. W_o is the dose of the drug in the matrix. If the drug completely dissolves on hydration of the matrix an analogue of equation 8.32 is used, namely

$$Q = \frac{2W_o}{V} \left\{ \frac{Dt}{\tau\pi} \right\}^{1/2} \tag{8.34}$$

In the initial stages of the process the rate of movement of water into the matrix may be important in determining release characteristics. When a homogeneous barrier wall is present diffusion through the walls has to take place and equations in section 8.4 apply.

8.5.2 Microencapsulation

This is a technique which, as its name suggests, involves the encapsulation of small particles of drug, or solution of drug, in a polymer coat. Microcapsules can be prepared by three main processes:

(A) *Coacervation* of macromolecules around the core material, this being induced by temperature change, solvent change or addition of a second macromolecule of appropriate physical properties.

(B) *Interfacial polymerisation of a monomer* around the core material by polymerisation at the interface of a liquid dispersion.

(C) *Spray coating and other methods* in which larger particles may be coated in suspension.

Any method which will cause a barrier to deposit itself on the surface of a liquid droplet or a solid particle of drug may be applied to the formation of micro-capsules. Many so-called microencapsulation procedures result in the formation of macroscopic 'beads' which are simply coated granules.

(A) Coacervation

Coacervation was the term used by Bungenberg de Jong and Kruyt in 1929 to describe the separation of macromolecular solutions into colloid-poor and colloid-rich (coacervate layers) when the macromolecules are desolvated. The liquid or solid to be encapsulated is dispersed in a solution of a macromolecule (such as gelatin, gum arabic, carboxymethylcellulose or poly(vinyl alcohol)) in which it is immiscible. A non-solvent, miscible with the continuous phase but a poor solvent for the polymer, under certain conditions will induce the polymer to form a coacervate (polymer-rich) layer around the disperse phase. This coating layer may then be treated to give a rigid coat or capsule wall. This is the process of *simple coacervation*. Successful application of the technique relies on the determination of the appropriate conditions for coacervate deposition, which can also be achieved not only by the addition of non-solvents such as ethanol and isopropanol and salts (sodium and ammonium sulphates) but also by the choice of macromolecules incompatible under selected conditions with the first macromolecular species. The latter technique involving two macromolecular species is termed *complex coacervation*. In both simple and complex coacerva-tion utilising hydrophilic macromolecules it is the decrease in solubility which results in deposition of the macromolecule layer at the particle–solution inter-face.

Desolvation of water-insoluble macromolecules in non-aqueous solvents would lead to the deposition of a coacervate layer around aqueous or solid disperse droplets. Table 8.11 lists both water-soluble and water-insoluble macromolecules which have been used in coacervation processes. Desolvation, and thus coacerva-tion, can be induced thermally and this is the basis of some preparative techni-ques. Conditions for phase separation are best obtained using phase diagrams.

table 8.11 **Materials used in coacervation microencapsulation**

Water-soluble macromolecules	Water-insoluble macromolecules
Arabinogalactan	Cellulose acetate phthalate
Carboxymethylcellulose	Cellulose nitrate
Gelatin	Ethylcellulose
Gum arabic (acacia)	Poly(ethylene vinyl acetate)
Hydroxyethylcellulose	Poly(methyl methacrylate)
Poly(acrylic acid)	
Poly(ethyleneimine)	
Poly(vinyl alcohol)	
Poly(vinylpyrrolidone)	
Methylcellulose	
Starch	

In the region of pH 6-8 gelatin will be positive and gum arabic negative. In admixture complex coacervates will form under the conditions described in figure 8.21 which shows the partial phase diagram of the ternary system gum arabic–gelatin–water. The hatched area at the top of the phase diagram is the restricted region in which coacervation occurs. At higher concentrations of the macromolecules, macroscopic precipitates of the complex will occur.

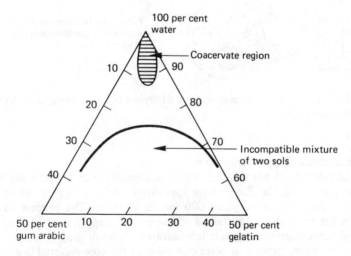

figure 8.21 Ternary diagram showing complex coacervation region for mixtures of gum arabic and gelatin at pH 4.5. Below the line the mixture separates into two sols. From US Pat. 2800457

(B) Interfacial reactions

Reactions between oil-soluble monomers and water-soluble monomers at the oil-water interface of w/o or o/w dispersions can lead to *interfacial polymerisation* resulting in the formation of polymeric microcapsules, the size of which is determined by the size of the emulsion droplets. Alternatively, reactive monomer can be dispersed in one of the phases and induced to polymerise at the interface, or to polymerise in the bulk disperse phase and to *precipitate* at the interface due to its insolubility in the continuous phase. There are many variations on this theme. Probably the most widely studied reaction has been the interfacial condensation of water-soluble alkyldiamines with oil-soluble acid dichlorides to form polyamides. Representative examples of other wall materials are polyurethanes, polysulphonamides, polyphthalamides and poly(phenyl esters). The selection of polymer is restricted to those that can be formed from monomers with the requisite preferential solubilities in one phase so that polymerisation takes place only at the interface. The process for the preparation of nylon 6,10 microcapsules is represented diagrammatically in figure 8.22.

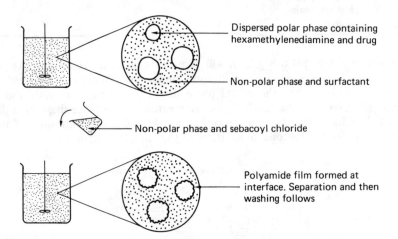

figure 8.22 Process for the preparation of nylon 6,10 microcapsules by interfacial polymerisation

(C) Physical methods of encapsulation

Various physical methods of preparing microcapsules such as spray drying and pan coating are available. *Pan coating* application of films to particles can only be used for particles greater than 600 µm in diameter. The process has been applied in the formation of sustained release beads by application of waxes such as glyceryl monostearate in organic solution to granules of drug.

The *spray drying* process involves dispersion of the core material in a solution of coating substance and spraying the mixture into an environment which causes the solvent to evaporate. In an analogous *spray congealing* process the coating material is congealed thermally or by introducing the core coat mixture into a non-solvent mixture. Both processes can produce microcapsules in the size range of 5000–6000 µm.

Spray polycondensation is a variant based on the polycondensation of surface-active monomers on a melamine–formaldehyde base on the surface of suspended particles during spray drying. A dispersion of the core material is prepared in a continuous phase containing aminoplast monomers or precondensates of relatively low molecular weight, in addition to other film-forming agents and catalyst. The reactive monomers derived from hexamethylol melamine derivatives are selectively adsorbed at the surface of the disperse phase. Spray drying at 150–200°C results in vaporising of the water and simultaneous polycondensation of the monomers and precondensates by acid catalysis.

Nanoencapsulation

Polymer microspheres with diameters of 50–500 nm have been described by Speiser[23] and termed nanoparts or nanocapsules. Speiser differentiated between solid polymeric spheres (nanoparts) and those spheres with thin polymeric walls (nanocapsules). The locus of polymerisation is not an emulsion droplet as in microencapsulation, but a micelle. The process involves the solubilisation of a water-soluble monomer such as acrylamide along with the drug or other agent such as antigen to be encapsulated. An organic liquid such as n-hexane serves as the outer phase. Polymerisation is induced by irradiation (γ-rays, X-rays, UV light), exposure to visible light or heating with an initiator.

The surfactants used to prepare the dispersions are not innocuous materials and can alter the properties of biological macromolecules. Thus in the encapsulation of antigens and other sensitive materials alternative techniques may have to be used. Influenza virus particles have thus been partly coated by poly(methyl methacrylate) by irradiation of monomeric methyl methacrylate dissolved in an aqueous virus suspension. Adsorption of substantial amounts of monomer onto the virus surface is thought to occur. Subsequent gamma-irradiation from a cobalt-60 source (500 krad) polymerised the monomer, and particles of 100–200 nm were produced.

Appropriate modification or control of the coacervation process has been shown to produce nanoparticles of gelatin. Gelatin nanoparticles have been prepared by desolvation (e.g. with sodium sulphate) of a gelatin solution containing drug bound to the gelatin, in a process which terminates the desolvation just before coacervation begins. In this manner colloidal particles rather than the larger coacervate droplets are obtained. The general method has been applied to yield nanoparticles of human and bovine serum albumin, ethylcellulose and casein. Salt and other low molecular weight species can be removed using Sephadex G80.

Hardening of the gelatin nanoparticles is achieved by glutaraldehyde which cross-links with gelatin, and is more efficient than formaldehyde. To achieve a high percentage incorporation of drug in the nanoparticles, the drug substance must have an affinity for the macromolecule used; thus water-soluble drugs are unlikely to be trapped in the embryo particles. Highly water-soluble drugs would be better incorporated in a non-aqueous system.

General considerations

In all of the techniques discussed there are several factors which are of importance in relation to the use of the product:

(1) the efficiency of encapsulation of the active ingredient
(2) the effect of the encapsulation process on the properties of the encapsulated agent
(3) the presence of potentially toxic residue (e.g. monomer, salts) in the final product

(4) the reproducibility of the process and ease of separation
(5) the biocompatibility of the encapsulating agent
(6) the biodegradability of the material in some cases
(7) the properties of the microcapsule in relation to use, size distribution, porosity and permeability of the wall.

The different processes and materials produce microcapsules in which optimum properties are not always obtained. Polyamide microcapsules are probably not biodegradable, hence the search for alternatives for forming microcapsules based on natural materials such as albumin. Biodegradable polymers cannot always be induced to form microcapsules and there is no guarantee that a polymer found to be degradable in solid or film form would be degradable when formed into microcapsules if cross-linking of the polymer chains occurs during formation. In their application in medicine the permeability of the capsule wall is probably the most important feature of the product. The effect of the various parameters relating to wall material, capsule and environment on permeability are outlined in table 8.12.

table 8.12 **General parameters affecting capsule wall permeability***

Parameter	For lower permeability
Properties of wall polymer	
Density	Increase
Crystallinity	Increase
Cross-linking	Increase
Plasticiser content	Decrease
Fillers	Increase
Solvents used in film formation	Use good solvents versus poor
Properties of capsule	
Size	Increase
Wall thickness	Increase
Treatment	Utilise (e.g. cross-linking, sintering)
Multiple coatings	Utilise
Environmental properties	
Temperature	Decrease

*From reference 24

Protein microspheres

Aqueous solutions of proteins such as albumin can be emulsified in an oil and induced to form microspheres either by cross-linking the protein molecules with glutaraldehyde or other agents or by coagulating the protein by heating. Incorporation of a drug within the initial protein solution results in drug-laden microspheres, which are biodegradable. The particle size of the microspheres (generally 0.2–300 μm diameter) is determined by the size distribution of the initial emulsion. Protein microspheres have been used for physical drug targeting, i.e. the entrapment of carrier and therefore drug in, for example, the capillaries of the lung. Microspheres greater than about 7 μm diameter will be physically trapped in capillary beds. On intra-arterial injection of large microspheres, the blood supply to an organ is reduced; the process of chemo-embolisation involves both blockage and delivery of drugs to the organ. External control over intravenously administered protein microspheres has been achieved by incorporation of magnetite (Fe_3O_4) particles into the microspheres, which then respond to an externally applied magnetic field.

8.5.3   Rate-limiting membranes and devices

The use of rate-limiting membranes to control the movement of drugs from a reservoir has been referred to above. Implants of silicone rubber or other appropriate polymeric material in which drug is embedded can be designed by choice of polymer, membrane thickness and porosity, to release drug at preselected rates. The Progestasert device (figure 8.23c) is designed for implantation into the uterine cavity and to release there 65 μg progesterone per day to provide contraceptive cover for one year. The Ocusert device and the Transiderm therapeutic system (figure 8.23a, b) which are discussed in chapter 9, are products of the Alza Corporation of the United States and rely on rate-limiting polymeric mem-

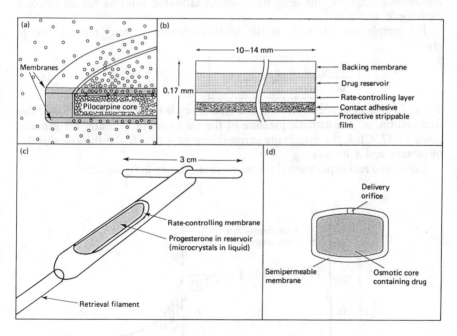

figure 8.23   Examples of drug delivery systems employing polymeric membrane. (a) Ocusert system for the eye with two rate-controlling membranes. (b) Transiderm system for medication with one rate-controlling layer. (c) The Progestasert device for intrauterine insertion in which the body of the device serves as the rate-controlling barrier. (d) The oral Oros device in which the membrane is a semipermeable membrane which forbids drug transport, allowing water ingress only

branes to control drug release. The opportunities for controlled release of drugs given by the oral route are less. One aims in oral dose forms for controlled rather than prolonged release, so that dosage frequency can be reduced or that side-effects resulting from fast dissolution of drug can be minimised. Ion-exchange resin–drug complexes have been used. Drugs may be embedded in hydrophilic or hydrophobic matrices. Potassium chloride has been formulated in this way, as this is a readily soluble but corrosive drug substance.

More precise control of release than is possible with matrices has recently been achieved by the application of several features of polymer physical chemistry.

Osmotic pump for oral administration

In the oral osmotic pump (Oros, figure 8.23*d*) the drug is mixed with a water-soluble core material. This core is surrounded by a water-insoluble semipermeable polymer membrane in which is drilled a small orifice. In water, water molecules can diffuse into the core through the outer membrane to form a concentrated solution inside. An osmotic gradient is set up across the semipermeable membrane with the result that drug is pushed out of the orifice.

The core may be a water-soluble polymer, an inert salt or, as in the case of metoprolol fumarate, the drug itself whose saturated solution has an osmotic pressure of 32.5 atm[25].

For simple osmotic systems the initial zero-order delivery rate (dm/dt) is given by

$$\frac{dm}{dt} = \frac{A}{H} k \left(\pi_f - \pi_e\right) S_d \qquad (8.35)$$

where S_d is drug concentration in the system, π_f is the osmotic pressure of the formulation, π_e the osmotic pressure of the environment (7.7 atm for isotonic saline at 37 °C), k the membrane permeability to water, H the thickness of the membrane and A its area.

Calculated and experimental release rates are shown in figure 8.24.

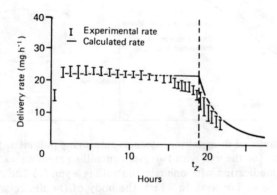

figure 8.24 *In vitro* release rate of potassium chloride from elementary osmotic pumps in water at 37 °C. From F. Theeuwes. *J. pharm. Sci.*, 64, 1987 (1975)

One problem is that of controlling the transit of the device down the gastro-intestinal tact, as individual subjects vary considerably in GI transit times. If the system is designed to release drug over a period of 10 h and total transit time in the gut is 5 h, then bioavailability will be reduced.

Transdermal delivery systems

Several transdermal systems dependent ostensibly on rate-controlling membranes are available for the delivery of nitroglycerin and other drugs. The word 'ostensibly' is used as there is current debate about whether the barrier membrane in

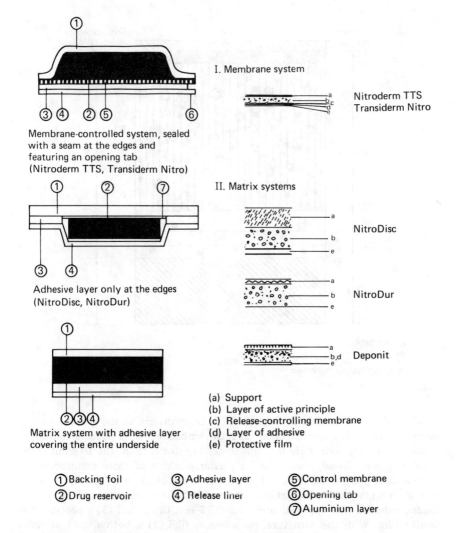

figure 8.25 The structure of some commercial transdermal membrane-controlled or matrix systems. (The Deponit TTS is shown in figure 8.26)

these devices is the rate-limiting step in absorption. However, the barrier properties of skin are so variable that one advantage of rate-controlling systems is that they prevent overdosing in patients with highly permeable skin. In those with less-permeable skin, the systems probably act only as reservoirs.

Some of the devices and the basis of their design are shown in figure 8.25. (The Transiderm system is shown in figure 8.23*b*.) There are two groups, membrane and matrix systems. Membrane systems generally consist of a reservoir, a rate-controlling membrane and an adhesive layer. Diffusion of the active principle through the controlling membrane governs release rate. This is usually present in suspended form; liquids and gels are used as dispersion media. In matrix systems

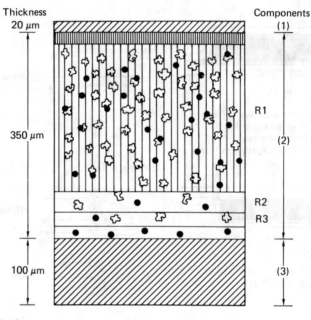

♋ Adsorbent
♋ Adsorbent plus nitroglycerin
● Dissolved nitroglycerin

figure 8.26 Cross section of Deponit TTS. Deponit, unlike the systems already described, contains nitroglycerin directly embedded in a lactose-containing adhesive film approximately 0.3 mm thick. In cross section, the Deponit transdermal system, closely resembling a plaster, consists of three components in macroscopic proportions, the ratios of which are as given in the figure: (1) a nitroglycerin-impermeable flexible carrier film; (2) a congruent nitroglycerin-loaded adhesive film approximately 0.3–0.35 mm thick; and (3) a peelable protective film. With this structure, the adhesive film (2) is both a store of active ingredient and the release-controlling matrix. From reference 26 the active principle is dispersed in a matrix which consists either of a gel or of an adhesive film.

In Transiderm Nitro, the rate-controlling membrane is composed of a poly-ethylene/vinyl acetate copolymer having a thin adhesive layer (membrane type). The reservoir contains nitroglycerin dispersed in the form of a lactose suspension in silicone oil. The NitroDur system consists of a hydrogel matrix, in which a nitroglycerin/lactose trituration is homogeneously dispersed. The hydrogel is composed of water, glycerin, poly(vinyl alcohol) and polyvinylpyrrolidone.

In NitroDisc, nitroglycerin is distributed between microscopically small liquid compartments and a cross-linked silicone matrix. In the approximately 10–200 μm sized microcompartments there is active ingredient also in the form of a lactose trituration, in an aqueous solution of PEG 400. The system is secured to the skin with the aid of a circular adhesive disc having a centrally located silicone matrix, not coated with adhesive. The system is termed a 'microsealed drug delivery system'. The Deponit system[26] is shown in detail in figure 8.26.

References

1. F. W. Billmeyer. *Textbook of Polymer Science*, 2nd Edition, Wiley, New York, 1971
2. J. G. Pritchard. *Poly(vinyl alcohol): Basic Properties and Uses*, Macdonald, London, 1970
3. A. G. Mattha. *Pharm. Acta helv.*, 52, 233 (1977)
4. R. L. Whistler. *Pure appl. Chem.*, 49, 1229 (1977)
5. W. P. T. James, *et al. Lancet*, i, 638 (1978)
6. R. H. Blythe, *et al. J. Am. pharm. Ass.*, 38, 59 (1949)
7. J. D. Ireson and G. B. Leslie. *Pharm. J.*, 205, 540 (1970)
8. R. Wasicky. *Planta Med.*, 9, 232 (1967)
9. C. Petty and N. L. Cunningham. *Anesthesiology*, 40, 400 (1974)
10. S. Weidenfeld, *et al. Diabetes*, 17, 766 (1968)
11. R. L. Whistler (ed.). *Industrial Gums*, 2nd Edition, Academic Press, New York, 1973
12. F. E. Windover. In *Water Soluble Resins* (ed. R. L. Davidson and M. Sittig), Reinhold, New York, 1962, p. 52 *et seq.*
13. B. Jirgensons. *Natural Organic Molecules*, Pergamon, Oxford, 1962
14. M. Wales, *et al. J. polymer Sci.*, 10, 279 (1953)
15. J. L. Azorlosa and A. J. Martinelli. In *Water Soluble Resins* (ed. R. L. Davidson and M. Sittig), Reinhold, New York, 1962, p. 110 *et seq.*
16. T. D. Turner. *Pharm. J.*, 228, 206 (1982)
17. J. A. Myers. *Pharm. J.*, 235, 270 (1985)
18. S. Aiba, N. Minoura, *et al. Biomaterials*, 6, 290 (1986)
19. R. C. Heal, *et al. Drugs*, 18, 89 (1979)
20. B. J. Munden, H. G. De Kay and G. S. Banker. *J. pharm. Sci.*, 53, 395 (1964)
21. W. A. Ritschel. *Drug Design*, vol. IV (ed. A. J. Ariens), Academic Press, New York, 1974

22. T. Higuchi. *J. pharm. Sci.*, 52, 1145 (1963)
23. P. Speiser. *Prog. colloid polymer Sci.*, 59, 48 (1976)
24. J. E. Vandegaer (ed.). *Microencapsulation, Processes and Applications*, Plenum Press, New York, 1974
25. F. Theeuwes, D. R. Swanson, *et al. Br. J. clin. Pharmacol.*, 19, 699 (1985)
26. M. Wolff, *et al. Pharm. Res.*, 2, 23 (1985)

9 Drug Absorption and Routes of Administration

We will deal here with the process of drug absorption from different sites in the body. Drug absorption, whether it be from the gastro-intestinal tract or from the buccal mucosa, requires the passage of the drug in a molecular form across the barrier membrane. Most drugs are presented to the body as solid or semisolid dosage forms and these must first release their drug content. Drug must then dissolve and, if it has the appropriate physical properties, will pass from a region of high concentration to a region of low concentration across the membrane surrounding the site of absorption into the blood (figure 9.1).

The special features of the different routes of administration are dealt with in separate sections, after a brief summary of the general properties of biological membranes and drug transport, a knowledge of which is important in understanding absorption processes. It is impossible to be comprehensive in a short space. Instead, we have attempted to bring out the matters unique to the routes discussed, such as the properties of the vehicle in topical therapy and the aerodynamic properties of aerosols in inhalation therapy. Where attempts have been made to quantify absorption, equations are presented, but the derivations of the equations have been omitted. Reference can be made to the original literature for these.

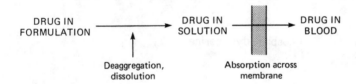

figure 9.1 Sequence of events in drug absorption from formulations

9.1 Biological membranes and drug transport

The main function of biological membranes is to contain the aqueous contents of cells and separate them from an aqueous exterior phase. To achieve this, membranes are lipoidal in nature. To allow nutrients to pass into the cell and waste products to move out, biological membranes are selectively permeable. Membranes thus have specialised transport systems to assist the passage of water-soluble materials and ions through their lipid interior. Lipid-soluble agents can pass by passive diffusion through the membrane from a region of high concentration to one of low concentration. Biological membranes differ from polymer membranes in that they are composed not of polymers, but of small amphipathic molecules, phospholipids (I) and cholesterol, which associate into lipoidal bilayers in aqueous media. Also associated with the lipid molecules are proteins which are generally hydrophobic in nature, embedded in the matrix of lipid molecules. Thus the membrane has a hydrophilic exterior and a hydrophobic

$$
\begin{array}{c}
\quad\quad\quad\quad\quad\quad\quad\quad O \\
\quad\quad\quad\quad\quad\quad\quad\quad \| \\
\quad\quad\quad\quad\quad CH_2-O-C-R_1 \\
\quad\quad\quad\quad\quad | \\
R_2-C-O-CH \quad\quad O \\
\quad\; \| \quad\quad\quad\quad | \quad\quad \| \\
\quad\; O \quad\quad\quad\quad CH_2-O-P-O-CH_2-CH_2-\overset{+}{N}-(CH_3)_3 \\
\quad\quad\quad\quad\quad\quad\quad\quad\; | \\
\quad\quad\quad\quad\quad\quad\quad\quad\; O
\end{array}
$$

Phosphatidyl choline (lecithin)

$$
\begin{array}{c}
\quad\quad\quad\quad\quad\quad\quad\quad O \\
\quad\quad\quad\quad\quad\quad\quad\quad \| \\
\quad\quad\quad\quad\quad CH_2-O-C-R_1 \\
\quad\quad\quad\quad\quad | \\
R_2-C-O-CH \quad\quad O \\
\quad\; \| \quad\quad\quad\quad | \quad\quad \| \\
\quad\; O \quad\quad\quad\quad CH_2-O-P-O-CH_2-CH_2-\overset{+}{N}H_3 \\
\quad\quad\quad\quad\quad\quad\quad\quad\; | \\
\quad\quad\quad\quad\quad\quad\quad\quad\; O
\end{array}
$$

Phosphatidyl ethanolamine

$$
\begin{array}{c}
\quad\quad\quad\quad\quad\quad\quad\quad O \\
\quad\quad\quad\quad\quad\quad\quad\quad \| \\
\quad\quad\quad\quad\quad CH_2-O-C-R_1 \\
\quad\quad\quad\quad\quad | \\
R_2-C-O-CH \quad\quad O \quad\quad\quad\quad \overset{+}{N}H_3 \\
\quad\; \| \quad\quad\quad\quad | \quad\quad \| \quad\quad\quad\quad\quad | \\
\quad\; O \quad\quad\quad\quad CH_2O-P-O-CH_2-CH-COOH \\
\quad\quad\quad\quad\quad\quad\quad\quad\; | \\
\quad\quad\quad\quad\quad\quad\quad\quad\; O
\end{array}
$$

Phosphatidyl serine

Phosphatidyl inositol

Structures of phospholipids found in membranes. R_1 and R_2 may vary in length; R_1 is usually saturated and R_2 usually unsaturated

(I)

interior. Cholesterol is a major component of most mammalian biological membranes; its removal causes the membrane to lose its structural integrity and to become highly permeable. Cholesterol complexes with phospholipids and its presence reduces the permeability of phospholipid membranes to water, cations, glycerol and glucose. The shape of the cholesterol molecule allows it to fit closely in bilayers with the hydrocarbon chains of unsaturated fatty acids. The present consensus of opinion is that cholesterol condenses and rigidifies membranes without solidifying them. The flexibility of biological membranes is an important feature; their ability to reform and to adapt to a changed environment is a vital part of their function. The dynamic characteristics of biological membranes are due to their unique construction from small molecules. On stretching (as with a soap film) the amphipathic molecules become less concentrated in the bilayer but are speedily replenished to maintain the original tension.

The details of membrane structure are still widely debated, and membrane 'models' which incorporate the results of latest investigations abound. Figure 9.2 shows the fluid mosaic model.

Although most of the data on permeation of non-electrolytes across biological membranes can be explained on the basis of the membrane behaving as a continuous hydrophobic phase, a fraction of the membrane may be composed of aqueous channels continuous across the membrane; that is, there are pores which offer a pathway parallel to the diffusion pathway in lipid. In the absence of bulk flow these pores play a minor part in the transfer of drugs, although in the case of ions the pore pathway must be important. The low electrical resistance of membranes compared with synthetic lipid membranes suggests that the ions move in the pores. Pores may be provided by the conjunction of hydrophilic faces of proteins or the polar heads of fatty acids and phospholipids orientated in the appropriate direction. The fluid mosaic model, in particular, allows the protein-lipid complexes to form either hydrophilic or hydrophobic 'gates' to allow transport of materials of different characteristics. It has been suggested that the permeability of the lipid bilayer is regulated by the density of hydrogen bonding in the outer polar layers of the membranes which contain phosphate, ammonium and carboxyl groups of phospholipids and hydroxyl groups of cholesterol. In layers of lipid without cholesterol the carboxyl groups bond to water, to cations or hydrogen bond donors such as glycerol or glucose. These solutes can thus pass through the membrane. Hydrogen bonding to cholesterol dehydrates and blocks the $C=O$ groups. When 50 per cent of them are involved the membrane 'closes'. The surplus of carboxyl groups in certain areas of the membrane mosaic could result in varying degrees of porosity of the lipid matrix. Thus membrane permeability is controlled by the nature of the membrane, its degree of internal

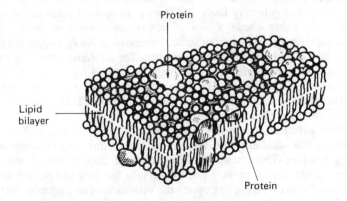

figure 9.2 Diagrammatic representation of the fluid mosaic model

bonding and rigidity, its surface charge and by the nature of the solute being transported.

There are some similarities between solute transport in biological membranes and in synthetic membranes. As we have discussed in chapter 8, the permeation of drugs and other molecules through hydrophobic membranes made of poly-dimethylsiloxane, for example, depends primarily on the solubility of the drug in the membrane. Drugs with little affinity towards the membrane are not likely to permeate through it. In porous membranes, for example of cellophane or collagen, drugs with little affinity for the polymer may be transported through the pores.

Most biological membranes bear a negative charge. Does this influence permeation? Molecular forms of solutes permeate faster than ionic forms through a membrane composed of collagen[1], which has been studied as a potential haemodialysis membrane or release controlling membrane for medication of the eye[2,3]. Of the ionic forms, the anionic solute permeates faster than the cationic form. With amphoteric drugs such as sulphasomidine and sulphamethizole, a similar order of permeation may be observed depending on the pH of the medium, namely: unionised molecule > anion > cation. Why? The basic groups in collagen which are mainly the basic amino acids lysine, arginine and histidine are positively charged in the conditions of the experiments as the pK_a of lysine and of arginine is about 10. In acidic media the membrane is positively charged and thus cationic drugs will be repelled from the surface. On the other hand, pH has little effect on the passage of drugs through cellophane membranes, a fact that can be rationalised by the lack of charge on the cellophane surface. In biological membranes, one might expect some preference for cationic drugs.

In chapter 5 we examined some relationships between the lipophilicity of drugs and their activity which was frequently controlled by their ability to pass across lipid membranes. Although of the same basic construction, membranes in different sites serve different functions and thus one might expect them to have different compositions and physicochemical properties, as indeed they do. Tissues derived from the ectoderm (the epidermis, the epithelium of nose and mouth and the anus and the tissues of the nervous system) have protective and sensory functions. Tissues evolved from the endoderm, such as the epithelium of the gastro-intestinal tract, are designed mainly for absorption.

If one studies the absorption of a sufficiently wide range of substances in homologous series one generally finds that there is an optimal point in the series for absorption. In other words, a plot of percentage absorption versus log P would be parabolic with the optimum value designated as log P_o (figure 9.3). By noting the values of optimal partition coefficient for different absorbing membranes and surfaces one can deduce something about their nature. Some values are given in table 9.1. It is because of these differences that we are considering separately absorption of drugs from various sites in the body. This subject is dealt with in greater detail by Lien in several papers[4-6].

The parabolic nature of activity–log P plots is due to a combination of factors: protein binding, low solubility and binding to extraneous sites of compounds with high log P values. This causes the more lipophilic drugs in some systems to have a lower activity than if it were possible to take the drug and place it at the receptor without the drug having to traverse the various lipid and aqueous hurdles that it usually finds on its way to the site of action. One reason for very low values of log P_o is, as discussed in chapter 10 (section 10.9), protein binding of

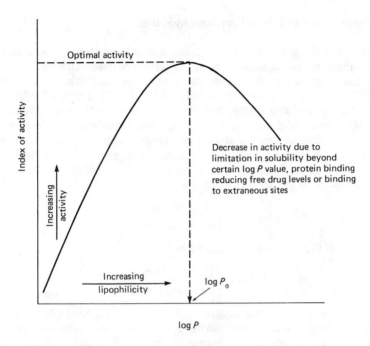

figure 9.3 Parabolic nature of many activity–log P plots. The decrease in activity beyond the optimal log P_0 probably is due to the factors listed, which limit the observed activity in intact animals

table 9.1 **Ideal lipophilic character (log P_0) in various transport systems**

System	Solute or drug	Log P_0 (octanol/water) for maximal transport
Buccal cavity (human)	Bases	5.52 (undissociated) 3.52 (dissociated)
	Acids	4.19 (undissociated)
Epidermis (human)	Steroids	3.34
Whole skin (rabbit)	Non-electrolytes	2.55
Small intestine (rat)	Sulphonamides	2.56–3.33
Stomach (rat)	Barbiturates	2.01
	Acids	1.97
Cornea (rabbit)	Steroids	2.8
Biliary excretion	Sulphathiazoles	0.60
Milk/plasma	Sulphonamides	0.53
Prostatic/plasma ratio	Sulphonamides	0.23

very lipophilic drugs which prevents entry into sites bounded by lipophilic membranes. Low values of log P_0 are especially evident for biliary excretion, diffusion into milk and diffusion into prostatic fluid (table 9.1).

9.1.1 Permeability and the pH-partition hypothesis

If, in the first instance, the plasma membrane is considered to be a strip of lipoidal substance, homogeneous in nature with a defined thickness, one must assume that only lipid-soluble agents will pass across this barrier. As most drugs are weak electrolytes it is to be expected that the unionised form of either acids or bases, the lipid-soluble species, will diffuse across while the ionised forms will be rejected. This is the basis of the pH-partition hypothesis in which the pH dependence of drug absorption and solute transport across membrane is treated. The equations of chapters 3 and 5 are relevant here, especially equations 3.65–3.68. For weakly acidic drugs such as aspirin and indomethacin the ratio of ionised to unionised species is given by the equation

$$pH - pK_a = \log \frac{[\text{ionised form}]}{[\text{unionised form}]} \tag{9.1}$$

For weak bases the equation takes the form

$$pK_a - pH = \log \frac{[\text{ionised form}]}{[\text{unionised form}]} \tag{9.2}$$

One can calculate from these equations (and equations 3.65–3.68) the relative amounts of absorbable and non-absorbable forms of a drug substance, given the prevailing pH conditions in the lumen of the gut. For several drugs such as aspirin, indomethacin, nalidixic acid, chlorpromazine and mecamylamine, one thus obtains the profiles in figure 9.4 for unionised drug (per cent) versus pH. In broad terms, therefore, one would expect acids to be absorbed from the stomach and bases from the intestine. The gradation of pK_a makes it difficult to state this categorically and there are several factors which cause the pH-partition hypothesis to be limited in its application.

A comparison of the intestinal absorption of several acids and bases at several pH values (table 9.2) indicates that the trend is that expected from the pH-partition hypothesis. However, it will be seen that salicylic acid is absorbed from the rat intestine at pH 8, although it is virtually completely ionised at this pH, as it has a pK_a of 3.0.

Experiments done in the rat stomach by instilling solutions of different pH directly into the lumen indicate the correct qualitative trends of absorption (61 per cent at pH 1 for salicylic acid and 13 per cent at pH 8). The data in table 9.2 taken at their face value suggest that the proportion of unionised to ionised molecules would be 1:5000 for acids and 1:15 for bases to allow rapid absorption, but this erroneously implies that the lipid boundary is more permeable to the unionised form of acids than it is to the unionised form of bases. Recalculation based on the 'virtual' pH gives the ratio of unionised drug to ionised species necessary for rapid absorption to be 1:300, the same for an acid of pK_a 2.8 and for a base of pK_a 7.8.

In table 9.2 absorption of acetylsalicylic acid at pH 4, 5, 7 and 8 is quoted. The amount of drug in the unionised form at these pH values is obtained from equation 9.1

$$pH - pK_a = \log \frac{[\text{ionised form}]}{[\text{unionised form}]}$$

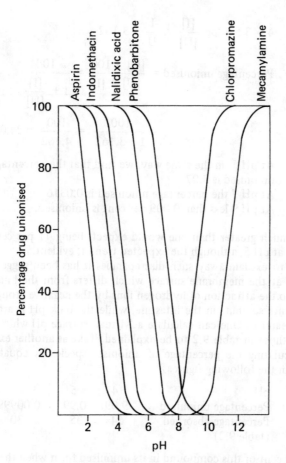

figure 9.4 Plot of percentage drug unionised (that is, in lipid-soluble form) as a function of pH, for aspirin, indomethacin, nalidixic acid, phenobarbitone, chlorpromazine and mecamylamine

table 9.2 Intestinal absorption of acids and bases in the rat at several pH values*

| | pK_a | Percentage absorption | | | |
		pH 4	pH 5	pH 7	pH 8
Acids					
Salicylic acid	3.0	64	35	30	10
Acetylsalicylic acid	3.5	41	27	–	–
Benzoic acid	4.2	62	36	35	5
Bases					
Amidopyrine	5.0	21	35	48	52
Quinine	8.4	9	11	41	54

*From B. B. Brodie. *Absorption and Distribution of Drugs,* (ed. T. Binns), Livingstone, Edinburgh, 1964

At pH 4

$$4 - 3.5 = \log \frac{[I]}{[U]} \quad \therefore \quad \frac{I}{U} = 3.162$$

$$\text{Percentage unionised} = \frac{[U] \times 100}{[U] + [I]} = \frac{100}{1 + \dfrac{[I]}{[U]}}$$

$$= \frac{100}{1 + 3.162} = \frac{100}{4.162} = 24.03 \text{ per cent}$$

At pH 5, in the same way, we find that the percentage unionised is 3.07
At pH 7 the percentage unionised is 0.0316
At pH 9 less than 0.009 per cent is unionised.

Absorption is much greater than one would expect, being 41 per cent at pH 4 and 27 per cent at pH 5, although the expected trend is evident.

In attempts to explain away such discrepancies it has been suggested that a local pH exists at the membrane surface which differs from the bulk pH. This local pH is due to the attraction of hydrogen ions by the negative groups of membrane components so that, in the intestine, while the bulk pH is around 7 the surface pH is nearer 6. One can calculate a 'virtual' surface pH which will allow results such as those in table 9.2 to be explained. Take as another example salicylic acid. Calculating the percentage of unionised species by equation 9.1, as above, we obtain the following figures

pH	4	5	7	8
Percentage unionised	9.09	0.99	0.009999	0.001
Percentage absorbed (table 9.2)	64	35	30	10

To have 30 per cent of this compound in its unionised form when the bulk pH is 7 would require a surface pH of about 3.4, which is too low. However, what one must remember is that the absorption and ionisation processes are dynamic processes. As the unionised species is absorbed so the level of [U] in the bulk falls and more of the unionised species appears in the bulk because of a shift in equilibrium. In fact, if a pH of 5.3 is taken as the pH of the absorbing surface the results in table 9.2 become more explicable. This is discussed in section 9.1.2.

9.1.2 Problems in the quantitative application of the pH-partition hypothesis

There are several reasons why the pH-partition hypothesis cannot be applied quantitatively in practical situations. Some are discussed here.

Variability in pH conditions in humans

The variation in the pH of the stomach of human subjects is remarkable bearing in mind that each pH unit represents a ten-fold change in hydrogen ion concentration. While the normally quoted range of stomach pH is 1–3, studies using a

table 9.3 **pH of blood and contents of alimentary tract***

Sample	pH
Blood	7.35–7.45
Buccal cavity	6.2–7.2
Stomach	1.0–3.0
Duodenum	4.8–8.2
Jejunum and ileum	7.5–8.0
Colon	7.0–7.5

*From W. C. Bowman, M. J. Rand and G. B. West. *Textbook of Pharmacology*, Blackwell, London, 1967

pH-sensitive radio-telemetric capsule have shown a greater spread of values, ranging up to pH 7 as seen in figure 10.2. This means that the dissolution rate of many drugs will vary markedly in individuals – this is indeed one of the reasons for individual to individual variation in drug availability.

The scope for variation in the small intestine is less, although in some states the pH of the duodenum may be quite low due to hypersecretion of acid in the stomach. Table 9.3 lists the pH of the blood and of different regions of the alimentary tract.

pH at membrane surfaces

Detailed observation of the relationship between absorption and pH of the intestinal contents on the one hand and the percentage of drug theoretically in the unionised state indicates a discrepancy between the two[7]. The absorption data can generally be rationalised if the apparent pH is reduced below the pH of the contents of the gut. This pH 'shift' is thought to be due to the existence of a pH at the membrane surface lower than that of the bulk pH. One might expect that the hydrogen ion concentration at the surface would be greater than that in the bulk, if the structure of the membrane is as outlined above; that is, if polyanionic chains protrude from the surface. Hydrogen ions would accumulate near the anionic groups. This can be quantified by the equation

$$[H^+]_{surface} = [H^+]_{bulk} \exp(-F\zeta/RT) \qquad (9.3)$$

where F is the Faraday, R the gas constant, T is absolute temperature and ζ is the zeta potential of the surface (see chapter 7).

Thus, in terms of pH

$$pH_{surface} = pH_{bulk} + \zeta/60 \qquad (9.4)$$

where ζ is expressed in millivolts. The secretion of acidic material in many parts of the gut wall is a complicating factor. The local pH in the region of the microvilli of the small intestine will undoubtedly influence the absorption of weak electrolytes. A drug molecule in the bulk will diffuse toward the membrane surface and will meet different pH conditions from those in the bulk phase. Whether or not this influences the extent of absorption would depend on the pH and the pK_a of the drug in question.

The negative charge on the membrane will attract small cations towards the surface and small anions will be repelled; one might thus expect some selectivity in the absorption process. The existence of this 'microclimate' has been questioned, but experimental evidence for its existence has been forthcoming[8,9] from the use of microelectrodes, which reveal the existence of a layer on the (rat) jejunum with a pH of 5.5 when the bathing buffer pH was 7.2. The existence of this acid layer has also been demonstrated on the surface of the human intestine[8,9]. Failure to maintain this acid microclimate has been quoted as a contributing factor to folate malabsorption in adult coeliac disease[10]. A more neutral pH would cause a greater percentage of folic acid to be in its ionised form and thus to be less well absorbed.

Convective flow

The movement of water molecules into and out of the alimentary canal will affect the passage of small molecules across the membrane: it can be appreciated that the absorption of water-soluble drugs will be increased if water flows from the lumen to the blood side across the mucosa, provided that drug and water are using the same route. Conversely, flow of water in the opposite direction will decrease absorption. Water movement is greatest within the jejunum.

The reasons for water flow will be differences in osmotic pressure between blood and lumen contents, and differences in hydrostatic pressure between lumen and perivascular tissue resulting from muscular contractions. Recently it has been shown in experimental animals that lipid-soluble drugs are affected by solvent flow induced by addition of salts to the lumen. Absorption of benzoic acid, salicylic acid, benzyl alcohol and digitoxin has been shown to be increased by efflux of water from the lumen and decreased by flow into the lumen (see figure 9.5). One likely explanation is that when water flows from the lumen the drug becomes concentrated and drug absorption is increased because of the more advantageous concentration gradient. Suggestions that water flow affects the 'unstirred' layers close to the membrane may also be valid in interpreting these data.

Unstirred layers

A layer of relatively unstirred water lies adjacent to all biological membranes[12]. The boundary between the bulk water and this unstirred layer is indistinct, but nevertheless it has a real thickness. During absorption drug molecules must diffuse across this water layer and then on through the lipid layer, and the overall rate of transfer is the result of the resistance in both water layer and lipid layer. The flux, J, for a substance across the unstirred layer is given by the expression

$$J = (C_1 - C_2)(D/\delta) \tag{9.5}$$

where C_1 and C_2 are the concentrations of the substance in the bulk water phase and in the unstirred water layer respectively, D is the diffusion coefficient and δ is the effective thickness of the unstirred layer. The flux of molecules which pass by passive diffusion through the lipid membrane can be written as

$$J = C_2 P_c \tag{9.6}$$

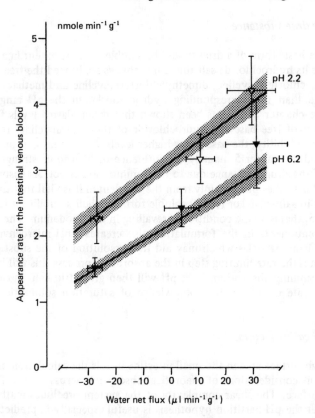

figure 9.5 The dependence of salicylic acid absorption on the net water flux (positive sign: directed towards the blood) in the rat jejunal loop perfused with hypo-, iso- and hypertonic solutions at pH 6.2 and 2.2. The curves with 95 per cent confidence limits (shaded area) were calculated by means of the parameters determined by a kinetic model with the following constants: concentration of salicylic acid in the perfusion solution 32.3 μmol ℓ^{-1}, wet tissue weight 0.453 g, perfusion rate 0.11 mℓ min^{-1}, intestinal blood flow 0.945 at pH 6.2 and 0.968 mℓ min^{-1} at pH 2.2. Mean values of experimental data with 95 per cent confidence intervals. From reference 11

where P_c is the permeability coefficient. The rate of absorption must equal the rate of transport across the unstirred layer; that is

$$J = (C_1 - C_2)(D/\delta) = C_2 P_c \qquad (9.7)$$

Neglect of the unstirred layer would introduce errors into the interpretation of experimental flux data. The rate of movement across the unstirred layer, as can be seen from the equations, is proportional to D/δ; the rate of absorption is proportional to P_c. Compounds with a large permeability coefficient may be able to penetrate across the cell membrane much faster than they can be transported through the unstirred layer. Under these circumstances diffusion through the water layer becomes the rate-limiting step in the absorption process.

Effect of the drug substance

Although the basic form of a drug should be soluble to some extent in a medium of low pH as its hydrochloride salt this is not always so. Indeed the free bases of, for example, chlortetracycline, dimethylchlortetracycline and methacycline are more soluble than the corresponding hydrochlorides in the pH range of the stomach (see chapter 5). It has been shown that mean plasma levels following administration of free base and hydrochloride of these tetracyclines reflect the differences in solubility, the base giving higher levels[13]. The reason is presumably that discussed in chapter 5, namely the influence of high ionic strength on the solubility of the drug substance due to the common ion effect. As absorption of the tetracyclines takes place mainly from the duodenum it is vital that they reach the intestine in a dissolved or readily soluble form, as their solubility is low in the pH range 4–5, that is, in the conditions prevailing in the duodenum. The presence of buffer components in the formulation also creates a pH microenvironment around dissolving particles which may aid the dissolution of the substance, and if dissolution is the rate-limiting step in the absorption process this will be significant in determining absorption. Bulk pH will then give little help in calculating the solution rate on the basis of knowledge of saturation solubilities in bulk conditions.

Other complicating factors

The very high surface area in the small intestine upsets the calculation of absorption based on considerations of theoretical absorption across identical areas of absorbing surface. The shear complexity of the system precludes mathematical precision, yet the pH-partition hypothesis is useful especially in predicting what follows from a change in bulk pH, for example on ingestion of antacids. The fact that aspirin, an acid with a pK_a of 3.5, is absorbed from the small intestine is due, it is believed, partly to the massive surface area available for absorption which allows significant absorption to occur even though the percentage of absorbable species is extremely low.

Drugs that are labile in the gastro-intestinal tract (for example, erythromycin), drugs that are metabolised on their passage through the gut wall, that are hydrolysed in the stomach to active forms (prodrugs), or drugs that bind to mucin or form complexes with bile salts may not always be absorbed in the manner expected.

Ion pairing

The interaction of drugs in the charged form with ions to form absorbable species with a high lipid solubility is a possible explanation for the ability of quaternary ammonium compounds, which are ionised under all pH conditions, to be absorbed readily. The origin of the ions which pair with drug ions is not clear but there is evidence that ion-pair formation will aid absorption.

One could assume that small organic ions are absorbed through water-filled pores or channels in the membrane, but the effective pore diameter of such pores means that large drug ions would be excluded from this route. If membranes are impermeable to large organic ions then ion-pair formation between a drug ion and an organic ion of opposite charge to form an absorbable neutral species is possible.

Two ion species, A^+ and B^-, may exist in solution in several states

$$A:B \rightleftharpoons A^+, B^- \rightleftharpoons A^+ B^- \rightleftharpoons A^+ + B^-$$
$$\text{'tight'} \quad \text{'loose'} \quad \text{free ions}$$
$$\text{ion pair} \quad \text{ion pair}$$

The formation of tight or loose ion pairs will depend on the solvent–ion interactions; hydrophobic ions might be encouraged to form ion pairs by the mechanism of 'water-structure enforced ion pairing' whereby the water attempts to minimise the disturbance on its structuring and, by reducing the polarity of the species in solution by ion-pair formation, achieves this end. Thus, ion pairing in highly structured solvents is due not to an electrostatic interaction but to a solvent-mediated effect. The importance of the phenomenon is that ion pairs have the property of being almost neutral species, so that the ion pair can partition into an oily phase when its parent ionic species cannot. This is important in drug absorption[14] and drug extraction procedures, and recently has been put to use in chromatographic techniques. The reactions below are examples of a quaternary amine pairing with a weak acid, and an alkyl sulphate with a weak base, both under pH conditions in which the solute is charged.

$$RCOO^- + N^+(C_4H_9)_4 \rightleftharpoons RCOON(C_4H_9)_4$$

$$RNH_3^+ + {}^-O_3SO(CH_2)_{11}CH_3 \rightleftharpoons RNH_3 O_3 SO(CH_2)_{11}CH_3$$

hydrophilic   ion-pair agent   hydrophobic ion pair
solute

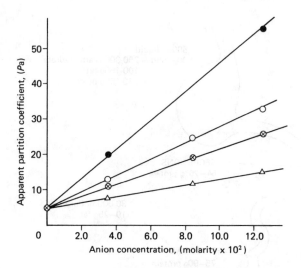

figure 9.6 Apparent partition coefficients for chlorpromazine between n-octanol and aqueous buffers. pH 3.9, in the presence of various anions at 30°C. ○, chloride: ●, propane sulphonate: ⊗, ethane sulphonate: △, methane sulphonate. From reference 15

In extraction procedures, bromothymol blue has been used as an ion-pairing counterion for amphetamine, and picrate ions for atropine, when the organic phase was methylene chloride. Tetrabutylammonium ion has been used for the extraction of penicillins into chloroform. Chloride ions have been used for desipramine extraction into this solvent. Figure 9.6 shows clearly the effect of chloride ion and other anions on the partition coefficient of chlorpromazine[15]. The nature of the anion significantly affects the partitioning of the drug. Ion pairing in the gastro-intestinal tract can thus influence absorption; it may be responsible for the absorption of charged drugs such as the quaternary ammonium ions and may be one further reason why it is difficult to apply the pH-partition hypothesis quantitatively to drug absorption.

9.2   The oral route

9.2.1   Drug absorption from the gastro-intestinal tract

The oral route is the most popular and convenient route of drug administration for those drugs that can survive the acid environment of the stomach and which are absorbed across the gastro-intestinal membranes. The functions of the gastro-intestinal tract are the digestion and absorption of foods and other nutrients and it is not easy to separate the two functions. Indeed the natural processes in the gut frequently influence the absorption of drug substance. This is not surprising when it is considered that approximately 500 g of solid and up to 2.5 litres of fluid are ingested on average each day. As well as this oral intake an estimated 30 litres of endogenous fluid are excreted each day into the intestine (figure 9.7).

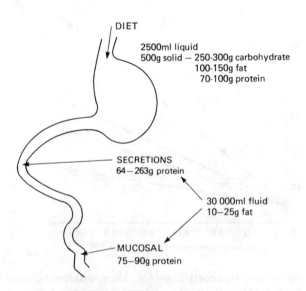

DIET

2500ml liquid
500g solid — 250-300g carbohydrate
100-150g fat
70-100g protein

SECRETIONS
64 – 263g protein

30 000ml fluid
10—25g fat

MUCOSAL
75—90g protein

figure 9.7 Representation of the approximate quantities of dietary and endogenous materials presented to the intestine each day. From reference 16

The pH of the gut contents, the presence of enzymes, foodstuffs, bile salts, fat and the microbial flora will all influence drug absorption. The complexity of the absorbing surfaces means that a simple physicochemical approach to drug absorption remains an *approach* to the problem and not the complete picture.

Whatever the limitations of theory — and one should not expect simple theories to hold in the complex and dynamic circumstances which are involved in drug absorption — theory provides a starting point in rationalising the behaviour of drugs in the gastro-intestinal tract.

Most drugs are absorbed by passive diffusion across the lipid membranes separating the gut contents from the rest of the body. A few drugs, among these molecules which resemble naturally occurring substances, are actively transported by special mechanisms. We first consider the physiological situation which might impinge on drug absorption.

9.2.2 Structural principles of the gastro-intestinal tract

Figure 9.8 diagrammatically represents the gastro-intestinal tract and some of the problems encountered in a consideration of drug absorption from this site. The stomach is not an organ specifically designed for absorption, the main site of absorption being in the small intestine. The stomach may be divided into its two main parts: (a) the body of the stomach (a receptacle or hopper), and (b) the pyloris (a churning chamber). Histologically these correspond to (a) the pepsin- and HCl-secreting areas, and (b) the mucus-secreting area of the gastric mucosa. The stomach varies its luminal volume with its content of food and this is one reason why food intake is important in relation to drug absorption; it may contain a few millilitres or a litre or more. Hydrochloric acid is liberated from the parietal cells at a concentration of 0.58 per cent, or 160 mmol ℓ^{-1}. The gastric glands produce some 1000–1500 mℓ of gastric juice per day.

The small intestine is divided anatomically into three sections: duodenum, jejunum and ileum. Histologically there is no clearly marked transition between these parts. All three are involved in the digestion and absorption of foodstuffs, absorbed material being removed by the blood and the lymph. The absorbing area is enlarged by surface folds in the intestinal lining which are macroscopically apparent; the surface of these folds possess villi, and the microvilli (figure 9.9). It has been calculated that with a maximum of 3000 microvilli per cell in the epithelial brush border (so called because of its physical appearance) the number of microvilli in the small intestine mucosa (of the rat!) is 200 000 000 per mm^2. Although one would expect organic acids to be absorbed only from the stomach where they will exist in the unionised lipid-soluble and membrane-diffusible form, the enormous surface area in the intestine allows significant absorption of acidic drugs from the intestine even though the fraction of unionised molecules is very small.

Over the entire length of the large and small intestines and the stomach the brush border has a uniform coating (3 nm thick) of mucopolysaccharides which consist of multibranched chains. This coating layer appears to act as a mechanical barrier to particles such as bacteria, cells or food particles. It may be that this layer acts as a filter; whatever its function, the weakly acidic, sulphated muco-

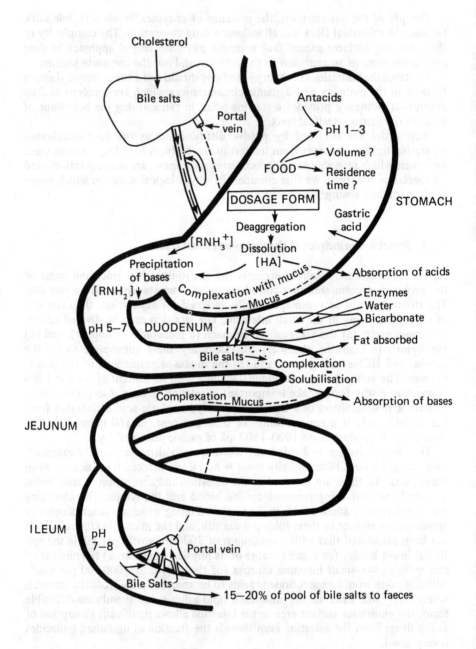

figure 9.8 Representation of the processes occurring along with drug absorption in the gastro-intestinal tract, and of the factors that must be taken into account in considering drug absorption

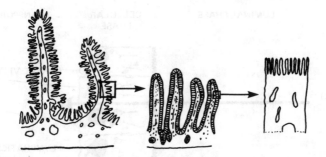

figure 9.9 Representation of the epithelium of the small intestine at different levels of magnification. From left to right: the intestinal villi and microvilli which comprise the brush border

polysaccharides influence the charge on the cell membrane and complicate the explanation of absorption.

The goblet cells form mucus; the secretion that is stored in granule form in the apical cell region is liquefied on contact with water to form mucus composed of protein and carbohydrate.

The large intestine is concerned primarily with the absorption of water and the secretion of mucus to aid the intestinal contents to slide down the intestinal tube. Villi are therefore completely absent from the large intestine, but there are deep crypts distributed over its surface.

Differences in the absorptive areas and volumes of gut contents in different animals are important when comparing experimental results on drug absorption in various species. The human small intestine has a calculated active surface area of approximately 100 m^2. No analogous calculations are available for most commonly used laboratory animals. The mucosal surface area of the small intestine of the rat is estimated to be 700 cm^2.

9.2.3 Bile salts and fat absorption pathways

Fat is absorbed by special mechanisms in the gut. The bile salts which are secreted into the jejunum are efficient emulsifiers and disperse fat globules to allow the action of lipase at the globule surface. Medium-chain triglycerides are thought to be directly absorbed. Long-chain triglycerides are hydrolysed, and the mono-glycerides and fatty acids produced form mixed micelles with the bile salts and are absorbed either directly in the micelle or, more likely, brought to the microvilli surface by the micelle and transferred directly to the mucosal cells, the bile salts remaining in the lumen[16] (figure 9.10). The bile salts are reabsorbed in the ileum and transported via the portal vein back into the bile salt pool.

There have been suggestions that lipid-soluble drugs may be absorbed by fat absorption pathways. Certainly administration of drugs in an oily vehicle can significantly affect absorption, increasing absorption in the case of griseofulvin[17,18], decreasing it in the case of vitamin D. Whether, in the former case, griseofulvin is absorbed by way of the bile salt micelles is not known, but it is likely that lipid-soluble drugs will at least be solubilised in the bile salt micelles.

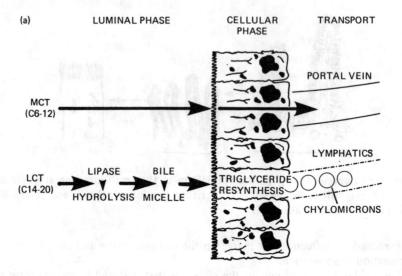

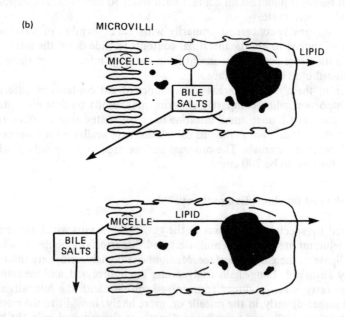

figure 9.10 (a) Comparison of the steps involved in absorption of medium-chain (C_6–C_{12}) (MCT) and long-chain (C_{14}–C_{20}) (LCT) triglycerides. MCT can be absorbed directly, while absorption of LCT involves intermediate steps. (b) Alternative hypotheses of penetration of micellar lipid into mucosal cells. The micelle may be absorbed intact and the bile salts returned to the lumen or, alternatively, the lipid alone may be absorbed, the bile salts remaining in the lumen. From reference 16

Cholesterol is thought to be transported into the lymphatic circulation by a simple diffusion process, depending on bile salt for its absorption. Cholesterol absorption in the presence of triglycerides is directly proportional to the chain length of the triglyceride fatty acid and is modified by the simultaneous absorption of the hydrolysis products of the triglycerides.

9.2.4 Gastric emptying, motility and volume of contents

The volume of the gastric contents will determine the concentration of a drug which finds itself in the stomach. The time the drug resides in the stomach will, if the drug is an acid, determine for how long absorption can occur and, if it is a drug absorbed lower down the gut, it will determine the delay before absorption begins.

The stomach empties liquids faster than solids. The rate of transfer of gastric contents to the small intestine is retarded by the activity of receptors sensitive to acid, fat, osmotic pressure and amino acids in the duodenum and the small bowel and stimulated by material that has arrived from the stomach[19]. Gastric emptying is a simple exponential or square-root function of the volume of a test meal — a pattern that holds for meals of variable viscosity. To explain the effect of a large range of substances on emptying, an osmoreceptor has been postulated which, like a red blood cell, shrinks in hypertonic solutions and swells in hypotonic solutions.

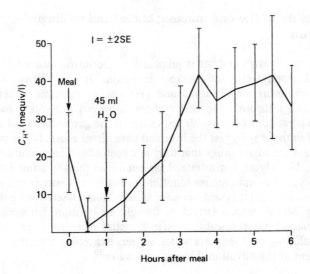

figure 9.11 Hydrogen ion concentrations (C_{H^+}) at intervals after a test meal. Mean results are shown ±2 s.e. The zero samples were taken just before the meal was begun, and the 1 hour sample just before 45 ml of water. After J. Fordtran *New Engl. J. Med.*, 274, 921 (1966)

Acids in test meals have been found to slow gastric emptying; acids with high molecular weights (for example, citric acid) are less effective than those, such as HCl, with low molecular weights. Natural triglycerides inhibit gastric motility, linseed and olive oils being the most effective[19]. The formulation of a drug may thus influence drug absorption through an indirect physiological effect. The nature of the dose form, whether solid or liquid, whether acid or alkaline, whether aqueous or oily, may thus influence gastric emptying†.

Drugs are rarely taken alone. Food and drugs are often taken at the same time, often inadvisedly. The effect of a meal on the hydrogen ion concentration of the stomach is shown in figure 9.11; the effect of two antacids on gastric volume and pH is shown in table 9.4. When considering the effect of an antacid, therefore, the effect of volume change and pH change and the effect on gastric emptying must all be considered.

table 9.4   **Effects of antacids on gastric volume and pH in the rat***

	Water	Maalox	Amphojel
Volume (ml)	0.29 ± 0.03	1.8 ± 0.3	4.0 ± 0.2
pH	2.2 ± 0.2	7.4 ± 0.1	4.7 ± 0.2

Ten rats per group. Measurements on gastric contents made 20 min after the third hourly dose of 1 ml water or antacid by gastric intubation.
*From M. Hava and A. Hurwitz. *Eur. J. Pharmacol.*, 22, 156 (1973)

9.3   Permeability of the oral mucosa; buccal and sublingual absorption

Mouth-washes, toothpastes and other preparations are introduced into the oral cavity for local prophylactic and therapeutic reasons. It is not known to what extent these substances are absorbed and give rise to systemic effects. The absorption of drugs through the oral mucosa, however, provides a route for systemic administration which avoids exposure to the gastro-intestinal system. Drugs absorbed in this way bypass the liver and have direct access to the systemic circulation. There are many drugs that have not been administered sublingually, although some, like glyceryl trinitrate, have been traditionally administered in this way. However, the oral mucosa functions primarily as a barrier and it is not a highly permeable tissue. It is said to resemble skin more closely than gut in this respect[22]. Dog buccal mucosa (which is thought to be more permeable than human oral mucosa) has permeability characteristics close to those of human dermis, both behaving as if they are water barriers exhibiting a permeability of about 30 per cent of the diffusion through pure water[23].

---

†The question of gastric emptying and transit down the gastro-intestinal tract has assumed further importance in relation to the design and performance of sustained-release preparations. Studies have been conducted[20,21] on the transit of disintegrating and non-disintegrating dosage forms containing pellets of different densities.

9.3.1 Mechanism of absorption

While cells of the oral epithelium and epidermis are capable of absorbing material by endocytosis it does not seem likely that drugs or other solutes would be transported by this mechanism across the entire stratified epithelium. It is also unlikely that active transport processes are operative in the oral mucosa. There is considerable evidence that most substances are absorbed by simple diffusion. For example, the buccal absorption of some basic drugs increases, and that of acidic drugs decreases, with increasing pH of their solutions[24] (figure 9.12).

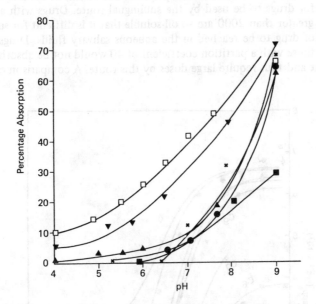

figure 9.12 Buccal absorption of some basic drugs. The drugs were dissolved in buffered solutions of different pH and placed in the mouth of human subjects. Absorption rates were determined from the decrease of drug concentration in expelled solutions. □, Chlorpheniramine (pK_a = 8.99); ▼, methadone (pK_a = 8.25); x, amphetamine (pK_a = 9.94); ▲, ephedrine (pK_a = 9.63); ●, pethidine (pK_a = 8.72); ■, nicotine (pK_a = 8.02). From reference 24

Nicotine in a gum vehicle when chewed is absorbed through the buccal mucosa; nicotine levels obtained are lower than those obtained by smoking cigarettes and the high peak levels are not seen. Buccal glyceryl trinitrate has been found to be an effective drug. The buccal route has the advantages of the sublingual route — buccal mucosa is similar to sublingual mucosal tissue — and a sustained release tablet can be held in the cheek pouch for several hours if necessary. A mucosal adhesive formulation has been devised[25].

Experiments with some analgesics showed that the highly lipid-soluble etorphine was absorbed several times more rapidly from the buccal cavity than the less lipid-soluble morphine[26]. Nitroglycerin exerts its pharmacological action

1-2 minutes after sublingual administration. The most striking demonstration of the importance of lipophilicity is the order of absorption of a series of n-alkyl fatty acids shown in figure 9.13, in which k_{app} is a function of amount absorbed.

Impressive absorption has been obtained with glyceryl trinitrate, methacholine chloride, isoprenaline, some steroids such as desoxycortisone acetate[27] and 17β-oestradiol at one-quarter of its oral dose, morphine, captopril and nifedipine. The only one of these that is regularly administered sublingually is glyceryl trinitrate; isoprenaline and oxytocin sublingual tablets are available but alternative routes of administration are preferred[28].

It has been claimed that an oil/water partition coefficient in the range 40-2000 is optimal for drugs to be used by the sublingual route. Drugs with a partition coefficient greater than 2000 are so oil-soluble that it is difficult for sufficiently high levels of drug to be reached in the aqueous salivary fluids. Drugs less lipophilic than those with a partition coefficient of 40 would not be absorbed to any great extent and thus require large doses by this route. A comparison of the sub-

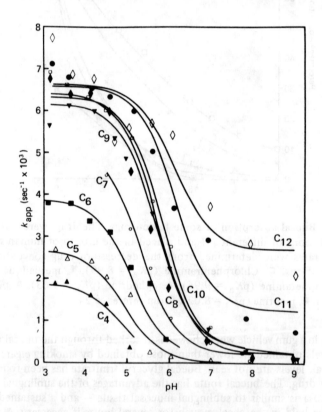

figure 9.13 Fit of Beckett and Moffat's data for the absorption of n-alkanoic acids through the buccal mucosa, to a model which quantifies the pH-partition hypothesis. In the diagram the original results for absorption in 5 minutes have been converted to $k_{app} = -\left\{ \ln\left[-\left(\% \text{ absorbed in 5 min}/100 \right) \right] \times 10^3 \right\} /(5 \times 60)$. From J. Wagner and A. Sedman. *J. Pharmacokinetics Biopharm.*, **1**, 23 (1973)

lingual/subcutaneous dose ratio serves as an index of the usefulness of the route[28]. Cocaine with $P \approx 28$ requires a sublingual/subcutaneous dosage ratio of 2 to obtain equal effects. Atropine with $P \approx 7$, requires eight times the subcutaneous dose, and for codeine ($P \approx 2$) over fifteen times the subcutaneous dose must be given sublingually.

9.4 Absorption of drugs from intramuscular and subcutaneous injections

The intramuscular and subcutaneous routes of administration have been regarded for a long time as efficient routes because they bypass the problems encountered in the stomach and intestine and it seems that all the drug injection must be bioavailable. Early views that subcutaneous (s.c.) and intramuscular (i.m.) administration of drugs results in only local action at the site were dispelled by Benjamin Bell, who wrote in the *Edinburgh Medical Journal* of 1858 that, 'absorption from the enfeebled stomach may not be counted on; we possess in subcutaneous injections a more direct, rapid and trustworthy mode of conveying our remedy in the desired quantity to the circulatory blood.' We now know that not all drugs are efficiently or uniformly released from i.m. or s.c. sites; relatively recent advances in analytical techniques have allowed closer attention to be paid to bioavailability from parenteral sites. Some of the drugs that are not fully absorbed from the i.m. site are listed in table 9.5.

A measure of the bioavailability of a drug from an intramuscular site is its plasma level profile compared to oral or intravenous (i.v.) administration. Where differences occur the clinical importance is most marked when the route of administration is changed and of course are of especial note in drugs with a low therapeutic index such as digoxin and phenytoin.

In order for the drug to have a systemic action following injection it must be released from the formulation and reach the site of action in sufficient amounts and at a sufficient rate to produce the desired pharmacological effect. The various regions into which injections are given are shown in figure 9.14. The subcutaneous region has a good supply of capillary and lymphatic vessels. Muscle tissue also has a rich supply of capillaries, although it is generally agreed that there are few, if any, lymph vessels in muscle proper. Drugs with the correct physicochemical characteristics can thus diffuse through the tissue and pass across the capillary walls and thus enter the circulation.

table 9.5 **Widely used drugs that may be incompletely absorbed after intramuscular injection**

Ampicillin	Digoxin*
Cephaloridine	Insulin*
Cephradine	Phenylbutazone
Chlordiazepoxide	Phenytoin*
Diazepam	Quinidine
Dicloxacillin	

*Clinically important problems have been demonstrated with these

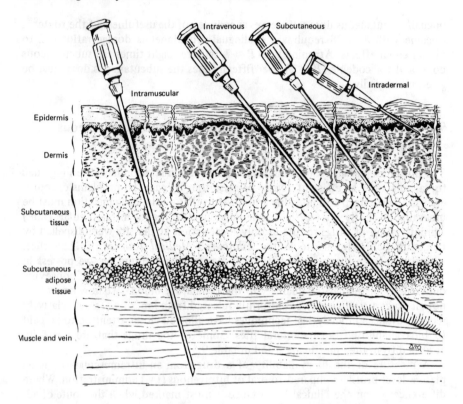

figure 9.14   Routes of parenteral medication, showing the tissues penetrated by intramuscular, intravenous, subcutaneous, and intradermal injections. The needles, with bevel up, penetrate the epidermis (cuticle) consisting of stratified epithelium with an outer horny layer, the corium (dermis or true skin) consisting of tough connective tissue, elastic fibres, lymphatic and blood vessels, and nerves, the subcutaneous tissue (tela subcutanea) consisting of loose connective tissue containing blood and lymphatic vessels, nerves, and fat-forming cells, the fascia (a thin sheet of fibrous connective tissue), and the veins, arteries, and muscle. Drawing by David S. Quackenbush. From E. W. Martin. *Techniques of Medication*, Lippincott, Philadelphia, 1969

If it is assumed that drug absorption proceeds by passive diffusion of the drug, the absorption can be considered to be a first-order process. Thus the rate of absorption is proportional to the concentration, C, of drug remaining at the injection site

$$dC/dt = -k_a C \qquad (9.8)$$

where k_a is the first-order rate constant. The half-life of the absorption process is

$$t_{1/2} = 0.693/k_a \qquad (9.9)$$

Drug absorption is 90 per cent complete when a time equivalent to three times the half-life has elapsed[29].

If dissolution and diffusion are important parameters it is easiest to isolate the important physicochemical parameters by considering soluble, neutral compounds and their dispersion from intramuscular sites. Table 9.6 shows that mannitol rapidly diffuses from the site of injection, insulin less rapidly and dextran of molecular weight 70 000, as might be expected, disperses more slowly.

Molecular size would be a factor in controlling release, but it is a minor one as the molecular weights of most drugs fall in the range 100–1000. Unless the drug is attached to a macromolecular backbone significant retardation cannot be achieved. Hydrophilic molecules such as those listed in table 9.6 will be transported to the blood after diffusing through muscle fibres and then through the pores in the capillary walls, being incapable of absorption through the lipid walls. The transport through the capillary wall is the rate-limiting step in most cases; the larger the molecule the more slowly it diffuses, as seen from equation 3.91, and the greater difficulty it has in traversing the aqueous pore in the capillary walls or the cell junction. The 'pores', of whatever character they are, account for only one per cent of the available surface of the capillary wall.

table 9.6 **Influence of molecular size on clearance from intramuscular sites**

Substance	Molecular weight	Fraction cleared (5 min)
Mannitol	182	0.7
Inulin	3500	0.2
Dextran	70000	0.07

With weak electrolytes, absorption across the capillary walls follows the expected patterns, absorption of more lipid-soluble agents being relatively fast. Hydrophobic drugs may bind to muscle protein leading to a diminution in free drug and perhaps leading to a prolongation of action. Dicloxacillin is 95 per cent bound to protein; ampicillin is bound to the extent of 20 per cent. Dicloxacillin as a consequence is absorbed more slowly from muscle than is ampicillin.

The significance of the diffusion phase through muscle tissue was demonstrated long ago by Duran Reynolds who, in 1928, observed that dye solutions injected together with hyaluronidase spread out more rapidly and over a greater distance in the tissue than in the absence of the enzyme. The enzyme achieves its effect by breaking down the hyaluronic acid in intercellular spaces, thus leading to a decrease in viscosity and so easing the passage of small molecules in the complex matrix.

The region into which the injection goes is complex, being composed of both aqueous and lipid components. Recent studies indicate that the muscle tissue is more acidic than normal physiological fluids. Measurement of percutaneous muscle surface pH (pH_m) by O'Donnell[30] using microelectrodes gave values of 7.38 ± 0.02; patients with peripheral vascular disease demonstrated a mean pH_m of 7.16 ± 0.04. pH_m reflects the intramuscular ion concentrations but is stated to be higher than the actual intramuscular pH, which has been said to be as low as 6.4. The pH of the region will determine whether or not drugs will dissolve in the tissue fluids or precipitate from formulations. In some cases precipitation will occur and may be the cause of the pain experienced on injection. The rate of solution here, as in many other routes, therefore determines how quickly the drug begins to act or, in some cases, for how long it acts. The deliberate reduction of the solubility of a drug to prolong its action by the i.m. or s.c. route is a ploy which has been resorted to on several occasions.

Many preparations of drugs for i.m. administration are formulated in water-miscible solvents such as polyethylene glycol 300 or 400, or propylene glycol or ethanol mixtures. Dilution by the tissue fluids may cause a drug to precipitate. Drugs formulated as aqueous solutions by adjustment of pH will alter only momentarily the pH of the injection site. Figure 9.15 shows diagrammatically the three main types of formulation used for i.m. and s.c. injections and their fate. Rapid removal of aqueous vehicles is to be expected. If the vehicle is non-aqueous and is in fact an oil such as sesame oil, or other vegetable or mineral oil, the oil phase disperses as droplets in the muscle and surrounding tissue, and is eventually metabolised. The rate-determining step in the absorption of drug esters such as fluphenazine decanoate (which has an aqueous solubility of about 1 part per million) is the hydrolysis of the drug at the surface of the oil droplet. Hydrolysis of the fluphenazine decanoate to its soluble alcohol therefore depends on the state of dispersion and surface area of the droplets. Dispersing the droplets by rubbing the site of injection or by violent exercise can result in excessive dosage resulting in toxic effects. Exercise also causes increased blood flow, and as absorption is a dynamic process requiring the sweeping away of the drug from the localised absorption site, this increased flow increases the rate of drug dispersal.

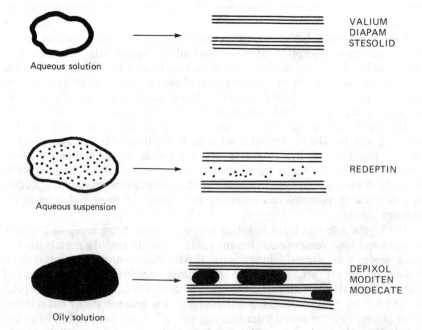

figure 9.15 Diagrammatic representation of three types of intramuscular formulation, and mode of dispersal. Examples of commercial formulations of drugs used in psychiatric medicine are shown

Different rates of blood flow in different muscles means that the site of i.m. injection can be crucial. Resting blood flow in the deltoid region is significantly greater than in the gluteus maximus muscle; flow in the vastus lateralis is intermediate. The difference between blood flow in the deltoid and gluteus is of the order of 20 per cent and is likely to be the cause of the fact that deltoid injection of lignocaine has been found to give higher blood levels than lateral thigh injection. In fact, therapeutic plasma levels for the prevention of arrhythmia with a dose of 4.5 mg kg^{-1} of a 10 per cent solution of lignocaine were achieved only following deltoid injection. Age should influence the behaviour of the injection as ageing will affect vascular blood flow, and fatty deposits, but age has not been a factor specificaly isolated in studies to date. In some disease states it is possible to predict that the outcome of an i.m. injection might be different from that in normal patients; for example, in patients with circulatory shock, hypotension, congestive heart failure and myxoedema, blood flow to skeletal muscle is decreased.

Formulation effects

The deliberate manipulation of formulations to produce special, often long-acting effects is discussed elsewhere. Crystalline suspensions of fluspirilene, steroids and procaine penicillin G can be prepared in different size ranges to produce different pharmacokinetic profiles. Variability in response to a drug, or difference in response to a drug from different manufacturers, can be the result of the formulation employed. Diazepam given as Valium (Roche) produced in one trial plasma levels ranging from about zero to 160 mg mℓ^{-1} 90 minutes after injection into muscle (see figure 9.16). The formulation contained diazepam dissolved in an ethyl alcohol–propylene glycol mixture. Upon injection this solvent is diluted out and the drug precipitates (see chapter 10, section 10.2). Injection of the formulation into the fatty tissue in the buttock can result in poor dispersal of drug, as the drug would have little opportunity for solution in that environment. Therefore the depth of the injection is significant. If, in addition, the blood supply to the region is limited there will be an additional restriction to rapid removal. A European formulation of diazepam (Stesolid) containing a surface-active agent as a solubiliser has a greater bioavailability than a non-aqueous solution form of the drug (Diapam). The additives can therefore influence dispersion, the solubiliser undoubtedly reducing precipitation at the site of injection and increasing the rate of solution. (Reports of marked difference in side-effects and adverse reactions to diazepam formulations are undoubtedly due to the additives present, either indirectly or directly.)

Insulin[31,32]

The classic example of controlled-release injections is that of insulin preparations. Modification of the crystallinity of the insulin allows control over solubility and duration of activity.

An acid-soluble formulation of insulin on amphoteric protein was introduced for clinical use in 1923. It had a short duration of action and attempts were made to prolong the action of the insulin. A castor oil vehicle was found to disperse too slowly but in 1936 Hagedorn and his colleagues found that insulin complexed

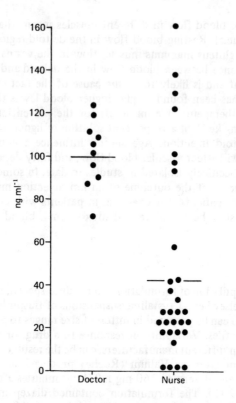

figure 9.16 Plasma diazepam levels 90 minutes after intramuscular injection by one doctor and several nurses, showing the importance of technique and site of injections, which was variable in the latter group. Individual values; horizontal lines denote average levels. From J. W. Dundee, J. A. Gamble and R. A. Assof *Lancet* ii, 1461 (1974) with permission

with zinc and protamine from the trout (*Salmo irideus*) formed an amorphous precipitate at neutral pH. When injected subcutaneously, the insulin slowly released from the complex into the blood was active for about 36 hours. The prolonged-acting insulins introduced since the advent of protamine–zinc insulin (PZI) have been designed to have intermediate duration of action (table 9.7).

The long-acting insulins in use today are mainly protamine insulin, zinc insulins and surfen insulins. Protamine insulins are the salt-like compounds found between the acid (insulin) polypeptide and the polypeptide protamine. They are used in the form of neutral suspensions of protamine insulin crystals (isophane insulin). Protamine consists primarily of arginine. In surfen insulins, which were developed by Hoechst, a complex is formed between the synthetic bis(4-amino-chinaldine-6)-*N*,*N*-urea hydrochloride (surfen) and insulin.

Isophane insulin is produced by titration of an acidic solution of insulin with a buffered solution of protamine at neutral pH until so-called isophane precipitation occurs; that is, no insulin or protamine is present in the supernatant. Under these conditions the precipitate consists of rod-shaped crystals. The dev-

table 9.7 Pharmaceutical injections of insulin BP*

Preparation	pH	Buffer	Description	Onset (hours)	Duration of effect (hours)
Insulin injection	3.0–3.5	–	Solution	~0.5–1	6–8
Neutral injection	6.6–7.7	Acetate	Solution	~0.5–1	6–9
Protamine zinc	6.9–7.4	Phosphate	Amorphous particles; rod-shaped crystals	~5–7	36
Globin zinc	3.0–3.5	–	Solution	~2	18–24
Isophane	7.1–7.4	Phosphate	Rod-shaped crystals (about 20 μm long)	~2	28
Zinc suspension (amorphous) 'Semilente'	7.0–7.5	Acetate	Amorphous particles (2 μm diameter)	~1	12–16
Zinc suspension (crystalline) 'Ultralente'	7.0–7.5	Acetate	Rhombohedral crystals (about 20 μm across)	~5–7	36
Zinc suspension 'Lente'	7.0–7.5	Acetate	Amorphous particles (30 per cent) Rhombohedral crystals (70 per cent)	~2	24
Biphasic	6.6–7.2	Acetate	Insulin in solution (25 per cent) Rhombohedral crystals (75 per cent)	~1	18–22

*Modified from Stewart. *Analyst*, 99, 911 (1974)

elopment of crystalline forms of insulin in the absence of foreign protein came in 1951. Under conditions of high zinc ion concentration (ten times that normally used to crystallise insulin) and in the absence of citrate and phosphate ions, rhombohedral crystals of insulin are formed in acetate buffer (IZS crystalline). By adjusting the pH during the crystallisation stage the insulin is produced as an amorphous precipitate (IZS amorphous) which has a duration of up to 16 hours. Various mixtures can be used providing modified durations of action, as in biphasic insulin. However, the insulin zinc suspensions are incompatible with PZI and isophane insulin, both of which act as a phosphate buffer which destroys the zinc complex. Variable insulin activity may result from the mixing of PZI and soluble insulin in a syringe prior to administration. It has been suggested that this variability is due to the combination of the soluble insulin with excess protamine present. As a general rule insulins of different pH should not be mixed.

Absorption of insulin is faster from injections made subcutaneously in the arms than in the thighs. There have been suggestions that insulin absorption in some people is poor because the subcutaneous tissue acts as a mechanical barrier and as an active site of degradation[32].

The use of insulins in solution obviates the potential source of error which arises when drawing a suspension into the syringe, but soluble insulins have the drawback that they must be stored at acid pH. Injected subcutaneously the insulin precipitates as amorphous particles. The hydrogen ion concentration of insulin preparations influences their stability, solubility and immunogenicity. After i.m. administration, short-acting insulin is absorbed about twice as rapidly as after s.c. injection, and therefore the i.m. route is used in the management of ketoacidosis in those cases where continuous i.v. infusion cannot be established. After s.c. injection absorption of short-acting insulin varies considerably depending in particular on the site of injection. Patient-to-patient variability is as great with these preparations as the variations in the absorption rate of intermediate acting insulins. This leads to difficulty in control. A self-regulating delivery system such as the artificial β-cell is of obvious clinical relevance. The artificial β-cell, now in developmental stage, consists of an implanted glucose electrode, computer, pump and reservoir. Signals from the glucose electrode cause insulin to be pumped from the reservoir when circulating levels of glucose are high. The reservoir can be topped up through a closure. The advantage of the system is that the dose administered is in direct response to the levels of glucose in the blood.

9.5   Topical preparations

It is difficult to write a succinct account of the pharmaceutics and biopharmaceutics of topical preparations. This section has therefore been deliberately restricted in scope to deal with the physicochemical principles involved in the process of medicating the skin or in systemic medication by the 'transdermal' or 'percutaneous' route. The bibliography at the end of the chapter will guide the interested reader to more comprehensive reviews of the topic[33]. Formulation of topical vehicles for the potent drugs now applied to the skin is an exact art. It is readily demonstrated that the vehicle in which the drug is applied influences the rate and extent of absorption. The factors responsible for the efficiency and efficacy of a topical formulation are discussed below.

Before any drug applied topically can act either locally or systemically it must penetrate the barrier layer of the skin, the stratum corneum. This layer is now widely acknowledged to be essentially uniformly impermeable. It behaves like a passive diffusion barrier with no evidence of metabolic transport processes. Penetration of water and low molecular weight non-electrolytes through the epidermis is proportional to their concentration, and to the partition coefficient of the solute measured between tissue and vehicle. A form of Fick's law similar to that expressed in equation 8.24 describes steady-state transport through the skin

$$J = \left(\frac{DP}{\delta}\right) \Delta C_v \qquad (9.10)$$

where J is the solute flux, D the solute diffusion coefficient in the stratum corneum, P the solute partition coefficient between vehicle and skin, and δ is the thickness of the stratum corneum. ΔC_v is the difference in solute concentration between vehicle and tissue. This relation is obtained as follows. Fick's law of diffusion shows that (for a given vehicle)

$$J \propto \Delta C_v \qquad (9.11)$$

A proportionality constant κ_p may be added. Thus

$$J = \kappa_p \Delta C_v \qquad (9.12)$$

where κ_p is the permeability constant which provides a means of expressing absorption measurements for comparing different vehicles and conditions. The units of a permeability constant are $m\ s^{-1}$, the concentration term being $mol\ m^{-3}$, so that J has the correct units of $mol\ m^{-2}\ s^{-1}$. It has been shown that

$$\kappa_p = \frac{PD}{\delta} \qquad (9.13)$$

so that equation 9.10 is readily obtained. Values of D range from 1×10^{-12} to $1 \times 10^{-17}\ m^2\ s^{-1}$ for human stratum corneum.

Before steady-state penetration is achieved, the rate builds up over a period of time and a lag phase will be apparent (figure 9.17). The lag time, τ, does not indicate the point at which steady-state is achieved but it is obtained, as shown in figure 9.17, by extrapolation. The linear portion of the graph can be described by an equation relating the total amount absorbed at time t, Q_t, to $\Delta C_v, P, \delta, t$ and D. From equation 9.10 as $Q_t = J \cdot t$

$$Q_t = \frac{DP \Delta C_v}{\delta} \cdot t \qquad (9.14)$$

But from figure 9.17 it is obvious that the time to be substituted is the time over which steady-state flux has been maintained; namely, $t - \tau$.

Thus we write

$$Q_t = \frac{DP \Delta C_v}{\delta} (t - \tau) \qquad (9.15)$$

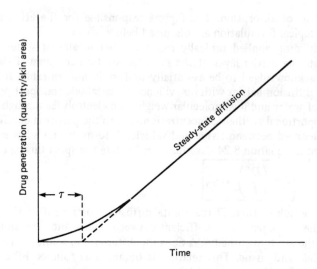

figure 9.17 Drug penetration–time profile for an idealised drug diffusing through human skin. Once steady-state diffusion occurs Q_t can be obtained using equation 9.14 where $t = $ (time elapsed $- \tau$)

The lag time, τ, has been shown to be equal to $\delta^2/6D$. Thus D is readily obtained from data such as those shown in figure 9.17. Values of τ range from a few minutes to several days, so lag times are of obvious clinical relevance.

9.5.1 Routes of penetration

Solute molecules may penetrate the skin not only through the stratum corneum but also by way of the hair follicles or through the sweat ducts, but these offer only a comparatively minor route because they represent such a small fraction of the surface area. Only in the case of molecules that move very slowly through the stratum corneum may absorption by these other routes predominate. Passage through damaged skin is increased over normal skin. Skin with a disrupted epidermal layer will allow up to 80 per cent of hydrocortisone to pass into the dermis but only 1 per cent is absorbed through intact skin.

The physicochemical factors that control drug penetration include the hydration of the stratum corneum, temperature, pH, drug concentration, the molecular characteristics of the penetrant and the vehicle[33]. Hydration of the stratum corneum is one determinant of the extent of absorption: increased hydration decreases the resistance of the layer, presumably by causing a swelling of the compact structures in the horny layer. Occlusive dressings increase the hydration of the stratum corneum by preventing water loss by perspiration; certain ointment bases are designed to be self-occluding (for example, Alphaderm). The use of occlusive films may increase the penetration of corticosteroids by a factor of 100 and of low molecular weight penetrants two- to five-fold.

As temperature varies little in the clinical use of a topical preparation this parameter will not be discussed here. A pH in excess of 11 will greatly increase skin permeability.

The stratum corneum is a heterogeneous structure containing about 40 per cent protein (keratin — a disulphide cross-linked linear polypeptide), about 40 per cent water and 18-20 per cent lipids (principally triglycerides and free fatty acids, cholesterol and phospholipids).

Occluded skin may absorb up to 5-6 times its dry weight of water. An idealised model of the stratum corneum (figure 9.18) has been used by Michaels *et al.*[34]. In this model L represents the lipid-rich interstitial phase and P the proteinaceous phases. If $\rho = P_L/P_P$ (the ratio of the partition coefficients of the drug between vehicle and the L and P 'phases'), and D_L and D_P are the diffusion coefficients of the drug in these phases, the flux through stratum corneum of average thickness (that is, 40 μm) is found to reduce to $J = 0.98\ \rho\ (D_L/D_P)$ (μg cm^{-2} h^{-1}) when $\rho(D_L/D_P)$ is very small. When $\rho(D_L/D_P)$ is very large, the flux becomes $J = 2.3 \times 10^{-4}\ \rho\ (D_L/D_P)$ (μg cm^{-1} h^{-1}), which emphasises the importance of the partition coefficient and diffusion coefficient of the drug in the absorption process.

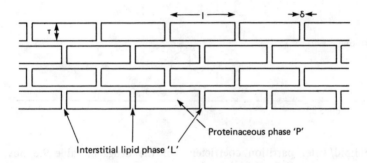

figure 9.18 Idealised model of the stratum corneum. Lipid (L) and proteinaceous (P) parts of the stratum corneum are represented; this model is used to derive model equations for drug transport across this layer (see text). From reference 34

9.5.2 Influence of drug

The diffusion coefficient of the drug in the skin will be determined by factors such as size, shape and charge; the partition coefficient will be determined not only by the properties of the drug but also by the vehicle as this represents the donor phase, the skin being the receptor phase. The quantity ρ can be approximated by experimentally determined oil/water partition coefficients. Thus, substances that have a very low oil solubility will display low rates of penetration.

The major pathway of transport for water-soluble molecules is transcellular, involving passage through cells and cell walls. The pathway for lipid-soluble molecules is presumably the endogenous lipid within the stratum corneum, the bulk of this being intercellular. Activation energies for the permeation of water and water-soluble alcohols have been found to be around 63 kJ mol^{-1}; for the corresponding lipid-soluble alkanols it is lower, at 42 kJ mol^{-1}. Increase in the polar character of the penetrant molecule decreases permeability, as seen from Scheuplein's data[35] in table 9.8.

table 9.8 **Permeability constants of steroids***

Steroid	Structure	Permeability constant (M_p, cm^{-2} h^{-1} × 10^{-6})
Progesterone	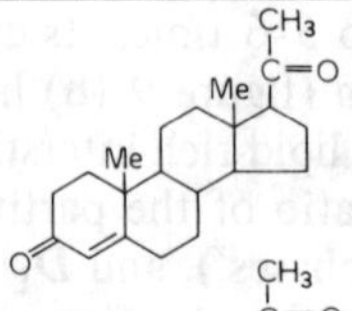	1500
Hydroxyprogesterone	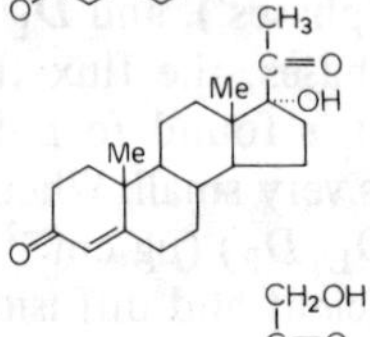	600
Cortexone	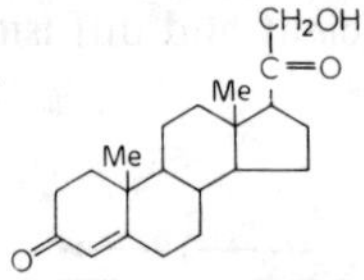	450
Cortexolone	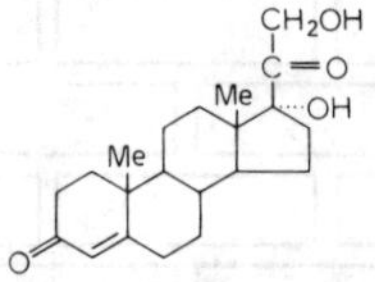	75

*From reference 35

The lipid/water partition coefficients of the drugs in table 9.8 obviously decrease as the number of hydroxyl groups increases, but a simple lipid/water partition coefficient is not an ideal guide, as the stratum corneum is a complex system as described above. However, if a substance is more soluble in the stratum corneum than in the vehicle in which it is presented, the concentration in the first layer of the skin may be higher than that in the vehicle. If depletion of the contact layers of the vehicle occurs then the nature of the formulation will dictate how readily these are replenished by diffusion and therefore will dictate the rate of absorption.

It has been found that the penetration rates of four steroids through intact abdominal autopsy skin were in the order of their physiological activity; namely, betamethasone 17-valerate (VI) > desonide > triamcinolone acetonide (III) > hydrocortisone (II). Triamcinolone is five times more active systemically than hydrocortisone, but has only one-tenth of its topical activity.

The acetonide of triamcinolone has a topical activity 1000 times that of the parent steroid because of its favourable lipid solubility. Betamethasone (V) is 30 times as active as hydrocortisone systemically but has only four times the topical potency. Of 23 esters of betamethasone the 17-valerate ester possesses the highest topical activity. The vasoconstrictor potency of betamethasone 17-valerate is 360 (fluocinolone acetonide = 100), that of betamethasone 0.8, its 17-acetate 114, the 17-propionate 190, and the 17-octanoate 10. The peak activity coincides with an optimal partition coefficient — one which favours neither lipid nor aqueous phase.

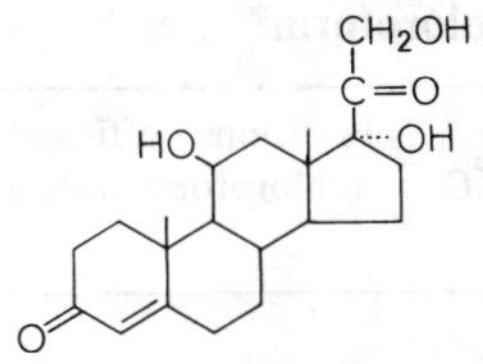

HYDROCORTISONE
(II)

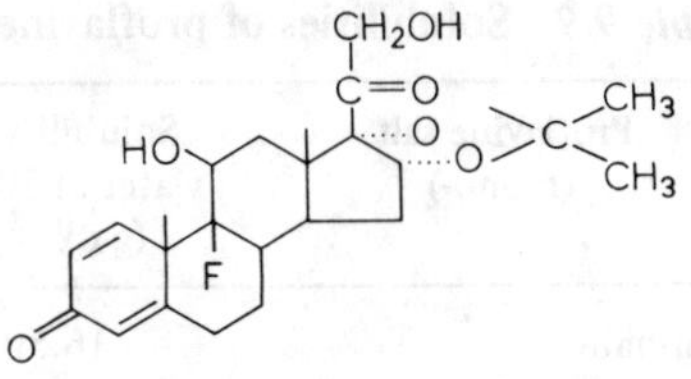

TRIAMCINOLONE ACETONIDE
(III)

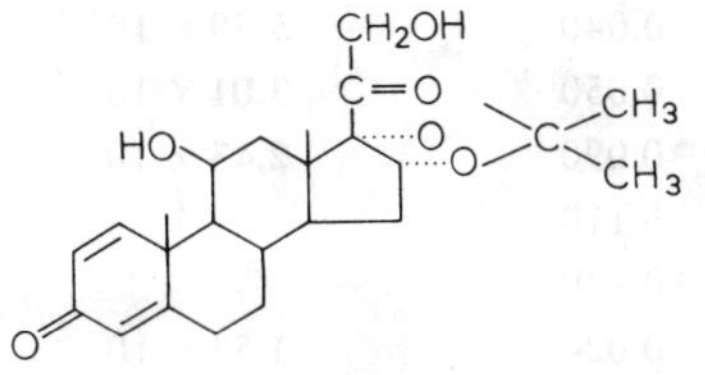

16-ALPHA HYDROXYPREDNISOLONE ACETONIDE
(IV)

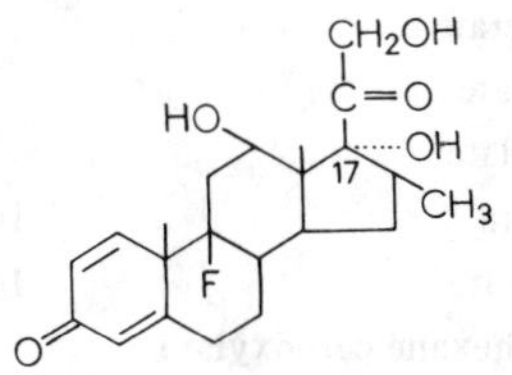

BETAMETHASONE
(V)

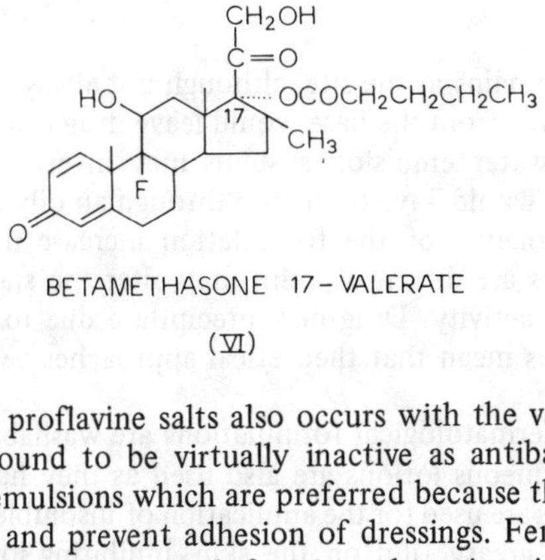

BETAMETHASONE 17-VALERATE
(VI)

Optimal activity of proflavine salts also occurs with the valerate. Acriflavine and proflavine were found to be virtually inactive as antibacterials when presented as water-in-oil emulsions which are preferred because they protect injured sites, act as lubricants and prevent adhesion of dressings. Fenton and Warren[36] found that the salts of proflavine with formic and acetic acid were only slowly released from a water-in-oil cream because of their high water and low oil solubility. With the higher acid salts such as n-valeric, capric, etc., diffusion increased, but even then release was less than from oil-in-water preparations. Higher acid salts are insoluble in water. The most effective cream was that made with proflavine n-valerate; a reasonably high aqueous concentration could be achieved with sufficient oil solubility to ensure release of the medicament from the aqueous phase (table 9.9).

9.5.3 Influence of vehicle

Consideration of equation 9.10 shows immediately that the vehicle has an influence on the absorption of the drug; if the vehicle is changed so that the drug becomes less soluble in it, P increases so that permeability increases. The vehicle is more dominant in topical therapy than in most routes of administration be-

table 9.9 **Solubilities of proflavine salts in water and in chloroform***

Proflavine salt (mono-)	Solubility in water at 20 °C ($g\ d\ell^{-1}$)	Solubility in chloroform at 20 °C ($g\ d\ell^{-1}$)	Partition coefficient (chloroform/water)
Formate	16.50	Almost insoluble	–
Acetate	15.50	0.008	5.16×10^{-4}
Propionate	13.75	0.012	8.73×10^{-4}
n-Butyrate	12.60	0.016	1.27×10^{-3}
n-Valerate	6.90	0.040	5.79×10^{-3}
Caproate	1.66	0.050	3.01×10^{-2}
Caprylate	0.37	0.090	2.43×10^{-1}
Laurate	Insoluble	0.110	–
Myristate	Insoluble	0.130	–
Cyclohexane carboxylate	1.57	0.024	1.53×10^{-2}
Hemisulphate	0.34	Insoluble	–

*From reference 36

cause the vehicle remains at the site, although not always in an unchanged form. Evaporation of water from the base would leave drug molecules immersed in the oily phase. Oil-in-water emulsion systems may invert to water-in-oil systems, such that the drug would have to diffuse through an oily layer to reach the skin. Non-volatile components of the formulation increase in concentration as the volatile components are driven off; this may alter the state of saturation of the drug and hence its activity. Drug may precipitate due to lack of remaining solvent. These changes mean that theoretical approaches very much represent the ideal cases.

Many modern dermatological formulations are washable oil-in-water systems. However, simple aqueous lotions are also used as they have a cooling effect on the skin. Ointments are used for the application of insoluble or oil-soluble medicaments and leave a greasy film on the skin, inhibiting loss of moisture and encouraging the hydration of the keratin layer. Aqueous creams combine the characteristics of the lotions and ointments. A classification of semisolid bases is given in figure 9.19. This is taken from a comprehensive account of vehicles for dermatological products by van Abbé et al.[37].

The descriptions 'ointment' and 'cream' have no universally accepted meaning and generally refer to the completed formulation. Ointments are generally composed of single-phase hydrophobic bases, for example of pharmaceutical grades of soft paraffin or microcrystalline paraffin wax.

The absorption bases have an alleged capacity to facilitate absorption by the skin but the term also alludes to their ability to take up considerable amounts of water to form water-in-oil emulsions. Lipogels are gels prepared by dispersion of long-chain fatty acid soaps such as aluminium monostearate in a hydrocarbon base. Hydrogels prepared from Carbopols or cellulose derivatives are discussed in chapter 8.

The complexity of many of the topical bases means that a physicochemical

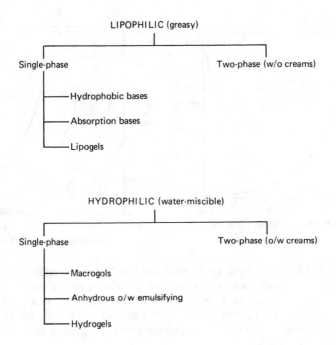

LIPOPHILIC (greasy)

Single-phase   Two-phase (w/o creams)

    — Hydrophobic bases
    — Absorption bases
    — Lipogels

HYDROPHILIC (water-miscible)

Single-phase   Two-phase (o/w creams)

    — Macrogols
    — Anhydrous o/w emulsifying
    — Hydrogels

figure 9.19 Classification of semisolid bases. Some of these systems are dealt with in more detail in chapter 7. After van Abbé *et al.*, reference 37

explanation of their influence on the release of medicament is not always possible. It is not difficult, however, to ascertain this influence and to highlight the problem. The effect of the base–drug relationship has been elucidated by Poulsen and his collaborators[38] who studied the effect of propylene glycol in simple bases. Figure 9.20 summarises the results. The thermodynamic activity of the drug is obviously the determinant of biological activity. If the solubility of the drug in the base is increased by addition of propylene glycol then its partition coefficient, P, towards the skin is reduced. On the other hand, increasing the amount which can be incorporated in the base is an advantage and the concentration (C_v) gradient is increased. It is apparent that there is an optimum amount of solubiliser. The optimum occurs at the level of additive which just solubilises the medicament. Addition of excess results in desaturation of the system, and therefore a decrease in thermodynamic activity. As the total amount of drug in the vehicle is increased from 0.5 to 1.0 mg g^{-1} (figure 9.20*b*) more propylene glycol has to be added to cause a decrease in C_vP. From figure 9.20*b* one could postulate that threefold dilution of a formulation containing 1.0 mg g^{-1} and 60 per cent solubiliser with a base with no solubiliser would cause the activity to *increase*. The vehicle affects penetration only when the release of drug is rate-limiting.

In situations where all of the activity gradient is in the applied phase, skin properties play no part. In these cases drug concentration in the vehicle, the

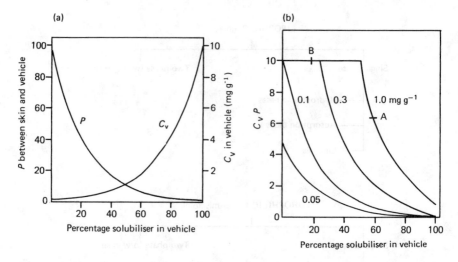

figure 9.20 (a) Drug partition coefficient and solubility dependence on vehicle composition for 'ideal' case. (b) Predicted relative steady-state penetration (C_vP) as a function of vehicle composition for 'ideal' case. Point A represents a formulation containing 1.0 mg g^{-1} and 60 per cent solubiliser. On threefold dilution with a base containing no solubiliser point B is arrived at where C_vP is higher. From reference 38

diffusion coefficient of the drug in the system and the solubility of the drug are the significant factors. For the cases in which these factors are not important the only significant factor involving the vehicle is the thermodynamic activity of the penetrating agent contained in it[39]. Using the simplest model, a solution of concentration C applied to an area A of the skin, the steady-state rate of penetration, dQ/dt, is given by

$$\frac{dQ}{dt} = \frac{PCDA}{\delta} \qquad (9.16)$$

δ being the thickness of the barrier phase. An equivalent form of this equation expresses release in terms of the thermodynamic activity, a, of the agent in the vehicle

$$\frac{dQ}{dt} = \frac{a}{\gamma}\frac{DA}{\delta} \qquad (9.17)$$

where γ is the activity coefficient of the agent in the skin barrier phase[39]. All ointments containing finely ground particles of the drug where the thermodynamic activity is equal to that of solid drug will have the same rate of penetration, provided that the passage of the drug in the barrier phase is rate-limiting. Some activity coefficients of the nerve gas sarin in various solvents are shown in table 9.10. The concept of thermodynamic activity is considered in more detail in chapter 3.

Solutes 'held firmly' in the vehicle exhibit low activity coefficients (low escaping tendencies) and thus low rates of penetration. Higuchi[39] cites the dif-

table 9.10 **Limiting activity coefficients of sarin in organic solvents and water***

Solvent	γ
Water	14
Diethylene glycol	2.4
Isoamyl alcohol	1.07
Benzyl alcohol	0.446

*From reference 39

ferences in activity of phenols in mineral oil and polypropylene glycol bases. The latter are bland, the former corrosive. Polyether–phenol complex formation decreases the thermodynamic activity of the phenols which are therefore less toxic.

In more complex vehicles the activity *a* is impossible to determine and other approaches must be adopted. For example, in emulsion systems the relative affinity of drug for the external and internal phases is an important factor. A drug dissolved in an internal aqueous phase must be able to diffuse through the oily layer to reach the skin. Three cases can be considered: solution, suspension and emulsion systems.

(1) Release from *solutions* is most readily understood and quantified by equations of the form

$$\frac{dQ}{dt} = C_0 \left(\frac{D}{\pi t}\right)^{1/2} \tag{9.18}$$

or

$$Q = 2 C_0 \left(\frac{Dt}{\pi}\right)^{1/2} \tag{9.19}$$

where C_0 is the initial medicament concentration in solution, D is the diffusion coefficient, and t is the time after application of the vehicle. As D is inversely proportional to the viscosity of the vehicle, one would expect that drug release would be slower from a viscous vehicle. There is evidence for this in section 9.11, where the rheological characteristics of rectal formulations are correlated with bioavailability.

(2) If a drug in *suspension* is to have any action it must have a degree of solubility in the base used. A suspension of a drug will therefore have a saturated solution of the drug present in the continuous phase. Release of medicament in these conditions is given by

$$\frac{dQ}{dt} = k_p \left(\frac{C_0 C_s D}{2 t}\right)^{1/2} \tag{9.20}$$

where C_s is the solubility and C_0 the concentration of the drug in the vehicle. This equation applies only when $C_s \ll C_0$. Only material in solution penetrates the stratum corneum and the depleted layer in the vehicle is replenished only by solution of distant particles and diffusion of the drug molecules to the depleted layer.

(3) For *emulsion*-type vehicles equations similar to those used to describe germicidal behaviour in two-phase systems can be applied. If D_1 and D_2 are the diffusion coefficients of the medicament in the continuous and disperse phases respectively, ϕ_1 and ϕ_2 are the volume fractions of these two phases, and P is the partition coefficient of the drug between the phases, the effective diffusion coefficient is given by

$$D_e = \frac{D_1}{\phi_1 + P\phi_2} \left[1 + 3\phi_2 \left\{ \frac{PD_2 - D_1}{PD_2 + D_1} \right\} \right] \tag{9.21}$$

If $\phi_2 \gg \phi_1$ and $D_2 \gg D_1$ (likely if phase 2 is water), then

$$D_e = \frac{D_1}{P\phi_2} (1 + 3\phi_2) \tag{9.22}$$

For a system containing 20 per cent water ($\phi_2 = 0.20$)

$$D_e = \frac{D_1 (1.6)}{0.2P}$$

Therefore

$$D_e = \frac{8D_1}{P} \tag{9.23}$$

The importance of P is seen in this equation. The value of D_e obtained here can be used in equations 9.18 and 9.19 above to obtain an approximation of effects.

In vitro assessment of topical formulations

Steroid formulations can be tested clinically using the vasoconstrictor activity of the steroid to quantitate the results. In a few cases it is feasible to measure quantities of drug reaching the systemic circulation after topical application. An *in vitro* system using isopropyl myristate as the 'solvent phase' or 'biophase' has been used by Ostrenga *et al.*[40] (see figure 9.21). The amount of drug released in 7 hours in this system correlates with the vasoconstriction index for four formulations of fluocinonide. Vasoconstriction evidenced by skin blanching correlates with steroid potency in topical preparations.

9.5.4 Dilution of topical steroid preparations

Busse[41] drew attention to the dangers of diluting topical steroid formulations. Inappropriate dilution of carefully formulated creams and ointments may result in changes in stability and bacteriological effectiveness. The biopharmaceutical considerations will be apparent from the discussion in this chapter. Busse gives the example of clobetasol propionate formulated in a base containing the optimum amount of propylene glycol. The solubilities of the steroid in propylene glycol are those shown in table 9.11. A 1 in 2 dilution of a 0.05 per cent cream, with a vehicle containing water and no propylene glycol, will precipitate a large proportion of the steroid. The same principles apply to steroids presented in fatty acid propylene glycol base.

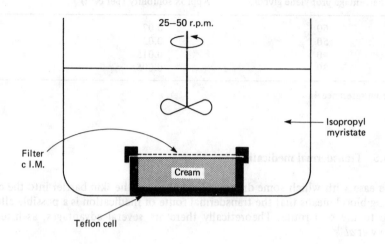

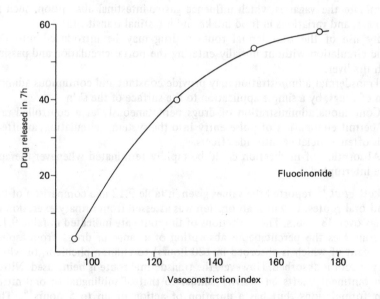

figure 9.21 An *in vitro* apparatus for the examination of the release of a steroid from a cream formulation. The cream is placed in a Teflon receptacle and covered with a membrane impregnated with isopropyl myristate. The concentration of fluocinonide is monitored in the bulk isopropyl myristate layer. In the lower figure release *in vitro* is compared with results of vasoconstrictor tests on four formulations. After Ostrenga *et al.*, reference 40

table 9.11 **Solubility of clobetasol propionate***

Percentage propylene glycol	Approx solubility (per cent)
60	0.07
50	0.03
40	0.015
30	0.005

*From reference 41

9.5.5 Transdermal medication

The ease with which some drugs can pass through the skin barrier into the circulating blood means that the transdermal route of medication is a possible alternative to the oral route. Theoretically there are several advantages, as listed by Shaw et al.[42]:

(1) For drugs that are normally taken orally, administration through the skin can eliminate the vagaries which influence gastro-intestinal absorption, such as pH changes, and variations in food intake and intestinal transit time.

(2) By use of the transdermal route a drug may be introduced into the systemic circulation without initially entering the portal circulation and passing through the liver.

(3) Transdermal administration may provide constant and continuous administration of drugs, by a simple application to the surface of the skin.

(4) Continuous administration of drugs percutaneously at a controlled rate should permit elimination of pulse entry into the systemic circulation, an effect which is often associated with side-effects.

(5) Absorption of medication could be rapidly terminated whenever therapy must be interrupted.

Beckett et al.[43] reported the values given in table 9.12 in a comparison of the skin and oral routes, in which absorption was assessed from urinary excretion of the drugs over 48 hours. The limitations of the route are indicated in table 9.13, which compares the percutaneous absorption of a range of drugs, from aspirin (22 per cent of which is absorbed in 120 hours) to chloramphenicol, of which only 2 per cent is absorbed. However, the transdermal route is being used. Nitroglycerin ointment exerts an activity similar to that of sublingual or oral nitroglycerin formulations, but has a duration of action of up to 5 hours[44]. The number of transdermal patches is increasing. Some of these are discussed in chapter 8.

In preliminary studies of a transdermal therapeutic system, workers at Alza have studied the absorption of ephedrine, scopolamine and chlorpheniramine. Maximum flux from a saturated aqueous system was 300 μg cm^{-2} h^{-1} and for a mineral oil system 250 μg cm^{-2} h^{-1}. Using a 'patch' with a rate-limiting polymeric membrane (see chapter 8) delivery is controlled to 40–50 μg cm^{-2} h^{-1}; that is, it is presented to the skin at that rate. A 10 cm^2 device will therefore deliver 0.4–0.5 mg h^{-1}. Control of release can be exercised by altering the mem-

table 9.12 **Relative urinary excretion (acidic pH) in 48 hours of ephedrine derivatives given by oral route and percutaneously***

	Partition coefficient of unionised drug†	Mean recovery of unchanged drug	
		Skin route	Oral route
Norephedrine	<0.01	44	100††
Ephedrine	0.17	63	83
Methylephedrine	2.11	50	63
Ethylephedrine	7.74	31	34

†Determined between n-heptane and 0.1 N NaOH
††Excretion of norephedrine by oral route taken to be 100
*From reference 43

table 9.13 **Percutaneous absorption of a range of drugs in man***

Drug	Percentage dose absorbed (120 hours)
Aspirin	22
Chloramphenicol	2
Hexachlorophene	3
Salicylic acid	23
Urea	6
Caffeine	48

*From reference 43

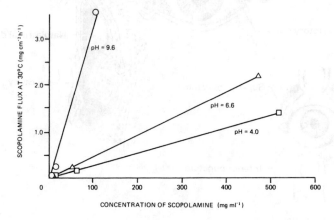

figure 9.22 The effect of scopolamine concentration and pH on its flux across a membrane. Increasing the pH increases the flux as ionisation is decreased. From reference 34

brane properties and by changing the pH of the reservoir solution for a drug such as scopolamine (figure 9.22). The Alza transdermal device is designed for affixing behind the ear.

9.6   Medication of the eye

The eye is not, of course, a general route for the administration of drugs to the body, but it is considered here because absorption of drugs does occur from medications applied to the eye, sometimes to produce toxic effects, sometimes the desired local effect on the eye or its component parts[45]. We must consider the factors affecting drug absorption from the eye, and those properties of formulations that affect drug performance. A wide range of drug types are placed in the eye, including antimicrobials, antihistamines, decongestants, mydriatics, miotics and cycloplegic agents[46].

The eye (figure 9.23) has two barrier systems: a blood–aqueous barrier and a blood–vitreous barrier. The former is composed of the ciliary epithelium, the epithelium of the posterior surface of the iris, and blood vessels within the iris. Solutes and drugs enter the aqueous humour at the ciliary epithelium and at blood vessels of the eye. The blood–vitreous barrier consists of the ciliary epithelium, the retinal pigmented epithelium and the endothelial layer which lines the retinal blood vessels. Many substances are transported out of the vitreous humour at the retinal surface. Solutes also leave the vitreous humour by diffusing to the aqueous humour of the posterior chamber.

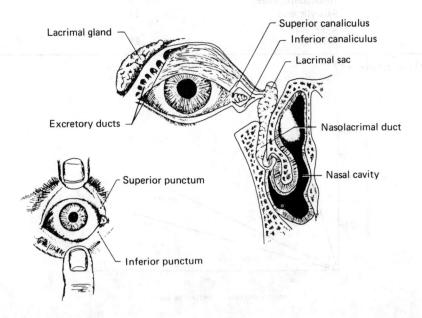

figure 9.23 Diagrams of parts of the eye of importance in medication. The superior and inferior punctae are the drainage ports for solutions and tear fluids. Medicaments can drain via the canaliculi into the nasolacrimal duct and then to the nasal cavity from whose surfaces absorption can occur. From J. R. Robinson (ed.), *Ophthalmic Drug Delivery Systems*, American Pharmaceutical Association, Washington DC, 1980

Drugs are usually applied to the eye in the form of drops or ointment for local action. The absorbing surface is the cornea. Drug that is absorbed by the conjunctiva enters the systemic circulation. It is useful to consider some of the properties of the absorbing surfaces and their environment.

9.6.1 The eye

Figure 9.23 is a diagrammatic representation of those parts of the eye involved in drug absorption. The cornea and the conjunctiva are covered with a thin fluid film, the tear film, which protects the cornea from dehydration and infection. Cleansed corneal epithelium is hydrophobic, physiological saline forming a contact angle of about 50° with it. It has, in this clean condition, a critical surface tension of 28 mN m^{-1}. However, normal eyes produce mucin and this adsorbs onto the corneal surface to convert the surface into a hydrophilic entity. The surface tension of the tear film is around 46 mN m^{-1}. The aqueous phase of tear fluid is secreted by blinking. Tears comprise inorganic electrolytes — sodium, potassium and some calcium ions, chloride and hydrogen carbonate counterions — as well as glucose. The macromolecular components include some albumin, globulins and lysozyme. Lipids which form a monolayer over the tear fluid surface derive from the Meibomian glands which open on to the edges of the upper and lower lids. This secretion consists mainly of cholesteryl esters with low melting points (35 °C) due to the predominance of double bonds and branched-chain structures. This fluid lies on the surface of the cornea (figure 9.24) and its importance in formulation lies in the possibility that components

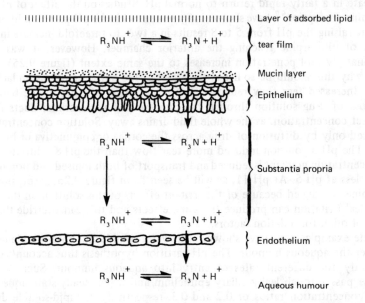

figure 9.24 Diagrammatic representation of the tear film and the penetration of a base through the cornea. In this example R$_3$N represents a drug such as homatropine. After R. A. Moses. *Adler's Physiology of the Eye*, 5th Edition, Mosby, St. Louis, 1970

of formulations or drugs themselves can so alter the properties of the corneal surface or interact with components of the tear fluid that tear coverage of the eye is disrupted. When this occurs the so-called dry-eye syndrome may arise, characterised by the premature break-up of tear fluids giving rise to dry spots.

9.6.2 Absorption of drugs applied to the eye

The cornea, which is the main barrier to absorption, consists of three parts: the epithelium, the stroma and the endothelium. Both the endothelium and the epithelium have a high lipid content and, as most membranes are, they are penetrated by drugs in their unionised lipid-soluble forms. However, the stroma lying between these two structures has a high water content and thus drugs which have to negotiate the corneal barrier successfully must be both lipid-soluble and water-soluble to some extent (figure 9.24). The tears have some buffering capacity so, as we noted before, the pH-partition hypothesis has to be applied with some circumspection. The acid neutralising power of the tears when 0.1 mℓ of a 1 per cent solution of a drug is instilled into the eye is approximately equivalent to 10 $\mu\ell$ of a 0.01 mol ℓ^{-1} strong base. The pH for maximum solubility (chapter 5) or maximum stability (see chapter 4) of a drug may well be below that of its optimum in relation to acceptability and activity. Under these conditions it is possible to use a buffer of low buffering capacity to maintain a low pH adequate to prevent change in pH due to alkalinity of glass or carbon dioxide ingress from the air. When such drops are instilled into the eye the tears will participate in a fairly rapid return to normal pH. Studies on the effect of pH on pilocarpine absorption[47] have shown, in agreement with the pH-partition hypothesis, that raising the pH from 5 to 8 results in a two- to threefold increase in the amount of pilocarpine reaching the anterior chamber. However, it was also found that glycerol penetration increases to the same extent (figure 9.25). The clue to why this should be so lies in the effect of the buffer solutions on lachrymation. Increased tear flow reduces the concentration of drug in the bathing fluid; loss of drug solution through the punctae and nasolacrimal ducts does not affect concentration, as the whole fluid drains away. Solution concentration is reduced only by diffusion of drug across the cornea or conjunctiva or by tear inflow. The pH 5 solution induced more tear flow than the pH 8 solution; thus the concentration gradient is reduced and transport of both ionised and non-ionic drugs is less at pH 5. At pH 11, as will be seen from figure 9.25, absorption of pilocarpine is reduced because of the irritant effects of this solution on the eye. Mechanical irritation can produce the same effects and this can override the influence of other formulation factors.

As the examples used have shown, both water-soluble and lipid-soluble drugs can enter the aqueous humour. The pH-partition hypothesis thus accounts only imperfectly for different rates of entry into aqueous humour. Sucrose and raffinose pass through leaky ciliary epithelium and reach steady-state aqueous/plasma concentration ratios of 0.2 and 0.3, respectively[48]. Lipid-soluble drugs including chloramphenicol and some tetracyclines can achieve higher concentrations as they can enter by both pathways. Penicillins, however, reach only low aqueous/plasma concentration ratios because they are removed from aqueous humour by absorption through the ciliary epithelium. Proteins are excluded and because this is so, protein binding of ophthalmic drugs limits their absorption[49].

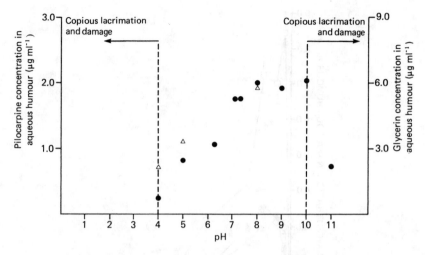

figure 9.25 The influence of vehicle pH on the aqueous humour concentration of pilocarpine and glycerol. ●, 10^{-2} mol ℓ^{-1} pilocarpine solutions; △, 0.14 mol ℓ^{-1} glycerol solutions. All samples obtained at 20 minutes post-instillation represent results from a minimum of six eyes. From reference 47

9.6.3 Influence of formulation

Pilocarpine activity has been compared in various formulations. Figure 9.26 shows some of the results on formulations including results on ointments designed to prolong the contact of the drug with the cornea. One of the most difficult problems is to design vehicles which will retard drainage and prolong contact. Viscous polymer vehicles help to some extent but are not the complete answer.

Chrai and Robinson[49] report that the rate of drainage decreased as the viscosity of the eye-drops increased and these factors contribute to an increased concentration of the drug in the precorneal film and aqueous humour. The magnitude of the concentration increase was small considering the 100-fold change in the viscosity, and it was concluded that the viscosity of the solution is not as important a factor in bioavailability as was previously thought.

The results of incorporating pilocarpine (VII), a relatively water-soluble drug, and fluorometholone (a lipophilic drug) into a water-in-oil ointment base can be compared in figure 9.27. Pilocarpine is thought to be released only when in contact with aqueous tear fluid, whereas the steroid, being soluble in the base, can diffuse through the base to replenish the surface concentrations and thus produce a sustained effect.

Pilocarpine (pK_{a1} = 7.05)
(VII)

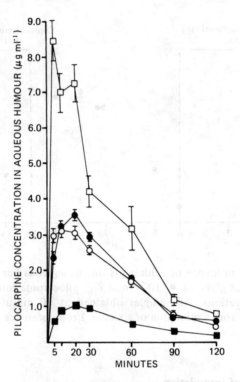

figure 9.26   Aqueous humour levels of pilocarpine after dosing with 10^{-2} mol ℓ^{-1} ointment and solution in intact and abraded eyes. ●, 25 mg of ointment, intact eyes; ○, 25 mg of ointment, abraded eyes; ■, 25 $\mu\ell$ of solution, intact eyes; □, 25 $\mu\ell$ of solution, abraded eyes. All points represent an average of 8–16 eyes. From reference 47

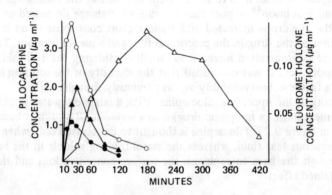

figure 9.27   Aqueous humour levels after dosing with pilocarpine and fluoro-metholone in ointment and aqueous solution. ●, 10^{-2} mol ℓ^{-1} pilocarpine solution, 25 $\mu\ell$; ○, 10^{-2} mol ℓ^{-1} pilocarpine ointment, 25 mg; ▲, saturated fluoro-metholone solution, 50 $\mu\ell$; △, 0.1 per cent fluorometholone ointment, 50 mg. From reference 47

The effect of the polymeric vehicles, poly(vinyl alcohol) and hydroxypropyl-methylcellulose, has been demonstrated by Wang and Hammerlund[50]. Poly(vinyl alcohol) increases the effectiveness of the drug substance. Hydroxypropylmethyl-cellulose (HPMC) has, by a similar mechanism, been found to reduce the effective dose of neomycin sulphate required to prevent infection of corneas of experimental animals (table 9.14).

table 9.14 **Calculation of effective dose of neomycin sulphate required to prevent infection in 50 per cent of rabbit corneas when incorporated into various vehicles* (ED$_{50}$)**

Vehicle	Derived values (mg base per mℓ)	
	ED$_{50}$	95 per cent confidence limits
HPMC, 0.5 per cent	0.50	0.403–0.620
PVA, 1.4 per cent	1.00	0.750–1.330
PVP, 1.4 per cent	1.10	0.89 –1.35
Distilled water	1.03	0.84 –1.27

*From F. C. Bach *et al. Am. J. Ophthalmol.*, 69, 659 (1970)

Eye-drops are often formulated to be isotonic with tear fluid but deviations from tonicity do not cause problems, although hypertonicity may cause stinging of the eye and hypotonicity may increase the permeability of the cornea[46]. Some ingredients of eye medications may increase the permeability of the cornea. Surface-active agents are known to interact with membranes to increase the permeability; benzalkonium chloride has surfactant properties and may well have some effect on corneal permeability, although its primary purpose is as a bacteriostat and bactericide. Chlorhexidine acetate and cetrimide, both of which are surface-active, are also used.

Prodrugs

Attempts have been made to improve the performance of drugs used in the eye. One approach has been to modify the drug substance to increase its ability to penetrate the corneal barrier. In figure 9.28 results of application of 0.5 per cent adrenaline and 0.16 per cent of the dipivoyl derivative of adrenaline are shown. The more hydrophobic derivative is absorbed to a greater extent and is then hydrolysed to the active parent molecule in the aqueous humour.

Reservoir systems

Since the advent of soft contact lenses attempts have been made to use these as drug reservoirs; drug is imbibed from solutions into the polymer matrix and when this is placed in the eye the drug leaches out, generally over a period of up to 24 hours. Pilocarpine has also been incorporated[51] into contact lenses. Levels of corticosteroid (prednisolone) applied in a copolymer of 2-hydroxyethyl methacrylate and *N*-vinyl-2-pyrrolidone were compared with levels in aqueous humour, cornea and iris after applications of solutions. When the drug was incor-

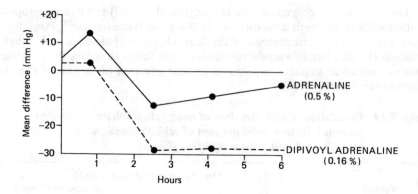

figure 9.28   The action of 0.5 per cent adrenaline (VIII) and 0.16 per cent dipivoyl adrenaline (IX) on intraocular pressurre. From D. A. McClure. *Prodrugs as Novel Drug Delivery Systems* (ed. T. Higuchi and V. Stella), American Chemical Society, Washington, 1975

porated into the lens, aqueous and corneal levels of the drug were maintained two- to threefold higher at four hours[52].

Dependence on leaching does not give sufficient control of drug release for prolonged action. The Alza Ocusert device releases controlled amounts of pilocarpine over a period of seven days and is designed for the treatment of glaucoma. The Ocusert (figure 9.29) comprises a drug reservoir in which pilocarpine is embedded in an alginic acid matrix, and this is bounded by two rate-controlling membranes of polymer (vinyl acetate–ethylene copolymer). The device has a raised rim to aid visibility and handling and it is inserted into the lower conjunctival sac of the eye. The rate-controlling membranes are subject to strict quality control[53] during manufacture and their thicknesses adjusted to give the appropriate flux of drug. The difference between the available 20 and 40 μg h^{-1}

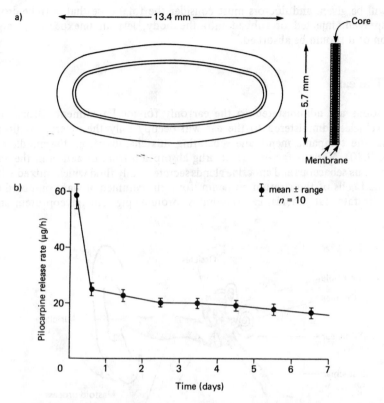

figure 9.29 Dimensions and structure of the Alza Ocusert device (a) and a release rate profile over one week (b). From J. W. Snell and R. W. Baker. *Ann. Ophthalmol.*, 6, 1037 (1974)

system is effected by altering the amount of plasticiser, di-(2-ethylhexyl)-phthalate, in the polymer; increasing the concentration of plasticiser increases the permeability to pilocarpine.

9.6.4 Systemic effects from eye-drops

Atropine toxicity resulting from the use of the drug in eye-drops to dilate the pupil has been reported[54], as has a rise in blood pressure in premature infants following the use of 10 per cent phenylephrine eye-drops in preparations for ophthalmoscopy[55]. In infants the dose which can be administered to the eye is relatively large compared to an adult. The mucosa in the eye, nose and mouth is much thinner and more permeable to a drug than is, say, the skin. Drugs placed in the eye, nose or mouth, moreover, may bypass the metabolic transformations which may inactivate the drug given orally. A commentator[56] has thus written, 'Drugs ought to be studied in both animals and man by all the routes by which

they will be given, and doctors must consider the total dose that is to be drop-
ped, sprayed, injected or rubbed into the body, lest an unexpectedly large
fraction of it should be absorbed.'

9.7 The ear

Medications are administered to the ear only for local treatment. Drops and
other vehicles administered to the ear will occupy only the external auditory
meatus, the tympanic membrane separating this chamber from the middle ear
(figure 9.30). Various factors affect drug absorption from or action in the ear.
The mucous sebaceous and apocrine glands secrete an oily fluid which, mixed with
exfoliated cells of the stratum corneum, form the cerumen or wax composed of
inter alia fats, fatty acids, carbohydrates, protein, pigment, glycoprotein and
water.

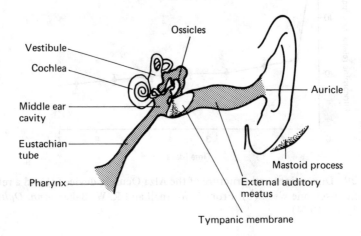

figure 9.30 The ear and its associated structures. Medications normally would
enter only the external auditory meatus

The acidic environment of the ear skin surface (around pH 6), sometimes
referred to as the acid mantle of the ear, is thought to be a defence against
invading micro-organisms.

Various ceruminolytic agents achieve their action by partially dissolving the
wax. Several commercial ear-drops contain poloxamers or sodium dioctyl sulphosuccinate[57]. In otitis media, infection of the Eustachian tube is involved; antibiotic treatment is indicated along with oral analgesics, there being little evidence
that topical analgesics give faster relief, suggesting poor absorption from the ear.

9.8 Absorption from the vagina

The vagina cannot be considered to be a route for the systemic administration of
drugs although oestrogens have been applied intravaginally[58]. However, a number

of medicaments applied to the vaginal epithelium are absorbed as it is permeable to a wide range of substances which include steroids, prostaglandins, iodine and some antibiotics[59]. Econazole and miconazole are two which are appreciably absorbed. The epithelial layer of the vagina comprises lamina propria and a surface epithelium of non-cornified, stratified squamous cells. The thickness of the epithelium increases after puberty and then again after menopause. The pH in the vagina decreases after puberty, varying between pH 4 and 5 depending on the menstrual cycle. There is, of course, little fluid in the vagina. The absorbing surface is under constant change, therefore absorption is variable. While the presence of mucus is likely to affect (i.e. retard) absorption, there is unlikely to be other material in the vagina which will inhibit absorption. The uterine and pudendal arteries are the main sources of blood to the vagina; the venous plexus which surrounds the vagina empties into the internal ileac veins. Lymph vessels drain the vagina.

9.9 Inhalation therapy

The respiratory system serves both as a mode of entry into the body for a variety of airborne substances and also as a route of medication. The large contact area of surfaces extends to more than 30 m^2. The surfaces have been described[60] as 'gossamer-thin membranes that separate the lung air from the blood, which courses through some 2000 km of capillaries in the lungs'. Hatch and Gross[61] refer to the consequent 'exquisite degree of intimacy between the lung tissue and blood and the atmospheric environment'. Even within the gastro-intestinal tract the possibilities for exchange of substances between its contents and other body tissues is markedly less than within the respiratory system.

Drugs administered by inhalation are mostly intended to have a direct effect on the lungs. However, the efficiency of inhalation therapy is often not high; for example, only about 8 per cent of the inhaled dose of sodium cromoglycate administered from a Spinhaler (see section 9.9.2) reaches the alveoli[60]. Crude inhalers have been used for medicinal purposes for at least two centuries. Solutions of volatile aromatic substances with a mild irritant action, inhaled as vapour arising from hot aqueous solutions, have been used for many years. Compounds such as menthol, thymol, eucalyptus, benzoin and creosote have been among the most frequently used, although their value has been attributed to the expectorant effect of hot moist air. Nebulisers designed to administer solutions of bronchodilators are of more recent origin. Adrenaline and isoprenaline have been administered in this way using simple nebulisers, the solution being atomised and inhaled. The particle size of the aerosol is so wide that many droplets are too large to penetrate the respiratory tract, for the particle size of the aerosolised drug is one of the prime determinants of efficiency. In the older nebulisers, particles would settle without reaching the patient's face: a 100 μm particle of unit density settles in still air at a velocity of 8 cm s^{-1}. The duration of existence of a suspension of particles of size around 10 μm is so brief that the upper limit of aerosols of therapeutic interest is well below this size. At the other end of the scale, particles that are too small can easily be exhaled without deposition in the respiratory tract.

Figure 9.31 shows the order of maximum size of particles that can penetrate

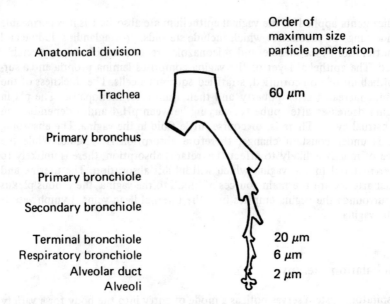

Anatomical division	Order of maximum size particle penetration
Trachea	60 μm
Primary bronchus	
Primary bronchiole	
Secondary bronchiole	
Terminal bronchiole	20 μm
Respiratory bronchiole	6 μm
Alveolar duct	2 μm
Alveoli	

figure 9.31   Deposition of particles in various regions of the respiratory tract according to particle size. From reference 62

to various parts of the respiratory tract, from trachea to alveoli. Three different physical forces operate within the respiratory system to remove particles from the inhaled air. These are inertial forces, gravity, and diffusion. As flowing air moves in and out, *inertial* forces within the nasopharyngeal chamber and at the points of branching of the airways, wherever the direction of flow changes, result in the collision of particles with surfaces. Along the finer airways particles are removed by *gravity*. Very fine particles (< 0.5 μm) are deposited on the walls of the finest airways by *diffusion*, the result of bombardment of the particles by gas molecules. Physical trapping by contact of particles with tube wall is of little practical significance in medical aerosols.

9.9.1   Physical factors affecting deposition of aerosols

Particles emanating from an aerosol valve or nebuliser and inhaled deposit in the respiratory tract by gravity, sedimentation, diffusion and inertial precipitation. Particle size, or particle size distribution, is important in several of these processes and will be affected by the nature of the aerosol-producing device and by the formulation. Growth of particles in a hygroscopic environment may occur, as may agglomeration and deagglomeration of aggregates. Hence appropriate sampling and measuring systems are of great importance (see section 9.9.3).

Gravitational settlement

Fine particles falling through the air under the force of gravity do so at a constant velocity such that the resistance of the air balances the mass of the particle. The following equation relates particle diameter, d, and density, ρ, to terminal velocity μ_t

$$\mu_t = \frac{\rho g d^2}{18\eta} \qquad (9.24)$$

where g is the gravitational constant and η the viscosity of the air. For air, $\eta = 1.9 \times 10^{-7}$ N m^{-2} s and therefore

$$\mu_t = 2.9 \times 10^6 \, \rho d^2 \ (\text{m s}^{-1})$$

Table 9.15 shows the terminal velocities for particles of a range of diameters from 40 to 0.04 μm, reinforcing the importance of small size in preventing removal of particles before entry into the lower reaches of the respiratory tract. In still air, a cloud of powder of about 20 μm diameter takes a few seconds to settle, whereas dust around 1 μm diameter takes about a minute.

table 9.15 **Terminal velocity (μ_t) of spherical particles of unit density in air***

Diameter (μm)	μ_t (m s^{-1} × 10^2)
40	4.8
20	1.2
10	2.9×10^{-1}
4	5×10^{-2}
1	3.5×10^{-3}
0.6	1.4×10^{-3}
0.1	8.6×10^{-5}
0.04	2.9×10^{-5}

*From reference 61. For particles smaller than 100 μm, resistance to fall decreases and a correction is applied to equation 9.24

Sedimentation

As the particles of drug move with the air in laminar flow in the airways, they fall, under the force of gravity, a distance equal to $\mu_t t$, where t is the time of travel. If the tube in which they move is of radius R and is inclined at an angle ψ to the horizontal, as most airways will be, the maximum distance of fall will no longer be $2R$, but $2R/\cos \psi$. The ratio, r, of the distance of fall to maximum distance for deposition to be achieved is thus

$$r = (\mu_t t \cos \psi)/2R \qquad (9.25)$$

The probability of deposition by sedimentation is proportional to this ratio; the closer $\mu_t t$ is to $2R/\cos \psi$, the higher the likelihood of deposition by this mechanism. If the particles are evenly distributed over the cross-section of an airstream, it is theoretically possible to calculate the probability of deposition in tubular airways and in a spherical alveolus. Individual tubes are of course randomly positioned with reference to the horizontal and therefore an average value of ψ is used; airflow is not always laminar, and orderly deposition will not always occur, but in spite of these problems the following calculations of percentage sedimentation (S) are quoted by Hatch and Gross[61]. If $S = 55$ per cent when

$d = 2$ μm and the deposition diameter is 1.0 μm, for 1 μm particles $S = 29$ per cent, and for 0.5 μm particles $S = 10$ per cent. That is, sedimentation reduces in importance as the particle size decreases.

Diffusion

The effectiveness of deposition by diffusion increases as particle size is reduced, which contrasts with the above. There must therefore be a particle diameter for which both processes have a combined minimum value; this occurs with particles of approximately 0.5 μm in diameter. Particles of this size have the minimum probability of respiratory deposition.

Inertial precipitation

When, during breathing, the airflow suddenly changes direction, a drug particle will continue in its original direction of flow owing to its inertia. In this way the particle may be precipitated on the channel wall. In curved tubes the particle in an airstream which experiences a sudden bend experiences a similar fate, and the effective stopping distance at right-angles to the direction of travel is

$$h_s = \frac{\mu_t \, \mu \sin \theta}{g} \tag{9.26}$$

where μ is the velocity of the airstream with particles approaching a bend of angle θ. $\mu \sin \theta$ is therefore a component of initial particle velocity at right-angles to the direction of airflow. The probability of inertial deposition, I, is proportional to the ratio of stopping distance, h_s, to the radius, R, of the airway; that is

$$I \propto \frac{h_s}{R} \propto \frac{\mu_t \, \mu \sin \theta}{gR} \tag{9.27}$$

from a calculation similar to that discussed for sedimentation above. Calculated inertial deposition shows a dependence on particle size as follows: 10 μm 50 per cent, 7 μm 33 per cent, 5 μm 20 per cent, 3 μm 10 per cent, and 1 μm 1 per cent.

9.9.2 Experimental observations

The complexity of the respiratory system prevents a precise mathematical approach to the problem. However, there are several clinical studies which demonstrate the importance of particle size. Work on the deposition of particles of different characteristics has also shown that particles of hygroscopic materials are removed in a higher percentage than are non-hygroscopic particles, because of the growth of these particles by uptake of water from the moist air in the respiratory system. Apart from its importance in determining the efficiency of an aerosol in reaching the alveoli, particle size may be critical in determining response. Proving that this is so is not easy. However, it may be inferred from studies aimed at explaining differences in toxicity of uranium oxide, in which particles of various sizes were injected intratracheally, and from our knowledge of the relationship between particle size, solubility and rates of solution (chapters 2 and 5). When the particle size of uranium oxide fell below 3 μm there was a striking increase in systemic response. Rate of solution, rather than absolute solubility, has been stated to be the probable rate-limiting step in the transfer of agent into the blood and the consequent dosage rate at critical sites beyond the lungs.

The swallowing of particles administered by aerosols complicates the study of systemic effects and Cohn *et al.*[63] have emphasised that the gastro-intestinal tract is a 'significant portal of entry of a dry-particle aerosol to the systemic circulation'. The effect of particle size on the fate of particles inhaled from an aerosol is shown in figures 9.32 and 9.33.

Figure 9.34 shows results for subjects receiving sodium cromoglycate (SCG) in the form of a powder of two particle size distributions from a Spinhaler. When used by patients, the Spinhaler delivers about 25 per cent by weight of the drug (SCG), which is normally dispersed as particles below 6 μm in size, about 5 per

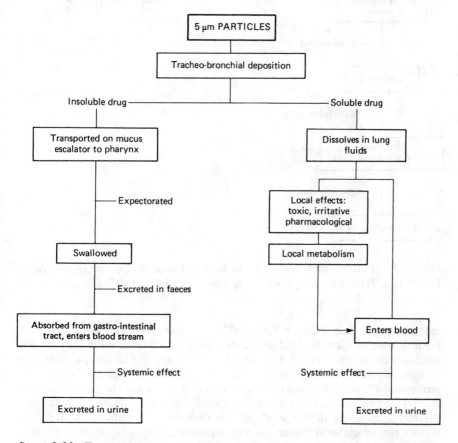

figure 9.32 Fate of particles of approximately 5 μm in diameter deposited on the ciliated epithelium of trachea. From D. G. Clark. *Proc. Eur. Soc. Study Drug Toxicity*, 15, 252 (1974)

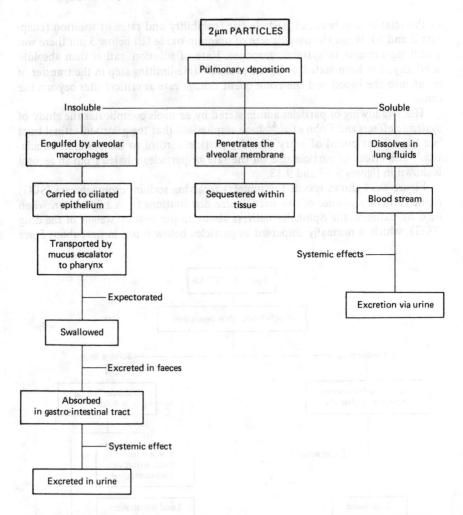

figure 9.33 Fate of particles of 2 μm in diameter deposited in alveoli. From D. G. Clark. *Proc. Eur. Soc. Study Drug Toxicity*, 15, 252 (1974)

cent being less than 2 μm in diameter[64]; in this experimental study, however, the mass median diameter (and geometric standard deviation) of the SCG particle batches were, respectively, 2 ± 1.2 and 11.7 ± 1.1 μm. The lactose particles used as a control were all in the range 1–20 μm. The results in figure 9.34 indicated provisionally that the SCG acts primarily on receptors in and around the terminal airways and alveoli, but final proof is required. There is no doubt, however, that the biological effect of the small particle material is dramatically greater than that of coarser material, hence the importance of storage conditions of the SCG cartridge capsules to prevent aggregation of the drug particles. Although the Spinhaler is designed to break up drug aggregates its efficiency will be reduced if moisture uptake is increased by storage in humid conditions, either in the

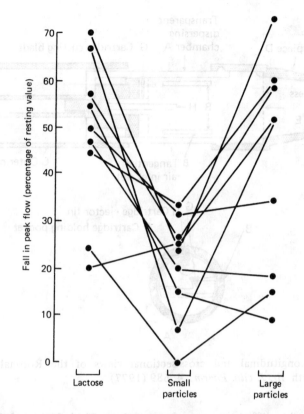

figure 9.34 The effect of placebo (lactose) and small and large particles of sodium cromoglycate on prevention of exercise-induced asthma, measured by the effect of peak airflow as a percentage of the resting value. From reference 64

pharmacy or in the home. Following administration of aerosols of this drug of large (11 μm diameter) size, 46 (± 20) per cent of the dose can be detected in the mouth. An alternative dry powder aerosol is illustrated in figure 9.35.

The Autohaler has been devised as a breath-activated pressurised inhaler system because of the difficulty experienced by some patients in coordinating manual operation of an aerosol with inhalation. The Autohaler is activated by the negative pressure created during the inhalation phase of respiration and is specifically designed to respond to shallow inhalation in those with restricted pulmonary capacity.

Orally administered corticosteroids are effective in the treatment of chronic bronchial asthma. The inhalation route has been widely used in attempts to avoid systemic side-effects, such as adrenal suppression. Evidence of a qualitative nature suggests that inhaled steroids are absorbed systemically to a significant extent. Studies by Burton and Schanker[65] showed that the respiratory tract epithelium has permeability characteristics similar to those of the classical biological membrane. Lipid-soluble compounds are absorbed more rapidly than lipid-insoluble molecules. Cortisone, hydrocortisone and dexamethasone are absorbed rapidly

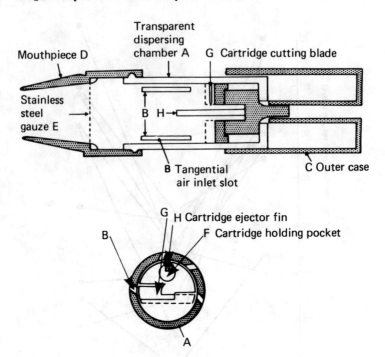

figure 9.35 Longitudinal and cross-sectional views of the Rotohaler. From G. W. Hallworth. *Br. J. clin. Pharm.*, 4, 689 (1977)

by a non-saturable diffusion process from the lung. The half-time of absorption for these steroids was of the order of 1–1.7 min. Quaternary ammonium compounds, hippurates and mannitol displayed absorption half-times, in contrast, of between 45 and 70 min.

Disodium cromoglycate BP (Cromolyn sodium, USAN)

$$(X)$$

Disodium cromoglycate (X) is a bischromone with two carboxylic acid groups and a pK_a of approximately 1.9. The molecule will be completely ionised at physiological pH. The free acid is very insoluble in both polar and non-polar solvents and has virtually no lipid solubility. Because of this, and the insolubility of the unionised form, very little of an oral dose of DSCG is absorbed[66]. Powder swallowed after inhalation therefore contributes little to the systemic dose and is

subsequently excreted in the urine and bile. None the less the drug is well absorbed from the lungs with a clearance rate of about 1 hour[67]. Relative to the gastro-
intestinal mucosa the pulmonary epithelium possesses a high permeability to
lipid-insoluble molecules and ions.

Nebulisers

Modern nebulisers for domestic and hospital use generate aerosols continuously
for chronic therapy of respiratory disorders. A Venturi-type system is shown in
figure 9.36a and an ultrasonic device in figure 9.36b. The particle size distribution and hence efficiency of such systems varies with the design and sometimes
with the model of use. Hence it is important that adequate monitoring of particle
size is carried out.

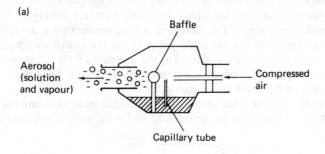

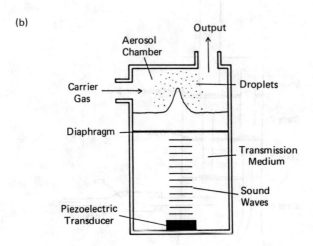

figure 9.36 Schematic diagrams of (a) a Venturi-type nebuliser and (b) an ultrasonic nebuliser. Adapted from D. F. Egan. *Fundamentals of Respiratory Therapy*,
3rd Edition, Mosby, New York, 1977

9.9.3   Analysis of particle size distribution in aerosols

Analysis of particle size distribution during formulation, development, clinical trial or after storage is of obvious clinical relevance. Aerosols are not easy to size, primarily because they are inherently unstable systems. Methods of sampling have been described by Bell[62]. These may be divided into techniques which utilise a cloud, and dynamic methods in which particles are carried in a stream of gas. In the former, sedimentation techniques based on Stokes' law are applied and the usual detection system is photometric. The Royco sizer is a commercially available instrument which measures individual particles in a cloud (it is used to monitor the air of 'clean rooms'). This instrument can be used to size particles in aerosol clouds provided that the particle size distribution does not change during the time of the analysis either by preferential settling of larger particles or by co-agulation. Dynamic methods depend on the properties of particles related to their mass. Instruments utilise both sedimentation and inertial forces. Probably the most widely used instrument in the sizing of airborne particles is the cascade impactor, in which large particles leave the airstream and are impinged on baffles or on glass microscope slides. The airstream is then accelerated at a nozzle, providing a second range of smaller sized particles on the next baffle and so on. Progressively finer particles are collected at the successive stages of impingement owing to increasing jet velocity and decreasing jet dimension. Shearing action of the jets may lead to the break-up of aggregates.

More useful for more routine examination of medicinal aerosols are the 'artificial throat' devices illustrated in figures 9.37 and 9.38, with which comparative

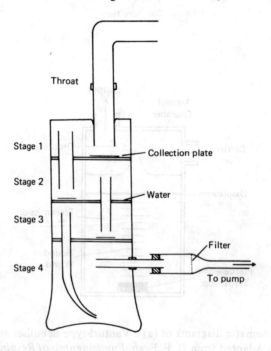

figure 9.37   The multistage liquid impinger. From G. W. Hallworth. *Br. J. clin. Pharm.*, 4, 689 (1977)

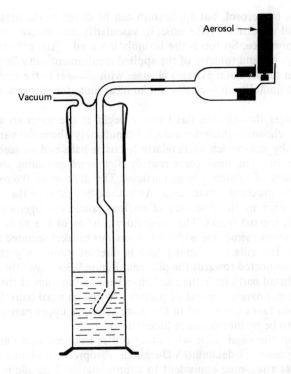

figure 9.38 A simple *in vitro* device with artificial mouth and throat devised in Riker's laboratories

studies of the behaviour of aerosols can be carried out. Segregation of particles according to size occurs in these devices. Analysis of the collecting layers at the several levels of the device in figure 9.37 allows the monitoring of changes in released particle size without rigorous particle size analysis. In the device in figure 9.38 the artificial mouth can be washed to reveal the extent of fall-out of large particles. The smallest particles reach the collecting solvent.

9.10 The intranasal route

Three classes of medicinal agents are applied by the nasal route: drugs for the alleviation of nasal symptoms; drugs that are inactivated in the gastro-intestinal tract following oral administration and where the route is an alternative to injection; and vaccines[68]. An intranasal vaccine for colds has recently been suggested, as there is a growing volume of evidence that antigens applied to the respiratory mucosa stimulate an immune response to the secretions[69]. Intranasal beclomethasone dipropionate in a dose as low as 200 μg daily is a useful addition to the therapy of perennial rhinitis[70]. Recently, interferon has been administered by this route.

What physical factors affect the utility of this route? The factors such as droplet size which affect deposition in the respiratory tract may be involved here if administration is by aerosol, but application can be direct to the nasal mucosa. The physiological condition of the nose, its vascularity, and mucus flow rate are therefore of importance. So too is the formulation used — the volume, concentration, viscosity, pH and tonicity of the applied medicament can affect activity. As the condition of the nasal passages changes with changes in the environment, temperature and humidity, it is clearly not an ideal route for absorption of drugs or vaccines.

The air passages through the nasal cavity begin at the nares (nostrils) and terminate at the choanae (posterior nares). Immediately above the nares are the vestibules, lined by skin which bears relatively coarse hairs and sebaceous glands in its lower portion. The hairs curve radially downward providing an effective barrier to the entry of relatively large particles. The division of the nasal cavity exposes the air to maximal surface area. As in the other parts of the respiratory tree sudden changes in the direction of airflow causes impingement of large particles through inertial forces. The respiratory portion of the nasal passage is covered by a mucous membrane which has a mucous blanket secreted in part by the goblet cells. The ciliary streaming here is directed posteriorly so that the nasal mucus is transported towards the pharynx. Figure 9.39 shows the fractional deposition of inhaled particles in the nasal chamber as a function of their particle size. In the external nares, removal of particles occurs on nasal hairs; further up inertial deposition takes place, and in the more tortuous upper passages deposition is assumed to be by inertia and sedimentation.

Comparison of the nasal route with other routes has been made in a few instances. Desmopressin (1-desamino-8-D-arginine vasopressin) administered as a 20 µg dose elicits a response equivalent to approximately 2 µg administered by i.v. injection. Also, a greater dose of virus is required to obtain an equivalent response to a nasal vaccine than when administered by other routes.

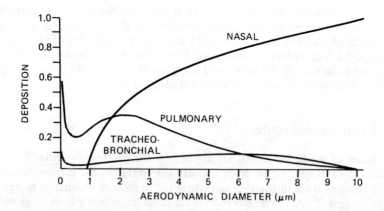

figure 9.39 Regional deposition of inhaled particulate matter as a function of particle size (µm). Nose breathing at fifteen respirations per minute and 730 cm^3 tidal volume. The pulmonary compartment refers to deposition beyond the terminal bronchiole. From C. D. F. Muir. *Clinical Aspects of Inhaled Particles*, Heinemann, London, 1972

In the treatment of nasal symptoms the patient adjusts the dose so that, perhaps, the theoretical bases of droplet and particle retention are less vital. However, formulation of the nasal drops, or sprays from plastic squeeze-bottles must obviously influence the efficiency of medication, yet little work has been carried out relating formulation to the effect of intranasal medicines.

9.11   Rectal absorption of drugs

Drugs administered by the rectal route in suppositories are placed in intimate contact with the rectal mucosa which behaves as a normal lipoidal barrier. The pH in the rectal region lies between 7.2 and 7.4, but the rectal fluids have little buffering capacity. As with topical medication, the formulation of the suppository can have marked effects on the activity of the drug. Other factors such as retention of the suppository for a sufficient duration of time also influence the outcome of therapy; the size and shape of the suppository and its melting point may well be factors in determining the drug bioavailability.

For many years, coconut butter (theobroma oil) was the only excipient used, but this is a variable natural product which undergoes a polymorphic transition on heating. It is primarily a triglyceride.

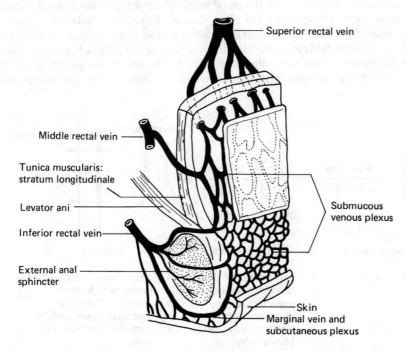

Superior rectal vein

Middle rectal vein

Tunica muscularis: stratum longitudinale

Levator ani

Inferior rectal vein

External anal sphincter

Submucous venous plexus

Skin
Marginal vein and subcutaneous plexus

figure 9.40   Blood supply to the rectum and anus. The rectum is the terminal 15–19 cm of the large intestine. The mucous membrane of the rectal ampulla, with which the suppositories and other rectal medications come into contact, is made up of a layer of cylindrical epithelial cells, without villi. From G. Tondury. *Topographical Anatomy*, Thieme, 1959

Four polymorphic forms exist: γ, m.p. 18.9 °C; α, m.p. 23 °C; β^1, m.p. 28 °C; and the stable β, m.p. 34.5 °C. Heating the mass above 38 °C converts the fat to a metastable mixture solidifying at 15-17 °C instead of 25 °C, and this subsequently melts at 24 °C instead of at 31-35 °C[71]. Reconversion to the stable β-form takes 1-4 days depending on storage conditions[72].

Modern bases include polyoxyethylene glycols of molecular weight 1000-6000 and semisynthetic vegetable fats[73]. Bases must be selected carefully for each substance. The important features of excipient materials are melting point, speed of crystallisation and emulsifying capacity. If the medicament dissolves in the base it is likely that the melting point of the base will be lowered, so that a base with a melting point higher than 36-37 °C has to be chosen. If the drug substance has a high density it is preferable that the base crystallises rapidly to prevent settling of the drug. Preservatives, hardening agents, emulsifiers, colouring agents and materials which modify the viscosity of the base after melting are common formulation additives.

The rectal cavity

Figure 9.40 shows the blood supply to the rectal area. The main artery to the rectum is the superior rectal (haemorrhoidal) artery. Veins of the inferior part of the submucous plexus become the rectal veins, which drain to the internal pudendal veins. Drug absorption takes place through this venous network. Superior haemorrhoidal veins connect with the portal vein and thus transport drugs to the liver; the inferior veins enter into the inferior vena cava and thus bypass the liver. The particular venous route the drug takes is affected by the extent to which the suppository migrates in its original or molten form up the gastro-intestinal

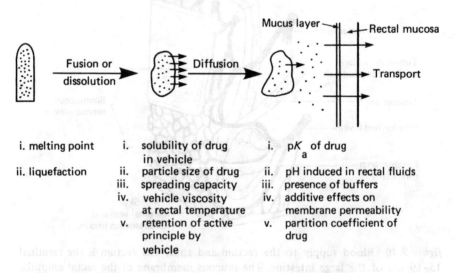

i. melting point	i. solubility of drug in vehicle	i. pK_a of drug
ii. liquefaction	ii. particle size of drug	ii. pH induced in rectal fluids
	iii. spreading capacity	iii. presence of buffers
	iv. vehicle viscosity at rectal temperature	iv. additive effects on membrane permeability
	v. retention of active principle by vehicle	v. partition coefficient of drug

figure 9.41 Schematic representation of rectal absorption of an active principle from a suppository, and the factors likely to affect the bioavailability of the drug. After Jaminet, reference 74

tract, and this may be variable. The rectal route therefore does not necessarily or reproducibly avoid the liver.

A schematic representation of the processes occurring following insertion of a suppository into the rectum is shown in figure 9.41. Cocoa butter suppositories usually liquefy within a few minutes, but the drug is not necessarily released from solution or suspension, as the fat in this case is not emulsified or absorbed. Surfactants may be required to aid dispersal of the fat and thus when this base is used the physicochemical properties of the drug are important. The rate-limiting step in drug absorption for suppositories made up from a fatty base is the partitioning of the dissolved drug from the molten base and not the rate of solution of the drug in the body fluids.[72]

The influence of the aqueous solubility on *in vitro* release from fat-based suppositories is shown in figure 9.42, the results being collated from the study of 35

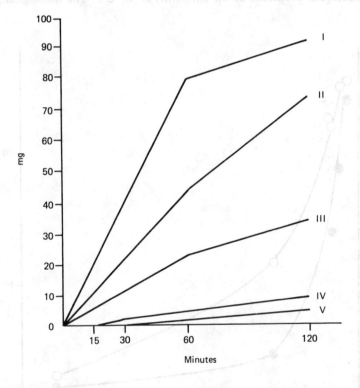

figure 9.42 Release of drugs of varying solubilities from fat-based suppositories of equal active agent content.

Type	Solubility in water
I	1 in 1
II	1 in 10
III	1 in 10 to 1 in 100
IV	1 in 100 to 1 in 1000
V	1 in 1000 to 1 in 10 000

After R. Voigt and G. Falk. *Pharmazie*, 23, 709 (1968)

drugs grouped into classes I–V in decreasing order of water solubility. The results may be explained as follows. The water-soluble active substances will be insoluble in the fatty base, while the less water-soluble material will tend to be soluble in the base, and will thus diffuse out of the base more slowly. According to published work, water-soluble barbiturates are better absorbed from a fatty excipient than from a water-soluble one. Conversely, ethyl nicotinate which is lipid-soluble is absorbed better from a water-soluble excipient.

Addition of substances that increase the viscosity of the base – for example, 2 per cent aluminium monostearate or 'Aerosil' – decreases the rate of release of soluble drugs but has little effect on the release of the less-soluble materials. (The action of aluminium stearate and Aerosil in oils is discussed in chapter 7.)

In a detailed study of the influence of the physical properties of excipients on the bioavailability of paracetamol, the most important parameter was found to be the rheological properties of the excipients at $37\,^{\circ}C$[75]. The rheological pro-

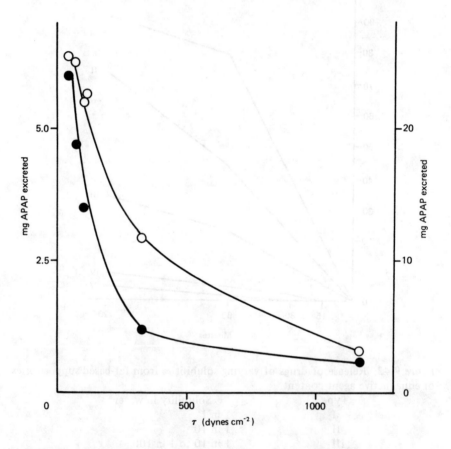

figure 9.43 Variation of the excretion of APAP in the urine after (●) 1 hour (left-hand axis) and (○) two hours (right-hand axis) with the limiting shear stress, τ, of the excipient drug-mixture at $37\,^{\circ}C$. From reference 75

files (η versus τ) of excipients vary widely; the relationship between the excretion of paracetamol (APAP; N-acetyl-p-aminophenol) and the rheology of the excipient drug suspension is shown in figure 9.43. The greater the limiting shear stress τ of the system, the lower the bioavailability of the drug.

Apparatus for studying the many variables in suppository formulation has been designed to measure rates of release *in vitro*. Both a circulating dissolution apparatus and a dialysis device utilising an aqueous and a non-aqueous phase have been described. In both these, the suppository is held in a cellophane 'rectum' which is placed in the water 15 minutes before introduction of the suppository. *In vitro* examination can be useful in determining the potential effects of the formulation of the suppository on bioavailability (chapter 11). As the suppository base is heated before moulding, certain effects can be noted which are unique for this type of medication. Jaminet[74] reports that it has been observed that testosterone dissolves when hot in the semisynthetic excipient Witepsol H, to give, on cooling, crystals of about 2–3 μm in diameter. After being dissolved in theobroma oil the drug does not crystallise on cooling but remains dissolved as a solid solution. In the former case, high absorption rates are obtained while, in the latter poor absorption is achieved. Because of the effect of particle size on viscosity of suspensions (see chapter 7) it is preferable to avoid the incorporation of ultrafine crystals as the resultant excipient becomes too viscous on melting.

The absorption of drug from the rectum depends, as in other sites, on the concentration of drug in absorbable form in the rectal cavity and, if the base is not emulsified, on the contact area between molten excipient and rectal mucosa. Addition of surfactants may increase the ability of the molten mass to spread and tends to increase the extent of absorption. Kakemi *et al.*[76] have extensively studied the effect of surfactants on drug absorption from theobroma oil suppositories. Significant increases in absorption can be obtained with polyoxyethylene sorbitan monostearate, sodium lauryl sulphate and cetyltrimethylammonium bromide. Surfactants increase wetting and spreading and may increase the permeability of the rectal mucosal membrane.

Study of the effect of polyoxyethylene glycols on the rectal absorption of unionised sulphafurazole demonstrated the importance of the affinity of the drug substance for base and for rectal mucosa. As the amount of Macrogol 4000 (polyoxyethylene glycol 4000) was increased, the partition coefficient (lipid/vehicle) fell; absorption fell correspondingly. Thus, as the affinity of the drug for the suppository base increased, its tendency to partition to the rectal lipids decreased. In a similar way incorporation of water-soluble drug into water-in-oil emulsified suppository base tends to decrease bioavailability as the drug is transferred in dispersed aqueous droplets in the molten base. The presence of ingredients likely to form water-in-oil emulsions in the rectum frequently results in poor bioavailability. Beeswax in alkaline conditions can form water-in-oil emulsions. Five per cent beeswax in a formulation of thiopentone sodium in Witepsol base reduces the availability of the drug, possibly because of emulsification[71].

Hygroscopicity of some hydrophilic bases such as the polyoxyethylene glycols results in the abstraction of water from the rectal mucosa. This causes a stinging sensation and discomfort and probably affects the passage of drugs across the rectal mucosa (see section 9.1.2). The hygroscopicity of the glycols decreases as the molecular weight increases. The problem, however, may be overcome by the incorporation of water into the base, although the presence of water may affect drug stability. Glycerogelatin bases are also hygroscopic. Deliberate incorporation of water into a formulation gives rise to the possibility that, on storage, water will be lost by evaporation and drug may crystallise out as a result. Reactions between components of the formulation are more likely to occur in the presence of moisture than in its absence.

Incompatibility between base and drug

Various incompatibilities have been noted with polyoxyethylene glycol bases. Phenolic substances complex with the glycol, probably by hydrogen bonding between the hydroxy group and the ether oxygens. The polyoxyethylene glycol bases are incompatible with tannic acid, ichthammol, aspirin, benzocaine, Vioform and sulphonamides[72]. High concentrations of salicylic acid alter the consistency of the bases to a more fluid state.

Glycerogelatin bases are prepared by heating together glycerin, gelatin and water. Although primarily used *per se* as an intestinal evacuant the glycerogelatin base may be used to deliver drugs to the body. For this purpose the USP XVIII specified two types of gelatin to avoid incompatibilities. Type A is acidic and cationic with an isoelectric point between pH 7 and 9; type B is less acidic and anionic with an isoelectric point between pH 4.7 and 5. Use of untreated gelatin renders the base incompatible with acidic and basic drugs.

References

1. M. Nakano, *et al. Chem. pharm. Bull.*, 24, 2345 (1977)
2. K. H. Sterzel, *et al. Trans. Am. Soc. artif. intern. Organs*, 17, 293 (1971)
3. A. L. Rubin, *et al. J. clin. Pharm.*, 13, 309 (1973)
4. E. J. Lien. *Drug Intell. clin. Pharm.*, 4, 7 (1970)
5. E. J. Lien. *ibid.*, 5, 38 (1971)
6. E. J. Lien. *ibid.*, 8, 470 (1974)
7. J. G. Wagner and A. J. Sedman. *J. Pharmacokin. Biopharm.*, 1, 23 (1973)
8. M. L. Lucas, W. Schneider, F. J. Haberich and J. A. Blair. *Proc. R. Soc. B*, 192, 39 (1976)
9. M. L. Lucas, J. A. Blair, *et al. Biochem. Soc. Trans.*, 4, 154 (1976)
10. A. Benn, *et al. Br. med. J.*, 1, 148 (1971)
11. H. Ochsenfahrt and D. Winne. *Arch. Pharmacol.* (NS), 281, 197 (1974)
12. J. M. Dietschy and H. Westergaard. *Intestinal Absorption and Malabsorption* (ed. T. Z. Csaky), Raven Press, New York, 1975
13. S. Miyazahi, *et al. Chem. pharm. Bull.*, 23, 2151 (1975)
14. M. Gibaldi and B. Grundhofer. *J. pharm. Sci.*, 62, 343 (1973)
15. L. S. Murthy and G. Zografi. *J. pharm. Sci.*, 59, 1281 (1970)
16. M. S. Losowsky, *et al. Malabsorption in Clinical Practice*, Longman, London, 1974
17. R. G. Crounse. *J. invest. Derm.*, 37, 520 (1961)
18. M. Kraml, *et al. Can. J. Biochem. Physiol.*, 40, 1449 (1962)
19. A. R. Cooke. *Gastroenterology*, 68, 804 (1975)
20. S. S. Davis, J. G. Hardy and J. W. Fara. *Gut*, 27, 886 (1986)

21. S. S. Davis, J. G. Hardy, M. J. Taylor, D. R. Whalley and C. G. Wilson. *Int. J. Pharmaceut.*, 21, 331 (1984)
22. C. A. Squier and N. W. Johnson. *Br. med. Bull.*, 31, 169 (1975)
23. W. R. Galey, *et al. J. invest. Derm.*, 67, 713 (1976)
24. A. H. Beckett and E. J. Triggs. *J. Pharm. Pharmacol.*, 19, 31S (1967)
25. J. M. Schar, S. S. Davis, A. Nigaleye and S. Bolton. *Drug Dev. ind. Pharm.*, 9, 1359 (1983)
26. E. H. Dobbs, G. F. Blane and A. L. A. Boura. *Eur. J. Pharmacol.*, 7, 328 (1969)
27. M. Gibaldi and J. L. Kanig. *J. oral Ther. Pharmacol.*, 1, 440 (1965)
28. C. F. Spiers. *Br. J. clin. Pharm.*, 4, 97 (1977)
29. D. Greenblatt and J. Koch-Weser. *New Engl. J. Med.*, 295, 542 (1976)
30. T. F. O'Donnell. *Lancet*, i, 533 (1975)
31. P. Turner. *Pharm. J.*, 217, 583 (1976)
32. R. Luft (ed.). *Insulin*, Nordisk, Denmark, 1976
33. B. W. Barry. *Pharm. J.*, 215, 322 (1975)
34. A. S. Michaels, *et al. A. I. Chem. J.*, 21, 985 (1975)
35. R. Scheuplein. *J. invest. Derm.*, 67, 672 (1976)
36. A. H. Fenton and M. Warren. *Pharm. J.*, 188, 5 (1962)
37. N. J. van Abbé, R. I. C. Spearman and A. Jarrett. *Pharmaceutical and Cosmetic Products for Topical Administration*, Heinemann, London, 1969
38. B. J. Poulsen. *Drug Design*, vol. IV (ed. E. J. Ariens), Academic Press, New York, 1973, p. 149 *et seq.*
39. T. Higuchi. *J. Soc. cosmetic Chem.*, 11, 85 (1960)
40. J. Ostrenga, J. Haleblian, *et al. J. invest. Derm.*, 56, 392 (1971)
41. M. Busse. *Pharm. J.*, 220, 25 (1978)
42. J. E. Shaw, S. K. Chandrasekaran and L. Taskovich. *Pharm. J.*, 215, 325 (1975)
43. A. H. Beckett, *et al. J. Pharm. Pharmacol.*, 24, 65P (1972)
44. Anon. *Lancet*, ii, 1287 (1976)
45. P. H. O'Connor Davies. *The Actions and Uses of Ophthalmic Drugs*, Barrie and Jenkins, London, 1972
46. J. Vale and B. J. Cox. *Drugs and the Eye*, Butterworths, London, 1978
47. J. W. Seig and J. R. Robinson. *J. pharm. Sci.*, 66, 1222 (1977)
48. S. I. Rapoport. *Blood Brain Barrier*, Raven Press, New York, 1976
49. S. S. Chrai and J. R. Robinson. *J. pharm. Sci.*, 65, 437 (1976)
50. E. S. Wang and E. R. Hammerlund. *ibid.*, 59, 1559 (1970)
51. S. R. Waltman and H. E. Jaufman. *J. invest. Ophthalmol.*, 9, 250 (1970)
52. D. S. Hull, H. F. Edelhauser and R. A. Hyndiuk. *Arch. Ophthalmol.*, 92, 413 (1974)
53. A. S. Michaels, W. J. Mader and C. R. Manning. *Quality Control of Medicines* (ed. P. B. Deasy and R. F. Timoney), Elsevier, Amsterdam, 1976, pp. 45-64
54. E. German and N. Siddiqui. *New Engl. J. Med.*, 282, 689 (1970)
55. V. Borromeo-McGrail, *et al. Paediatrics*, 51, 1032 (1973)
56. Anon. *Br. med. J.*, 1, 2 (1974)
57. A Li Wan Po. *OTC Medicines*, Blackwells, Oxford, 1983
58. I. Schiff, D. Tulchinsky and K. J. Ryan. *Fertil. Steril.*, 28, 1063 (1977)
59. D. P. Danziger and J. Edelson. *Drug Metab. Ther.*, 14, 137 (1983)

60. J. S. G. Cox, *et al. Adv. Drug Res.*, 5, 115 (1970)
61. T. F. Hatch and P. Gross. *Pulmonary Disposition and Retention of Inhaled Aerosols*, Academic Press, New York, 1964
62. J. H. Bell. *Mfg. Chem.*, 38 (9), 37 (1967)
63. S. M. Cohn, *et al. Inhaled Particles and Vapours* (ed. C. N. Davies), Pergamon, London, 1961, p. 178
64. S. Godfrey, E. Zeidiford, K. Brown and J. H. Bell. *Clin. Sci. mol. Med.*, 46, 265 (1974)
65. J. A. Burton and L. S. Schanker. *Steroids*, 23, 617 (1974)
66. K. M. Jones and G. F. Moss. *Excerpta Medica International Congress Series* No. 311, vol. XV, 1974, p. 269
67. G. F. Moss, *et al. Toxicol. appl. Pharmacol.*, 17, 691 (1970)
68. D. S. Freestone and A. L. Weinberg. *Br. J. clin. Pharm.*, 3, 827 (1976)
69. G. Thomas. *Lancet*, i, 400 (1975)
70. G. J. Gibson, *et al. Br. med. J.*, 4, 503 (1974)
71. N. Senior. *Adv. pharm. Sci.*, 4, 363 (1974)
72. J. Anschel and H. A. Lieberman. In *Theory and Practice of Industrial Pharmacy*, 2nd Edition (ed. L. Lachman, *et al.*), Lea and Febiger, Philadelphia, 1976, p. 245 *et seq.*
73. B. R. Guillot and A. P. Lombard (ed.). *The Suppository*, 1st English Edition, Maloine, Paris, 1973
74. F. Jaminet. *The Suppository* (ed. B. R. Guillot and A. P. Lombard), Maloine, Paris, 1973, p. 111 *et seq.*
75. A. Moes. *J. Pharm. belg.*, 29, 319 (1974)
76. K Kakemi, H. Sezaki, S. Muranishi and H. Matsui. *Chem. pharm. Bull.*, 15, 172 (1967)

10 Drug Interactions and Incompatibilities— A Physicochemical View

This chapter deals with one aspect of the topic of drug interactions, and contains a discussion of the problem from a physicochemical viewpoint. Many drug interactions *in vitro* are, not surprisingly, readily explained by resorting to the physical chemistry discussed in previous chapters of this book. There is no reason why the same forces and phenomena that operate *in vitro* cannot explain many of the observed interactions that occur *in vivo*, although of course the interplay of physicochemical and physiological forces makes simple interpretations a little hazardous. Interactions such as protein binding, whether as a result of hydrophobic or electrostatic interactions, adsorption of drugs onto solids, chelation and complexation all occur in physiological conditions and are predictable to a large degree.

Drug–drug or drug–adjuvant interactions can take place before administration of a drug. These may result in precipitation of the drug from solution, loss of potency, or instability. With the decline in extemporaneous dispensing this aspect of pharmaceutical incompatibility has diminished somewhat in importance. Subtler forms of extemporaneous preparation occur today; for example, in the form of addition of drugs to intravenous fluids. This is a practice which is carried out often without the pharmaceutical oversight needed to avoid incompatibilities and instabilities arising. The first part of this chapter is devoted to a discussion of some of the possible interactions that are most likely to occur before a drug is administered.

An incompatibility occurs when one drug is mixed with other drugs or agents to produce a product unsuitable for administration either because of some modification of the effect of the active drug, such as increase in toxicity, or because of some physical change such as decrease in solubility. Some drugs that are designed to be administered by the intravenous route cannot safely be mixed with available intravenous fluids. At times, as discussed in chapter 5, the solubility of a drug in a particular infusion fluid will be low so that when drug and fluid are mixed crystallisation may occur, often very slowly; microcrystals may be formed which are not immediately visible and which when infused will have potentially serious consequences. The mechanism of crystallisation will often involve a change in pH; the problem is a real one because the pH of commercially available infusion fluids can vary within a pH range of perhaps one or two units and therefore a drug may be compatible with one batch and not another. The proper application of the equations relating pH and pK_a to solubility should allow additions to be safely made or to be avoided.

We now discuss, in turn, pH effects *in vitro* and *in vivo*, cation–anion interactions, electrolyte effects, complex formation, ion-exchange interactions, adsorption and protein binding.

10.1 pH effects *in vitro* and *in vivo*

10.1.1 *In vitro*

Chemical, as well as physical, instability may result from changes in pH, buffering capacity, salt formation or complexation. Chemical instability may give rise to the formation of inactive or toxic products. Although infusion times are generally not greater than two hours, chemical changes following a change in pH may occur rapidly. pH changes may follow from the addition of a drug substance or solution to an infusion fluid. Some measured examples of pH changes are listed in table 10.1. This increase or decrease in pH may then produce physical or chemical changes in the system.

table 10.1 Changes in pH of 5 per cent dextrose (1000 mℓ) following drug additions*

Drug	Quantity	ΔpH	Final pH
Aminophylline	250 mg	+ 4.2	8.5
	500 mg	+ 4.2	8.5
Cephalothin sodium	1 g	+ 0.1	4.2
	2 g	+ 0.2	4.3
Oxytetracycline	500 mg	− 1.25	2.9
hydrochloride	1 g	− 1.45	2.7

*From reference 1

It has been pointed out[2] that pH *per se* is not so important in determining compatibility and stability as the titratable acidity or alkalinity of the mixture. An autoclaved solution of dextrose may have a pH of 4.0, but the titratable acidity in such an unbuffered solution is low, and thus the addition of benzylpenicillin or a soluble acidic drug may not be contraindicated. As pointed out in table 10.1, the additive may change the pH. Addition of as little as 500 mg of ampicillin sodium may raise the pH of 500 mℓ of some fluids to over 8. Carbenicillin or benzylpenicillin may raise the pH of 5 per cent dextrose or dextrose saline to 5–5.6 or even higher, both being stable in these conditions[3].

The solubility of calcium and phosphate in total parenteral nutrition (TPN) solutions is dependent on the pH of the solution, and TPN solutions are of course clinically acceptable only when precipitation can be guaranteed under all conditions. Dibasic calcium phosphate, for example, is soluble only to the extent of 300 mg ℓ$^{-1}$ whereas monobasic calcium phosphate is relatively soluble, having a solubility of 18 g ℓ$^{-1}$. At low pH the monobasic form predominates while at higher pH values the dibasic form becomes available to bind with calcium; precipitates tend to form[4].

Solubility curves for TPN solutions containing 1.5 g dℓ$^{-1}$ amino acid and 10 per cent dextrose at pH 5.5 are shown in figure 10.1. The broken lines show the calcium and phosphate concentrations as a reference at 3:1 and 2:1 ratios. The dotted curve for Aminosyn solutions shows the concentrations at which precipitation occurs after 18 h at 25 °C followed by 30 min in a water bath at 37 °C.

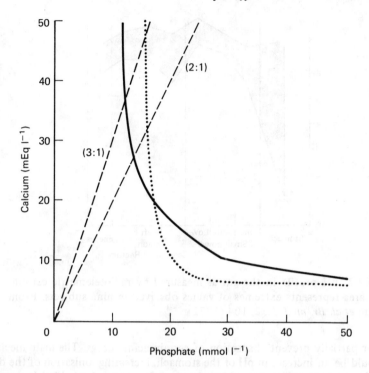

figure 10.1 Solubility curves for TPN solutions containing amino acid (1.5 g $d\ell^{-1}$) and 10 per cent dextrose at pH 5.5. Aminosyn, ...; Troph-Amine, ————; relative calcium to phosphate ratios, — — — — — —. From reference 4 with permission

The full curve is for TrophAmine solutions, and represents calcium or phosphate concentrations at which visual or microscopic precipitation or crystallisation occur. Compositions to the left of the curve represent physically compatible solutions.

10.1.2 *In vivo*

The sensitivity of the properties of most drugs to changes in the pH of their environment means that the hydrogen-ion concentration will be an important determinant of solubility, crystallisation and partitioning. The gastric pH is 1–3 in normal subjects but the measured range of pH values in the human stomach is wide[5]. Figure 10.2 shows the changes in pH that occur as the gastro-intestinal tract is traversed. As pH is on a logarithmic scale the order of the change in the environment and its effect on aqueous and lipid solubility in particular may readily be observed. Changes in the acid–base balance may thus have a marked influence on the absorption of drugs and thus on their activity. Antacids, food, and weak electrolytes will all change the pH of the stomach. Weakly acidic drugs, being unionised in the stomach, will be absorbed from the stomach by passive diffusion. One might expect, therefore, that concomitant antacid therapy would

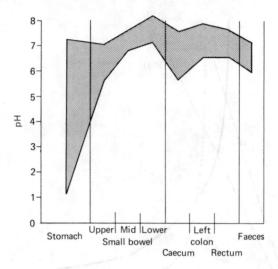

figure 10.2   pH profile in the gut as measured by radiotelemetric capsule. The shaded area represents extremes of values observed in nine subjects. From S. J. Meldrum *et al. Br. med. J.*, 2, 104 (1972)

delay or partially prevent absorption of certain acidic drugs. The main mechanism would be an increase in pH of the stomach, increasing ionisation of the drug and reducing absorption. A problem in generalisations of this kind is that the acid-neutralising capacity of antacids is very variable, as the results quoted in table 10.2 show. Some acidic drugs, listed in table 10.3, are also known to be absorbed in the intestine, in which case the co-administration of antacid is not necessarily prohibited.

There are contradictory reports of the effect of antacids on the absorption of levodopa. Levodopa is metabolised within the gastro-intestinal tract and more rapidly degraded in the stomach than in the intestine. The rate at which the drug is emptied from the stomach can therefore affect its availability. It has been suggested that when an antacid is administered prior to the drug, an increase in serum levodopa occurs. This finding has not always been confirmed by other workers.

The use of cimetidine and ranitidine has given rise to the possibility of a drug interaction involving the alteration of gastric pH. Both drugs inhibit gastric acid secretion; a few subjects have transient achlorhydria after oral cimetidine, for example, thus increased absorption of acid-labile drugs is a predictable side-effect. In one out of five subjects studied by Fairfax *et al.*[6] the absorption of benzylpenicillin was markedly increased after cimetidine, but remained unaltered in the other patients. (Gastric juice at pH 2 rapidly destroys benzylpenicillin.)

Sodium bicarbonate is one of the most effective antacids in terms of neutralising capacity. It can greatly depress the absorption of tetracycline – the mean amount of drug appearing in the urine of patients receiving only drug was 114 mg at 48 hours and 53 mg for those also given sodium bicarbonate. Chelation is not possible with monovalent Na^+, as it is with the components of other antacids. If the drug was dissolved prior to administration the antacid did not affect excretion

table 10.2   **Amount of various antacids required to neutralise 50 mEq HCl***

Antacid	Neutralising capacity of 1 g or 1 ml		Dose required to neutralise 50 mEq HCl	Weight of tablet (g)	No. of tablets
	ml 0.1 N HCl	mEq HCl			
Powders					
NaHCO₃	115	11.5	4.4 g		
MgO	85	8.5	5.9 g		
CaCO₃	110	11.0	4.5 g		
Magnesium trisilicate	10	1.0	50 g		
MgCO₃	8	0.8	63 g		
Solutions					
Al(OH)₃ gel	0.7	0.07	715 ml		
Milk of Magnesia	27.7	2.8	17.8 ml		
Titralac	24	2.4	20.6 ml		
Aludrox	1.7	0.17	294 ml		
Oxaine	2.4	0.24	208 ml		
Mucaine	1.7	0.17	294 ml		
Kolantyl gel	3.4	0.34	147 ml		
Tablets					
Gastrogel	5.0	0.5	100 g	1.08	93
Gastrobrom	15.0	1.5	33.3 g	1.48	23
Glyzinal	2.5	0.25	200 g	0.72	278
Actal	7.7	0.77	65 g	0.60	109
Amphotab	2.5	0.25	200 g	1.04	192
Gelusil	2.5	0.25	200 g	1.36	147
Nulacin	10.0	1.0	50 g	3.12	17
Kolantyl wafer	5.0	0.5	100 g	1.64	61
Sebella	2.5	0.25	200 g	0.53	379
Titralac	42.5	4.25	11.8 g	0.65	18
Almacarb	3.0	0.3	167 g	1.28	130
Dijex	4.6	0.46	109 g	1.65	66

*From D. W. Piper and B. H. Fenton. *Gut*, 5, 585 (1964). Note that proprietary preparations available in several countries may not have the same formulation

table 10.3   **Drugs* whose absorption may be affected by antacid administration**

Drugs whose activity would be reduced	Drugs whose activity would be potentiated
Tetracyclines	Theophylline
Nalidixic acid	Chloroquine
Nitrofurantoin	Mecamylamine
Penicillin G	Amphetamine
Sulphonamides	L-Dopa

*Structures of some of these compounds are shown in structures (I) on next page

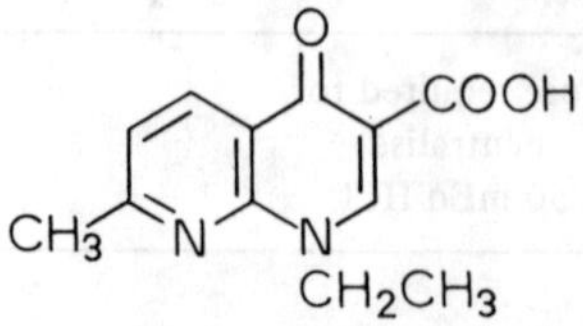

Nalidixic acid
(pK$_a$ = 6.0)

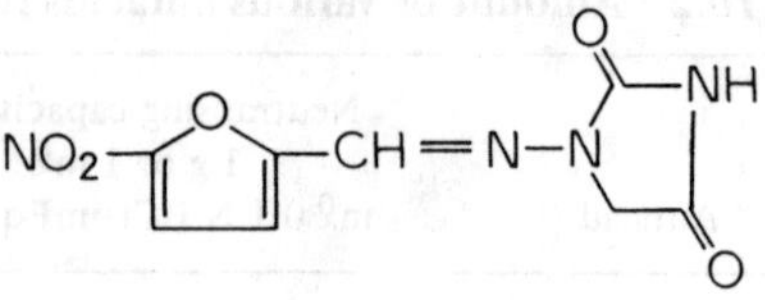

Nitrofurantoin
(pK$_a$ = 7.2)

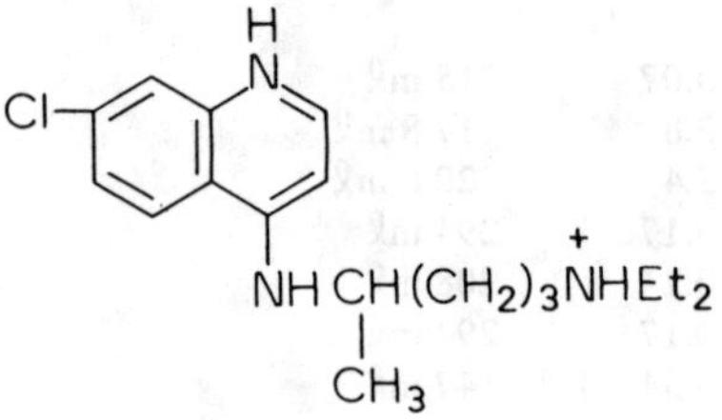

Chloroquine
(pK$_{a1}$ = 8.10; pK$_{a2}$ = 9.94)

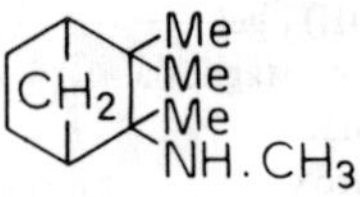

Mecamylamine

Structures of nalidixic acid, nitrofurantoin, chloroquine, and mecamylamine

(I)

of the antibiotic, suggesting that in normal dosage forms the dissolution of the drug is affected by the antacid.

The aqueous solubility of tetracycline at pH 1–3 is one hundred times greater than at pH 5.6. Consequently its rate of solution, dc/dt, at this pH is greatly reduced according to equation 10.1 (see also section 2.6)

$$\frac{dc}{dt} = kc_s \qquad (10.1)$$

where c_s is solubility. A 2 g dose of NaHCO$_3$ will increase the intragastric pH above 4 for a period of 20–30 minutes, sufficient time for 20–50 per cent of the undissolved tetracycline particles to be emptied into the duodenum where the pH (at 5–6) is even less favourable for solution to occur, so the fraction of drug absorbed is decreased.

The effect of antacids on gastric emptying rate is a factor which obviates direct physicochemical analysis of the problem. The difficulty in predicting the effect of antacids is clearly brought out by studies with naproxen, a weakly acidic non-steroidal anti-inflammatory. Several textbooks of drug interactions state that antacids decrease the absorption of acidic drugs such as nalidixic acid, nitrofurantoin and penicillin G. Antacids both increase and decrease the absorption of naproxen; magnesium carbonate, magnesium oxide and aluminium hydroxide decrease absorption and adsorption effects are suspected (see figure 10.3).

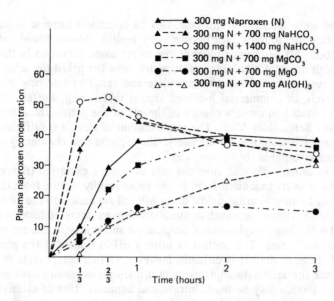

figure 10.3 Mean plasma concentrations of naproxen in fourteen male volunteers with and without intake of sodium bicarbonate, magnesium carbonate, magnesium oxide or aluminium hydroxide. From E. J. Segre, H. Sevelius and J. Varady. *New Engl. J. Med.*, 291, 582 (1974)

Intake of Maalox which contains magnesium and aluminium hydroxides, however, slightly increased the area under the curve.

Even though gastric emptying tends to become faster as the gastric pH is raised, antacid preparations containing aluminium or calcium are prone to retard, and magnesium preparations to promote, gastric emptying. However, the caution about the co-administration of antacids containing divalent or trivalent metals and tetracyclines should be extended to antacids containing sodium bicarbonate or any substance capable of increasing intragastric pH. Similar principles, of course, can be applied to intestinal absorption, but here the pH gradients between the contents of the intestinal lumen and capillary blood are smaller. Sudden changes in the acid–base balance will, nonetheless, change the concentration of drugs able to enter cells, providing that the pH change does not alter binding of the drug to protein, or drug excretion, which of course it invariably does. Phenobarbitone owes its biological effect to the concentration of drug available inside cells. Breathing CO_2 decreases the concentration of phenobarbitone in the plasma and causes an increased affinity for the cell phase and therefore increases its potency. Ingestion of large doses of sodium bicarbonate increases the alkalinity of the plasma and hence decreases the effective partition coefficient of the drug, thereby reducing its potency. Conversely, the hypotensive activity of the ganglion blocker mecamylamine is increased by inhalation of CO_2, as its activity is dependent on plasma levels. Thus, to predict accurately the effect of change in acid–base balance on the activity of any drug requires a knowledge of its site of action and the potential effect of pH changes in excretion and biotransformation; that is, it requires clarification of the extent of pH changes throughout the body.

Ingestion of some antacids over a period of 24 hours will increase urinary pH. Administration of sodium bicarbonate with aspirin reduces blood salicylate levels by about 50 per cent, probably due to increased excretion in the urine. Although high doses of alkalising agents which raise the pH of the urine will increase the renal excretion of free salicylate and result in lowering of plasma salicylate levels, in commercial buffered aspirin tablets (e.g. Bufferin) there is insufficient antacid to cause a change in the pH of the gastric fluids (as is also the case with Butazolidin-Alka). The small amount of antacid is sufficient, however, to aid the dissolution of the acetylsalicylic acid (see chapter 5) and this leads to more favourable absorption rates[7].

Change in urinary pH will alter the rate of urinary excretion (figure 10.4). When a drug is in its unionised form it will more readily diffuse from the urine to the blood. In an acid urine, acidic drugs will diffuse back into the blood from the urine. Acidic compounds such as nitrofurantoin are excreted faster when the urinary pH is alkaline. Amphetamine, imipramine and amitryptyline are excreted faster in an acid urine. The control of urinary pH in studies of the pharmacokinetics of drugs is vital. It is difficult, however, to find compounds to use by the oral route for deliberate adjustment of urinary pH. Sodium bicarbonate and ammonium chloride may be used. Intravenous administration of acidifying salt solutions presents one approach to the forced diuresis of basic drugs in cases of poisoning.

Urinary pH can be important in determining drug toxicity more directly. A preparation containing hexamine mandelate and sulphamethizole caused turbidity in the urine of 9 out of 32 patients. The turbidity was higher in acid urine, and was caused by precipitation of an amorphous sulphonamide derivative containing 63 per cent of sulphamethizole. *In vitro*, hexamine causes the precipitation of the sulphonamide at pH values from 5 to 6. The efficacy of both agents is reduced by precipitation and the danger of renal blockade is of course increased[8].

Pain on injection is frequently a result of precipitation of a drug at the site of injection brought about by either solvent dilution or by alteration in pH. Precipitation of drugs from formulations used intravenously can, of course, lead to thromboembolism. The kinetics of precipitation under realistic conditions must be appreciated as a sufficiently slow rate of infusion can obviate problems of this sort. In the treatment of Jusko *et al.*[9], of the precipitation of diazepam, the flow rate of normal saline (Q) required to maintain the drug in solution during its addition to an intravenous infusion was calculated from

$$Q = R/S_m \tag{10.2}$$

where R is the rate of injection of diazepam in mg min^{-1}, and S_m is the apparent maximum solubility of diazepam when mixed with normal saline (~0.3 mg mℓ^{-1}). If the value of R is chosen as 5 mg min^{-1}, Q would have to exceed 17 mℓ min^{-1} to prevent *observable* precipitation. As this is a high rate of infusion it is evident that the administration of diazepam through the tubing of an i.v. drip is likely to result in precipitate formation. Too rapid injection of the preparation directly into the venous supply might result in precipitation; slow venous blood flow would have the same adverse effect. This would perhaps explain a finding that thrombophlebitis occurs less frequently when smaller veins are avoided and when injection of diazepam is followed by rigorous flushing of the infusion system with normal saline[10].

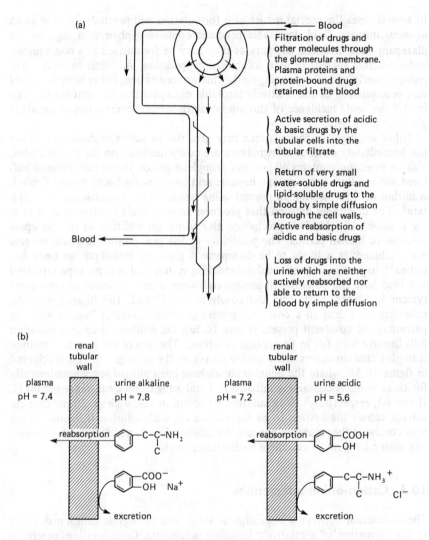

figure 10.4 (a) A highly simplified diagram of a kidney tubule to illustrate the filtration and secretion of drugs from the blood into the tubular filtrate, and their subsequent reabsorption or loss in the urine. From I. H. Stockley. *Pharm. J.*, 206 377 (1971)

(b) Schematic representation of the influences of urinary pH on the passive reabsorption of a weak acid and a weak base from the urine in the renal tubules. At a high pH the passive reabsorption of the weak base and the excretion of the weak acid are enhanced while at a low pH value the reabsorption of the weak acid and the excretion of the weak base are enhanced

10.2   Effects of dilution of mixed solvent systems

In several cases the special nature of a formulation will preclude dilution by an aqueous infusion fluid. Injectable products containing phenytoin, digoxin and diazepam come into this category as they may be formulated in a non-aqueous water-miscible solvent or as a solubilised preparation. Addition to water may result in precipitation, depending on the final concentration. It has been suggested that precipitation of the relatively insoluble diazepam may account for the high (~ 3.5 per cent) incidence of thrombophlebitis which occurs when diazepam is given intravenously.

Other additives in formulations may give rise to subtle problems which are not immediately obvious as ingredients are rarely disclosed on the product label. Valium injection contains 40 per cent propylene glycol, 10 per cent ethanol buffered with sodium benzoate and benzoic acid and preserved with benzyl alcohol. Addition of this injection to normal saline results in the formation of a precipitate[9]. The maximum dilution that produces an observable precipitate after mixing is about 15-fold. This precipitate also forms on addition of the diazepam solution to human plasma. The possibility of the precipitate being benzoic acid was excluded; it is likely to be diazepam. A graphical technique has been described[11] to predict whether a solubilised drug system will become supersaturated and thus have the potential to precipitate. When a drug dissolved in a cosolvent system is diluted, both drug and cosolvent are diluted. The logarithm of the solubility of a drug in a cosolvent system generally increases linearly with the percentage of cosolvent present (figure 10.5a). On dilution, drug concentration falls linearly with fall in percentage cosolvent. The aim of the graphical method is to plot dilution curves and solubility curves on the same graph. This is achieved in figure 10.5b, where the dilution curves have been plotted semilogarithmically for three systems containing initially 1, 2 and 3 mg of drug substance per mℓ (I, II and III, respectively). With solution III, dilution below about 30 per cent cosolvent causes the system to be supersaturated; with solution II, below 20 per cent cosolvent the solubility line and the dilution line touch. Only with solutions containing 1 mg mℓ^{-1} can there be dilution at will.

10.3   Cation-anion interactions

The interaction between a large organic anion and an organic cation may result in the formation of a relatively insoluble precipitate. Complexation, precipitation or phase separation can occur in these circumstances, the product being affected by changes in ionic strength, temperature and pH. Examples of cation-anion interactions include procaine amide and diphenylhydantoin sodium, procaine and thiopentone sodium, hydroxyzine hydrochloride and penicillin G. The nature of many of these interactions has not been studied in detail. In the absence of such work it is necessary to predict possible incompatibilities from a knowledge of the physical properties of the drug and other components in the formulation. The incompatibility when chlorpromazine and morphine injections are mixed[12] has been found to be due to the chlorocresol contained in the morphine injection prepared by heating with a bactericide. The chlorocresol caused the precipitation of the chlorpromazine, possibly by an anion-cation

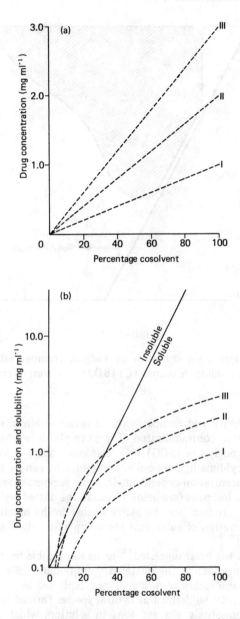

figure 10.5 (a) Dilution profiles for three solutions (I, II and III) in pure co-solvent containing 1, 2 and 3 mg drug per mℓ, respectively. (b) Curves from (a) plotted on semilog scale along with typical solubility line. From reference 11

interaction. Nitrofurantoin sodium must be diluted prior to use with 5 per cent dextrose or with sterile water for injection without alkyl *p*-hydroxy benzoates (parabens), phenol or cresol which tend to precipitate the nitrofurantoin[1].

Phase diagrams such as that shown in figure 10.6 are useful in determining

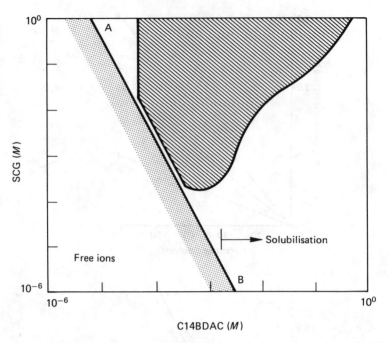

figure 10.6 Phase diagram for mixtures of sodium cromoglycate (SCG) and tetradecyldimethylammonium bromide (C14BDAC). From reference 13

regions of incompatibility and compatibility in cation–anion mixtures because admixture is not always contraindicated. The example shown here is for mixtures of sodium cromoglycate (SCG) – a dianionic drug – with a cationic surfactant – tetradecyldimethylammonium bromide. As can be seen the interaction is strongly concentration-dependent. In some regions, below the line AB, the two ions co-exist. Ion pairs (see below) form in the dotted region below AB. Above this solubility product line, turbidity occurs in the hatched region. On increasing the concentration of surfactant the complex is solubilised so that the interaction is masked[13].

Ion-pair formation has been suggested[14] to be responsible for the absorption of highly charged drugs such as the quaternary ammonium salts and sulphonic acid derivatives, the absorption of which is not explained by the pH-partition hypothesis. Ion pairs may be defined as neutral species formed by electrostatic attraction between oppositely charged ions in solution, which are often sufficiently lipophilic to dissolve in non-aqueous solvents. The formation of an ion pair (figure 10.7) results in the 'burying' of charges and alteration to the physical properties of the drug. This is discussed more fully at the end of section 9.1.2. Interactions between charged drug species and appropriate lipophilic ions of opposite charge may constitute a drug interaction and may occur *in vitro* or *in vivo*.

In a study of the incompatibility of organic iodide contrast media and antihistaminic drugs (added in attempts to reduce anaphylactic reactions) it was

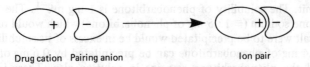

figure 10.7 Diagrammatic representation of ion-pair formation

found[15] that the acidity of the antihistamine solutions used caused the precipitation of the organic iodide. One of the antihistamines, promethazine, reacted strongly with all the contrast media, probably because its solution had the lowest pH of the drugs studied. It is not only with intravenous fluids that such interactions may occur. Examples have been quoted of the inadvisable mixture of syrups; for example, immediate precipitation in a prescribed mixture of Orbenin syrup, containing cloxacillin sodium, and Phensedyl syrup, containing three bases. Precipitation was followed by a 20 per cent loss in antibiotic activity in 5 hours and 99 per cent loss in five days. The double-decomposition reactions involved are likely to be those shown in figure 10.8.

Complexes which form are not always fully active. A well-known example is the complex between neomycin sulphate and sodium lauryl sulphate which will form when Aqueous Cream BP is used as a vehicle for neomycin sulphate. The formation of visible precipitates depends to a large extent on the insolubility of the two combining species in the particular mixture. One might assume that in an atropine and phenobarbitone mixture, a barbiturate–atropine complex may precipitate. Atropine base is soluble to the extent of 1 part in 460 parts of water. There is only 0.6 mg atropine in a 5 ml dose and it is therefore well within its

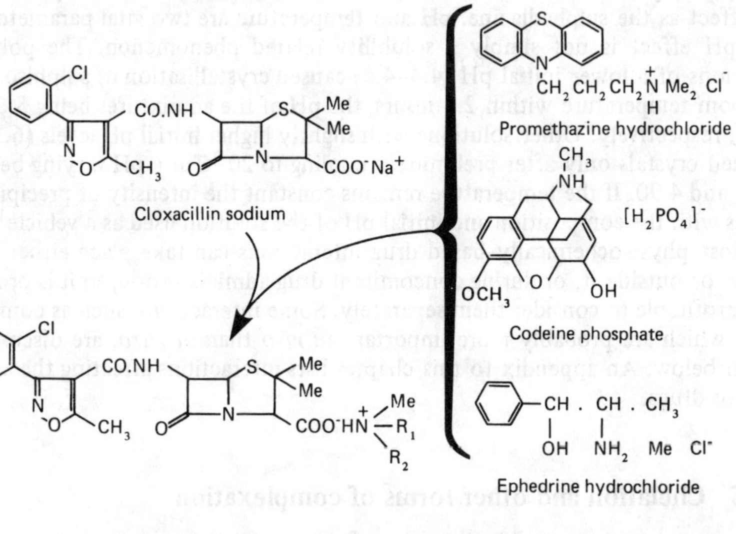

figure 10.8 Possible interaction of cloxacillin with promethazine, codeine or ephedrine. After J. B. Stenlake, *Pharm. J.*, 219, 533 (1975)

solubility limit. The solubility of phenobarbitone is 1 mg mℓ^{-1}. The 15 mg of phenobarbitone sodium ($\equiv$ 13.5 mg of phenobarbitone) which would result if all the sodium salt were to be precipitated would be in excess of its solubility. However, only 0.4 mg of phenobarbitone can be precipitated by 0.6 mg of atropine sulphate and the phenobarbitone remains in solution also. There is thus no precipitation of a complex.

Interactions between drugs and ionic macromolecules are another source of problems. Heparin sodium and erythromycin lactobionate are contraindicated in admixture as are heparin sodium and chlorpromazine hydrochloride or gentamicin sulphate. Tables of incompatibilities abound with such examples. Heparin is a polysulphuric ester of mucoitin. Interference with the sulphuric groups reduces its anticoagulant activity. The activity of phenoxymethylpenicillin against *Staphylococcus aureus* is reduced in the presence of various macromolecules such as acacia, gelatin, sodium alginate and tragacanth[16]. The interaction will not, in each case, be electrostatic in origin but may involve binding through hydrophobic interactions.

10.4   Simple electrolytes and their effect on drug solutions

The extensive clinical use of polyionic solutions for intravenous therapy means that drugs are frequently added to systems of complex ionic nature. Reduction in the solubility of both weak electrolytes and non-electrolytes can occur through salting out, a phenomenon discussed in chapter 5.

Sodium sulphadiazine and sulphisoxazole diethanolamine (Gantrisin) in therapeutic doses (1 g) added to 5 per cent dextrose and 5 per cent dextrose and saline solution have been found to be compatible, yet added to commerical polyionic solutions (such as Abbott Ionosol B, Baxter electrolyte No. 2) formed rapid and heavy precipitates, although the sulphisoxazole did not produce as pronounced an effect as the sulphadiazine. pH and temperature are two vital parameters but the pH effect is not simply a solubility related phenomenon. The polyionic solutions of a lower initial pH (4.4–4.6) caused crystallisation of sulphisoxazole at room temperature within 2.5 hours, the pH of the admixtures being 5.65 and 5.75, respectively. Other solutions with slightly higher initial pH levels (6.1–6.6) formed crystals only after preliminary cooling to 20 °C at a pH varying between 4.25 and 4.90. If the temperature remains constant the intensity of precipitation varies with the composition and initial pH of the solution used as a vehicle[17].

Most physicochemically based drug interactions can take place either in the body, or outside it, or during concomitant drug administration, so it is probably not profitable to consider them separately. Some interactions, such as complexation, which are probably more important *in vivo* than *in vitro*, are discussed in detail below. An appendix to this chapter lists interactions affecting the absorption of drugs.

10.5   Chelation and other forms of complexation

The term *chelation* relates to the interaction between a metal atom or ion and another species, known as the ligand, by which a heteroatomic ring is formed.

Chelation changes the physical and chemical characteristics of the metal ion, and the ligand. It is simplest to consider that the ligand is the electron-pair donor and the metal the electron-pair acceptor, the donation establishing a coordinate bond. Many chelating agents act in the form of anions which coordinate to the metal ion[18]. For chelation to occur there must be at least two donor atoms capable of binding to the same metal ion and ring formation must be sterically possible. For a fuller account Bell's book on metal chelation should be consulted[18]. When a drug forms a metal chelate the solubility and absorption of both drug and metal ion may be affected.

Probably the most widely quoted example of complex formation leading to decreased drug absorption is that of tetracycline chelation with metal ions. Polyvalent cations such as Fe^{2+} and Mg^{2+} and anions such as the trichloroacetate or phosphate ion interfere with absorption in both model and real systems[19]. Figure 10.9 shows the effect of a dose of 40 mg ferrous ion on the serum levels following 300 mg of tetracycline. As can be seen from this figure, the nature of the iron salt ingested is important. In the study, the iron salts were packed in readily dissolved gelatin capsules. Ferrous sulphate has the greatest inhibitory effect on tetracycline absorption; it dissolves in water more quickly than organic iron compounds. The ability of the various iron compounds to liberate ferrous or ferric ions in the upper part of the gastro-intestinal tract before tetracycline is absorbed would seem to be an essential step in the interaction. It has been pointed out that the order of activity of the different iron salts in the chelation process *in vivo* is the same as the order of the intestinal absorption of these iron compounds.

All the active tetracyclines form stable chelates with Ca^{2+}, Mg^{2+} and Al^{3+}. In fact, it has been suggested that the antibacterial action of the compounds depends on their metal binding activity, as their main site of action is on the ribosomes which are rich in magnesium. The tetracyclines have an avidity for divalent metals similar to that of glycine (II) but they have a greater affinity for the trivalent metals with which they form 3:1 drug–metal chelates. Doluisio and Martin[20] showed that therapeutically active tetracyclines formed 2:1 complexes with cupric, nickel and zinc ions while inactive analogues formed only 1:1 complexes.

$$CH_2-NH_2$$

The 1:1 copper–glycine chelate

(II)

The complexing of tetracyclines with calcium is an interaction which is a problem in paediatric medicine. Discolouration of teeth results from the formation of a coloured complex with the calcium in the teeth and the deposition of drug in the bones of growing babies can lead to problems in bone formation. Examination of table 10.4 will show that there is no correlation between the binding capacity of a tetracycline with iron and with calcium, suggesting different modes of complexation. The *in vitro* data are simpler to interpret; the serum levels are the result of two processes, the chelation of the tetracycline and the partitioning of

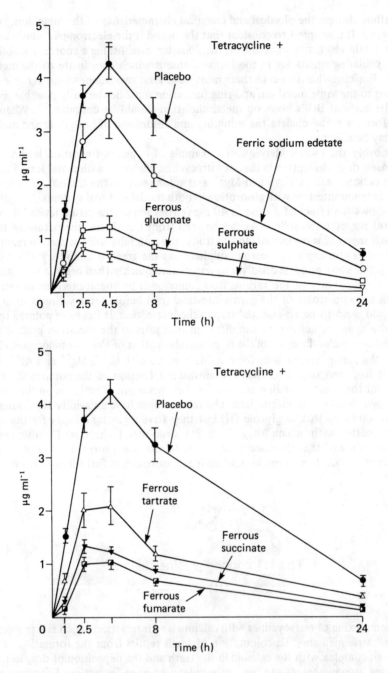

figure 10.9 The effect of simultaneous ingestion of various iron salts (dose equivalent to 40 mg elemental iron) and tetracycline hydrochloride (500 mg) on serum levels of tetracycline (mean levels in six patients). From P. J. Neuvonen and H. Turakka. *Eur. J. clin. Pharm.*, 7, 357 (1974)

table 10.4 **Relative calcium binding capacities of tetracycline derivatives and decreases in serum levels after 200 mg ferrous sulphate**

Tetracycline	Calcium binding (per cent)*	Decrease in serum concentration (per cent) after 200 mg ferrous sulphate
Demethylchlortetracycline	74.5	n.s.**
Chlortetracycline	52.7	n.s.
Tetracycline	39.5	40–50
Methacycline	39.5	80–85
Oxytetracycline	36.0	50–60
Doxycycline	19–22	80–90

*Percentage of dissolved antibiotic (5 mg/50 mℓ H_2O) bound to calcium phosphate after overnight shaking. (Source: *US Dispensatory*, 27th edition p. 1155 *et seq.*)
**n.s. = not studied

the chelate. The lipophilicity of the complex may or may not be readily correlated with known parameters.

Table 10.4 demonstrates the different affinities of the tetracyclines for metallic ions. Clinical studies have shown that the absorption of doxycycline is not significantly affected by milk in conditions where the absorption of tetracycline is reduced. Which portion of the tetracycline molecule is involved? Experiments with simpler compounds suggest that the

grouping is important as in the 1-hydroxyanthraquinone chelate (III). The structures of some tetracyclines are reproduced in table 10.5, together with their pK_a values; from these data it should be possible to determine something of the relative affinities of the tetracyclines for metal ions. One piece of evidence relating to the site of chelation is that isochlortetracycline, which lacks the C_{11}, C_{12} enolic system, does not chelate with Ca^{2+} ions.

The highly coloured nature of the tetracycline complexes may be utilised in analytical procedures. The uranyl ion-tetracycline complex has been used[21]. The possibility of mixed complexes, especially *in vitro*, has been suggested by Kohn[22], who found extractable mixed complexes in the presence of calcium ion and barbiturates.

1–Hydroxyanthraquinone chelate

(III)

table 10.5 **Structures and pK$_a$ values for some tetracyclines**

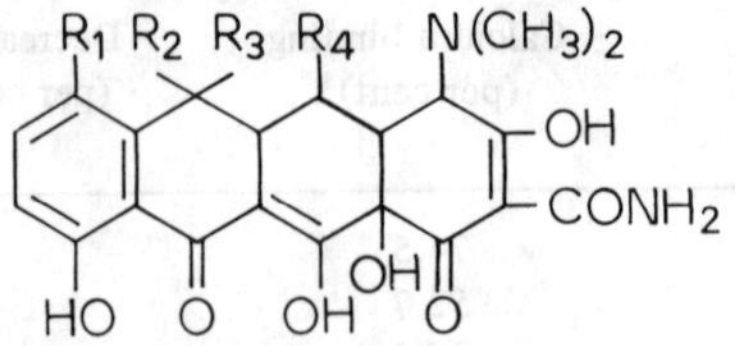

	R$_1$	R$_2$	R$_3$	R$_4$	pK$_{a1}$	pK$_{a2}$	pK$_{a3}$
Chlortetracycline	Cl	Me	OH	H	3.27	7.43	9.33
Oxytetracycline	H	Me	OH	OH	3.5	7.6	9.2
Tetracycline	H	Me	OH	H	3.33	7.7	9.5
Demeclocycline*	Cl	H	OH	H	3.3	7.16	9.25
Doxycycline	H	Me	H	OH	3.4	7.7	9.7

*Demethylchlortetracycline

The former practice of administering chlortetracycline and oxytetracycline with milk to reduce the intensity and incidence of the gastro-intestinal discomfort that frequently accompanied their ingestion was effective because of the bonding of the antibiotic with the calcium in the milk. Of course this decreased the amount of drug available for absorption. Much of the early work on tetracycline absorption is invalid because of the use of tetracycline dihydrochloride formulated with dicalcium phosphate as a control. The results showed enhanced levels with some formulations simply because of the decreased performance of the control brought about by chelation.

Martell and Calvin[23] have described the four criteria for determining chelate formation: coloured product, decreased aqueous solubility, a drop in pH during

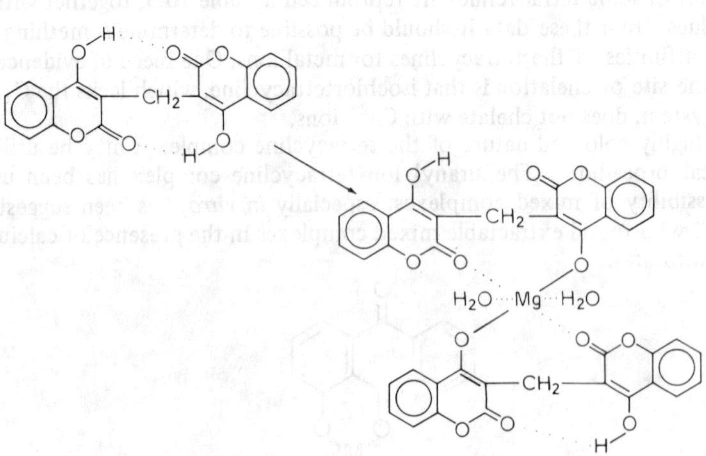

Suggested form of bishydroxycoumarin — magnesium chelate
From L. D. Bighley and R. J. Spirey. J. pharm. Sci., 66 1124 (1977)

(IV)

chelate formation, and the absence of metal ions in solution after chelate formation. The decreased aqueous solubility of chelates suggests increased lipophilicity, but, in the case of the tetracycline chelates, the precipitation of drug would decrease biological activity, as drug would not be available for transport across membranes; the size of the chelate would also prevent easy absorption in the intact form.

There is evidence for complex formation of a coumarin derivative with magnesium ions present in antacid formulations. In this case the formation of a more absorbable species is indicated since plasma levels of bishydroxycoumarin are elevated in the presence of magnesium hydroxide, but unaffected by aluminium hydroxide; neither antacid influenced warfarin absorption[24]. A magnesium-bishydroxycoumarin chelate having the structure (IV) has been isolated. The adsorptive properties of aluminium hydroxide may be responsible for the lack of effect of this antacid on the absorption of the drug. Adsorption may counteract a chelate-mediated increase in absorption. Results are shown in figure 10.10. It has been suggested that the more rapid and complete absorption of warfarin makes it less susceptible to interactions of this type[24].

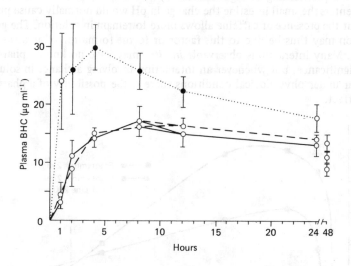

figure 10.10 Plasma levels of bishydroxycoumarin (BHC) in six subjects after a 300 mg oral dose with water (○——○), magnesium hydroxide (○. . . . .○), or aluminium hydroxide (○− − − ○). Closed data points represent a significant difference from control. From reference 24. Bishydroxycoumarin is also known as dicoumarol

Following treatment of acrodermatitis enteropathica with diodoquin (V) it was noted that the absorption and retention of dietary zinc and other trace metals was greater than in subjects not receiving the drug. Chelation of zinc and other trace metals by the drug might result in the increased absorption from the intestinal lumen.

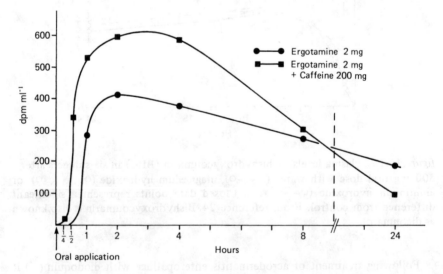

Suggested form of chelate of zinc–diodoquin
After H.T. Delves et al. Lancet, ii, 929 (1975)

($\underline{V}$)

Ergotamine tartrate given parenterally is still the most efficient treatment for severe migraine, but the drug given orally is less effective. Simultaneous intake of caffeine, however, improves the efficiency of ergotamine in aborting migraine attacks (figure 10.11). Caffeine enhances the dissolution rate of ergotamine by a factor of three in gastric juice; and at intestinal pH caffeine has been found to increase the partitioning of ergotamine (VI) into a lipid phase *in vitro*[25]. As the alkaloid enters the small intestine the change in pH would normally cause precipitation but the presence of caffeine allows more to remain in solution. The greater absorption may thus be due to this factor or to the formation of an absorbable complex. Many interactions observable *in vitro* may have no known pharmacological significance, but whenever an interaction involving a change in solubility can occur under physiological conditions there is the possibility of a pharmacological effect.

figure 10.11 The effect of caffeine on the enteral absorption of ergotamine in man. Plasma radioactivity after administration of [3H]-ergotamine tartrate alone or in combination with caffeine to six subjects is shown. From R. Schmidt and A. Fonchamps. *Eur. J. clin. Pharm.*, 7, 213 (1974)

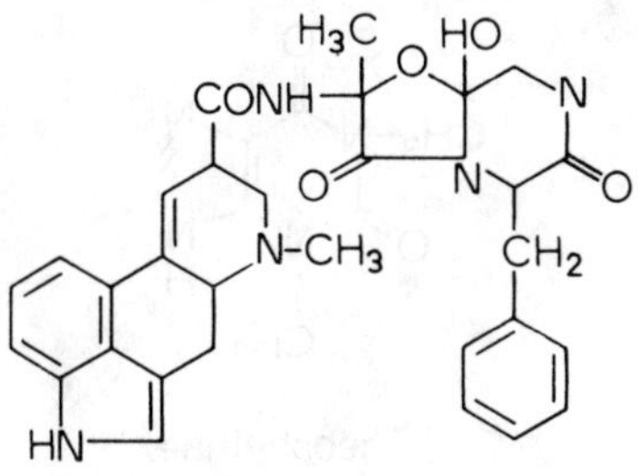

Ergotamine

(VI)

10.6 Other complexes

Molecular complexes of many types may be observed in systems containing two or more drug molecules. Generally, association follows from attractive interactions (hydrophobic, electrostatic or charge-transfer interactions) between two molecules. In the charge transfer system one component is usually an aromatic compound; the second may be a saturated moeity containing a lone electron pair (donor atom) or a weakly acidic hydrogen (acceptor atom). The interaction therefore takes place between electron-rich donors and electron-poor acceptors. Interactions between aromatic rings in which there is a parallel overlap of π-systems fall into this category, as in the following example involving nicotinamide (acceptor) and the indole moiety of tryptophan (donor) (VII).

The imidazole moiety (VIII) is involved in many interactions; it is regarded as aromatic and the molecule is planar. Not surprisingly caffeine (IX) and theophylline (X) are frequently implicated in interactions with aromatic species. Caffeine, for example, increases the solubility of benzoic acid. The marked difference in the solubilising properties of caffeine and dimethyluracil (XI) towards substances such as benzoic acid suggests that the imidazole ring of the xanthine nucleus is the portion involved in the interaction. In the main, 1:1 complexes are formed but 2:1 drug–caffeine complexes may also be found. Hydrophobic interactions

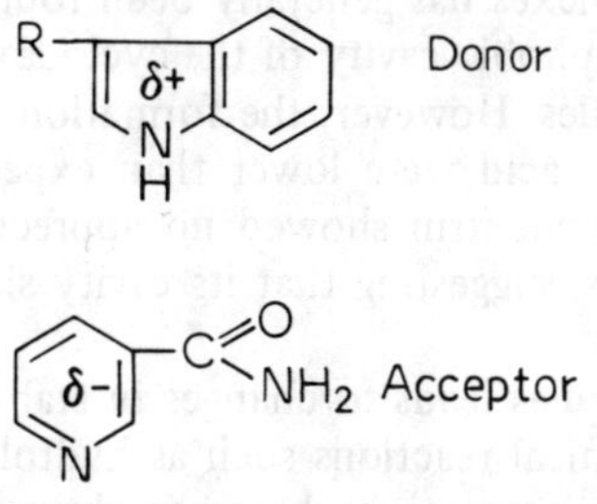

Example of donor acceptor interaction between tryptophan (donor) and nicotinamide

(VII)

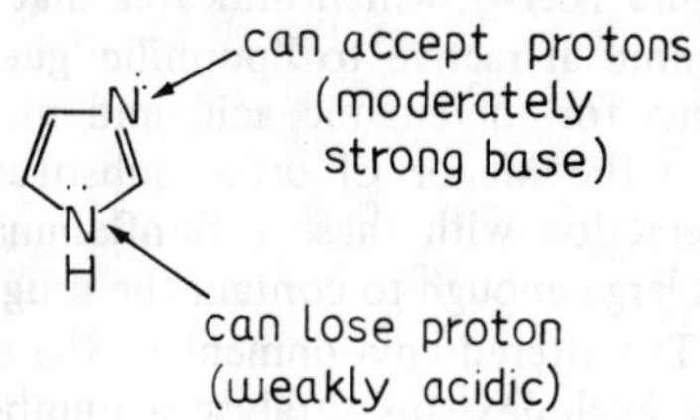

(VIII)

Caffeine

(IX)

Theophylline

(X)

Dimethyluracil

(XI)

are also implicated in the interaction since caffeine has a greater solubilising capacity than theophylline, which lacks the $N-CH_3$ group in the imidazole nucleus.

10.6.1 Interaction of drugs with β-cyclodextrin

Considerable interest has been shown in the interaction of drugs with cyclodextrins, which are cyclic β-1,4-linked D-glucose oligomers. Cyclodextrins form a number of crystalline adducts with aromatic compounds, alkanes and alkyl halides. The cyclodextrin molecules in solution form structures with internal diameters of 0.6–1 nm. The interior cavity of the cyclodextrin ring is hydrophobic in nature and binds a hydrophobic portion of the 'guest' molecule, usually forming a 1:1 complex. The assumed structure of β-cyclodextrin is shown in figure 10.12. This shows both equatorial and axial modes of incorporation of a guest molecule. A possible structure of β-cyclodextrin complexes of phenothiazines is shown in figure 10.13.

Inclusion complexes of various drugs have been successfully used to enhance solubility, chemical stability and the absorption of drugs. The solubilising effect of β-cyclodextrin on flufenamic acid suggests that a 1:1 complex is formed in aqueous solution. A relationship between the oil/water partition coefficient of a drug and its tendency to form inclusion complexes has generally been found[26] (figure 10.14), which indicates that the hydrophobic cavity of the cyclodextrin is more attractive to lipophilic 'guest' molecules. However, the formation constants for mefenamic acid and meclofenamic acid were lower than expected from the Me or Cl *ortho* substituent. α-Cyclodextrin showed no appreciable interaction with these anti-inflammatory acids, suggesting that its cavity size is not large enough to contain the drug molecules[27].

The altered environment of the drug molecules leads to changes in stability. The cyclodextrins catalyse a number of chemical reactions such as hydrolysis, oxidation and decarboxylation[28]. Although interactions between drugs and cyclodextrins have been mostly the result of deliberate attempts to modify the behaviour or properties of the drug, they are included here to illustrate an additional mode of interaction with other components which may well occur inside the body as well as in the laboratory.

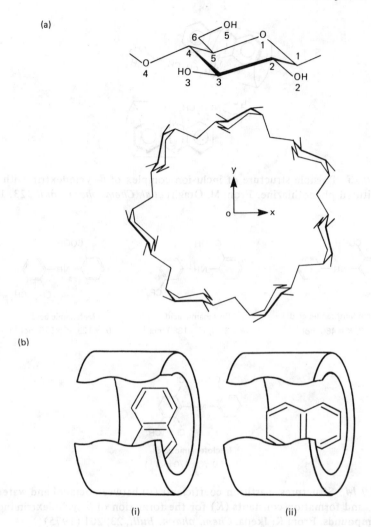

figure 10.12 (a) The assumed structure of β-cyclodextrin. (b) The two models of the complex; equatorial inclusion (i) and axial inclusion (ii). From K. Harata and H. Uedaira. *Bull. chem. Soc. Jap.*, 48, 375 (1975)

10.6.2 Ion-exchange interactions

Ion-exchange resins are now being used medicinally. Cholestyramine and colesti-pol are insoluble quaternary ammonium anion-exchange resins which, when administered orally, bind bile acids and increase their elimination because the high molecular weight complex is not absorbed. As bile acids are converted *in vivo* into cholesterol, cholestyramine is used as a hypocholesteraemic agent. When given to patients receiving other drugs as well, the resin would conceivably bind anionic compounds and reduce their absorption. In *in vitro* experiments,

figure 10.13   Possible structure of inclusion complex of β-cyclodextrin with an N-substituted phenothiazine. From M. Otagiri *et al. Chem. pharm. Bull.*, 23, 188 (1975)

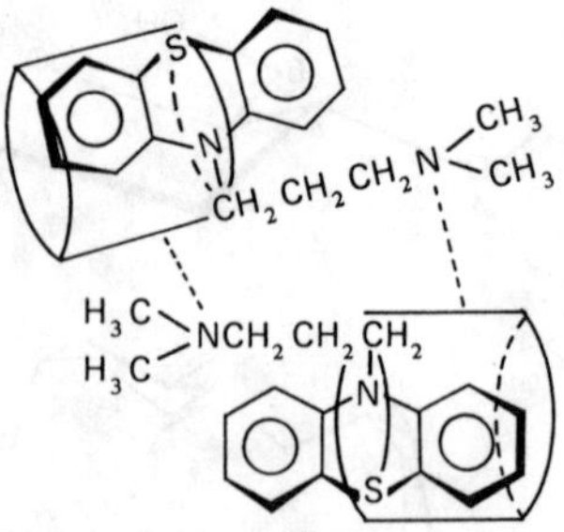

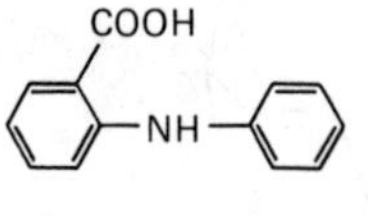

N-Phenylanthranilic acid

($P = 22$, $K = 460$ mol^{-1})

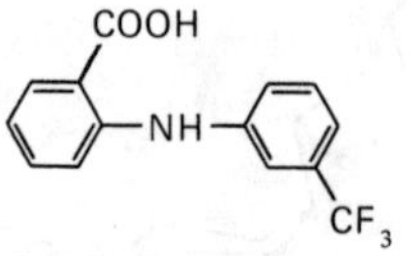

Flufenamic acid

($P = 39.0$, $K = 1380$ mol^{-1})

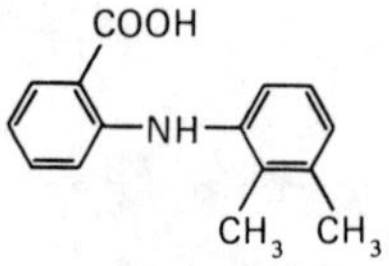

Mefenamic acid

($P = 123$, $K = 570$ mol^{-1})

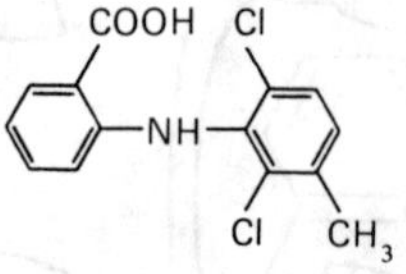

Meclofenamic acid

($P = 60$, $K = 470$ mol^{-1})

figure 10.14   Structures, partition coefficients (P) between octanol and water at pH 7.0, and formation constants (K) for the formation of β-cyclodextrin inclusion compounds. From K. Ikeda. *Chem. pharm. Bull.*, 23, 201 (1975)

phenylbutazone (XII), warfarin (XIII), chlorothiazide (XIV) and hydrochlorothiazide were bound strongly even after washing several times with buffer. Some drugs, such as tetracycline, were initially bound but were then removed by washing.

Cholestyramine had no effect in animal experiments on the blood levels of these drugs. However, 95 per cent of warfarin and of phenylbutazone were bound and peak blood levels of the latter were halved, although the same total amount eventually was absorbed. A single dose of resin did not significantly reduce the pharmacological effect of warfarin. Nevertheless, it is prudent to administer drugs orally a short time before cholestyramine to preclude delay in the action of the drug through adsorption and slow leaching. Table 10.6 shows the effect of very high resin levels on the serum concentration of sodium fusidate. Since the bind-

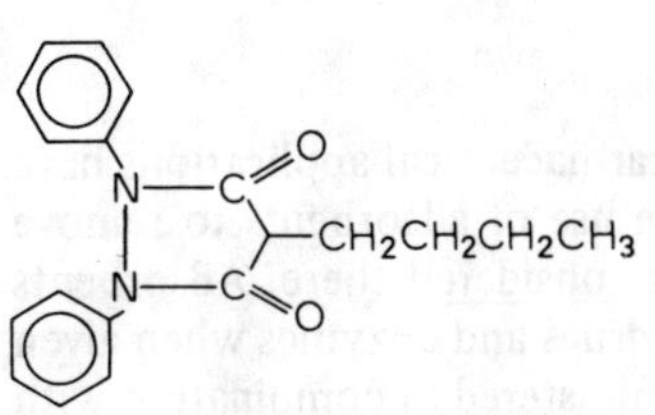

Phenylbutazone

(XII)

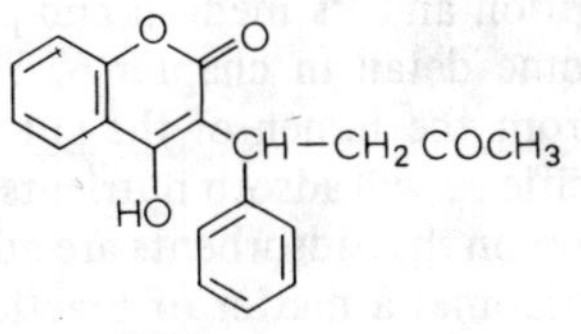

Warfarin

(XIII)

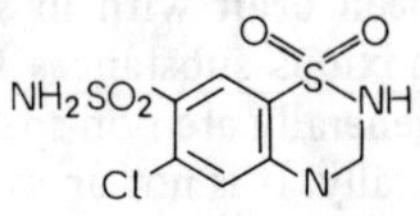

Chlorothiazide

(XIV)

ing is pH-dependent (table 10.7) an ion-exchange resin can bind any drug with an appropriate charge under pH conditions in which both species are ionised. Decreased drug absorption caused by cholestyramine or colestipol in the main has been reported with thyroxine, aspirin, phenprocoumon, warfarin, chlorothiazide, cardiac glycosides, and ferrous sulphate as a consequence of bonding. Vitamin K and vitamin B_{12} absorption is decreased indirectly by competition of the resin for binding sites on intrinsic factor molecules[29,30].

table 10.6 **The effect of cholestyramine administration on sodium fusidate levels***

Resin dose (mg kg^{-1})	Mean serum conc. (μg mℓ^{-1})
0	3.7
72	2.5
215	1.9
356	0.8

*From Johns and Bates. *J. pharm. Sci.*, 61, 735 (1972)

table 10.7 **Binding of acetylsalicylic acid *in vitro* to cholestyramine***

	Percentage binding			
	pH 2	pH 3	pH 5	pH 7.5
Buffer	5	–	6	6
Water	51	95	98	97

*From K.-J. Halin *et al. Eur. J. clin. Pharm.*, 4, 142 (1972)

The influence of buffer ions on the binding of the drug in table 10.7 indicates one of the problems of making *in vitro* measurements and attempting to assess the biological significance of the results. Binding of cephalexin, clindamycin and the components of co-trimoxazole to cholestyramine has been measured; *in vivo* absorption of these substances in the presence of the resin is delayed and somewhat reduced, but the changed pattern of absorption is unlikely to affect therapeutic efficacy, except perhaps with cephalexin where it is more pronounced.

10.7 Adsorption of drugs

The process of adsorption and its medical and pharmaceutical applications have been dealt with in some detail in chapter 6. The use of adsorbents to remove noxious substances from the lumen of the gut is considered there. Adsorbents generally are non-specific so will adsorb nutrients, drugs and enzymes when given orally. It is not uncommon that adsorbents are administered in combination with various drugs and it becomes a matter of practical importance to determine the extent to which adsorbents will interfere with the absorption of the drug sub-stance from the gastro-intestinal tract. Several consequences of adsorption are possible. If the drug remains adsorbed until the preparation reaches the general area of the absorption site, the concentration of the drug presented to the absor-bing surfaces will be much reduced. The driving force for absorption would then be reduced, resulting in a slower rate of absorption. During the course of absorp-tion of the drug, it is probable that the adsorbate will dissociate in an attempt to re-establish equilibrium with drug in its immediate environment, particularly if there is competition for absorption sites from other substances in the gastro-intestinal tract. As a consequence, the concentration of free drug in solution at the absorption site would be maintained at a low level and the absorption rate would be slowed. Alternatively, the release of drug from the adsorbent might be complete before reaching the absorption site, possibly hastened by the presence of electrolytes in the gastro-intestinal tract; in which case absorption rates would be virtually identical to those in the absence of adsorbent. Figure 10.15 shows the reduced amounts of promazine excreted in the urine when the promazine was administered with activated charcoal – evidence of a reduced absorption rate of this drug in this situation.

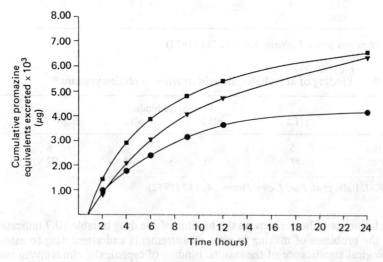

figure 10.15 Cumulative amounts of promazine equivalents excreted in the urine following administration of various dosage forms to humans. (■), Promazine in simple aqueous solution; (▼) promazine plus activated attapulgite; (●) pro-mazine plus activated charcoal. From D. L. Sorby. *J. pharm. Sci.*, 54, 677 (1965)

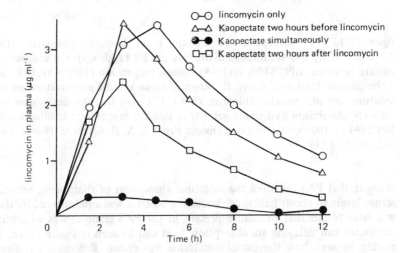

figure 10.16 The influence of Kaopectate (kaolin and pectinic acid) on the absorption of lincomycin in man after oral application. Note the decrease in plasma levels caused by the simultaneous administration of the adsorbent. From J. G. Wagner. *Can. J. Pharm. Sci.*, 1, 55 (1966)

A further example is the delayed absorption of lincomycin (XV) when administered with kaolin and pectinic acid in the form of Kaopectate (figure 10.16). The dramatic effect of several antacids on the dissolution *in vitro* of tablets containing digoxin and on the adsorption of digoxin from elixirs is shown in figure 10.17. In view of the relatively low doses of these glycosides, these adsorption effects are likely to have a significant impairing effect on digoxin bioavailability. In an *in vitro* study of drug adsorption onto a series of antacids, Khalil and Moustafa[31] demonstrated significant adsorption of some anticholinergics, tranquillisers, ataractics and mild sedatives, particularly by magnesium trisilicate and magnesium oxide.

If the possibility of adsorption onto formulation ingredients is not kept in mind, erroneous conclusions may be drawn from bioavailability studies, as was the case in the suspected *in vivo* interaction between PAS and rifampicin. It was

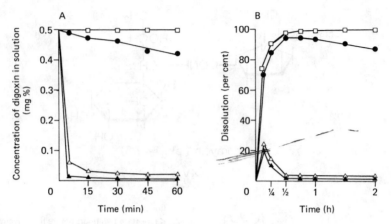

figure 10.17 A. Adsorption of digoxin from Lanoxin paediatric elixir at 37 ± 0.1 °C by (●) aluminium hydroxide gel BP (10% v/v); (△) magnesium trisilicate mixture, BPC (10% v/v); (▲) Gelusil suspension (10% v/v); (□) Lanoxin elixir diluted 1:10 with water. B. Effect of some antacid preparations on the dissolution rate of Lanoxin tablets at 37 ± 0.1 °C. (□) Dissolution in water; (●) in 10% v/v aluminium hydroxide gel; (△) in 10% v/v magnesium trisilicate mixture, BPC; (▲) in 10% v/v Gelusil suspension. From S. A. H. Khalil. *J. Pharm. Pharmacol.*, 26, 961 (1974)

thought that PAS impaired the intestinal absorption of rifampicin, reducing its serum levels to about half those occurring when it was administered by itself. It was later found that bentonite present in the PAS granules was adsorbing the antibiotic and delaying its absorption[32], as can be seen in figure 10.18. It can readily be seen how the initial conclusion was drawn. Bentonite has also been shown to be capable of binding brucine and methapyrilene by adsorption or ion-exchange reactions. Bentonite is a naturally occurring mineral (montmorillonite), consisting chiefly of hydrated aluminium silicate, and is thus related to kaolin.

McCarthy[33] suggests that a loss of activity of preservatives could arise from their adsorption onto solids commonly used as medicaments. Such an explanation was proposed for the spoilage of the BPC mixture of sulphadimidine for infants[34]. This possibility was investigated further by Khalil and Nasipuri[35], who reported that benzoic acid could be adsorbed to the extent of 94 per cent by the sulphadimidine. The adsorption could, however, be suppressed with hydrophilic polymers, the suppressive efficiency of these polymers following the sequence polyvinylpyrrolidone > methylcellulose > sodium carboxymethylcellulose.

El-Masry and Khalil[36] report that in the BPC mixture of magnesium trisilicate and belladonna, almost complete (approximately 93 per cent) adsorption of hyoscyamine by the magnesium trisilicate occurred. Studies on the effect of pH on the release of the bound drug showed that, because of the presence of both sodium bicarbonate and magnesium carbonate in the mixture, release could only occur when the pH value was less than 2, a condition which is unlikely to be fulfilled after the administration of the BPC mixture.

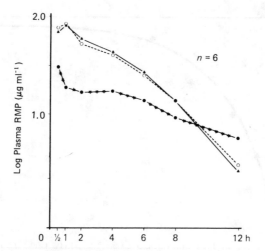

figure 10.18 Plasma concentrations of rifampicin (RMP) after oral administration in solution (10 mg kg^{-1}), given alone (▲——▲), with *p*-aminosalicylic acid (PAS) granules (●——●), or with Na-PAS tablets (○--------○). Means and s.d. are indicated. From reference 32

Similar problems may arise in solid dosage forms. Talc, which is commonly used as a lubricant for tablet making, has been reported to adsorb cyanocobalamin and consequently to interfere with intestinal absorption of this vitamin[37,38]. Other vitamins, notably pyridoxine and thiamine, are often formulated with the cyanocobalamin, and competitive adsorption of these three vitamins, leading to suppression of the adsorption of cyanocobalamin by talc, has been noted[39].

Protein and peptide adsorption

There has been increasing interest in the physical properties of peptides and proteins in solution because of the larger number of such molecules now used therapeutically. We have previously discussed the adsorptive loss of insulin from solution; adsorption can be a problem because of the amphipathic nature of many peptides, and becomes pharmaceutically important when the agent is present in low concentration in solution. The adsorption of peptides onto glass has been ascribed to bonding between their amino groups and the silanol groups of the glass. D-Nal(2)^{6} LHRH is a decapeptide derivative of the natural luteinising hormone releasing hormone. It has two basic amino groups which will be positively charged at low pH and hence provide an opportunity for binding to silanol groups[40]. The adsorption isotherm of the compound from aqueous solution at pH 7 is shown in figure 10.19. Siliconisation of the glass did not inhibit adsorption completely, suggesting that ionic binding was not the only mechanism of interaction. Phosphate buffer at 0.1 *M* concentration and acetate ions at 0.16 *M* concentration (both at pH 5) were most effective in preventing adsorption[40].

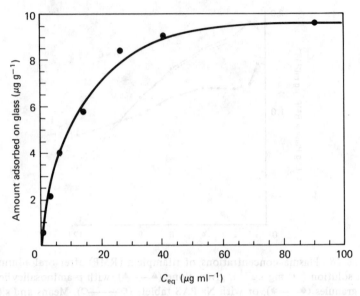

figure 10.19 Adsorption isotherm for D-Nal(2)^6 LHRH from aqueous solution at pH 7 onto glass beads. From reference 40 with permission

10.8 Drug interactions with plastics

In section 8.2.8 we discussed the adsorption of insulin onto glass and plastic materials used in syringes and giving sets. The plastic tubes and connections used in intravenous containers and giving sets can adsorb or absorb a number of drugs, leading to significant losses in some cases. D'Arcy[41] in a review of the topic lists those drugs which show a significant loss when exposed to plastic, in particular poly(vinyl chloride) (PVC). These include insulin, nitroglycerin, diazepam, chlormethiazole, vitamin A acetate, isosorbide dinitrate and a miscellaneous group of drugs such as phenothiazines, hydralazine hydrochloride and thiopentone sodium.

A theoretical treatment to account for the loss of nitroglycerin from solution to plastic containers has been developed[42]. It is derived on the basis of the model representing the two-stage loss of drug:

$$
\begin{array}{ccc}
\text{Nitroglycerin in} & \xrightleftharpoons[k_2]{k_1} & \text{Adsorbed} & \xrightarrow{k_3} & \begin{array}{c}\text{Nitroglycerin}\\ \text{dissolved in}\end{array} \\
\text{aqueous solution} & & \text{nitroglycerin} & & \text{the matrix} \\
(A) & & (B) & & (C)
\end{array}
$$

Therefore

$$\frac{\mathrm{d}A}{\mathrm{d}t} = -k_1 A + k_2 B \qquad\qquad (10.3)$$

$$\frac{\mathrm{d}B}{\mathrm{d}t} = k_1 A - k_2 B - k_3 B \qquad\qquad (10.4)$$

and
$$\frac{dC}{dt} = k_3 B \qquad (10.5)$$

It is found that $dB/dt \gg dC/dt$. At steady state $dB/dt = 0$, and at $t = 0$, $A = A_0$, $B = 0$ and $C = 0$. Rate constant k_1 is a function of the amount of drug in solution, the surface area available for absorption and the nature of the plastic, and k_3 is a function of the volume of the plastic matrix and the solubility of the nitroglycerin in the plastic matrix. The ratio k_3/k_2 describes the partitioning of the drug between the plastic and the aqueous phase. Where P is the partition coefficient

$$k_3/k_2 = aP \qquad (10.6)$$

a being a proportionality constant related to the mass of the plastic, the volume of the solution and other parameters. The ratio k_1/k_2 can be related to a Langmuir type of adsorption constant. The final form of the equation accurately predicts the nitroglycerin remaining in solution:

$$A = 8.957e^{-0.028t} + 14.943e^{-0.235t} \qquad (10.7)$$

Some idea of the rate and extent of disappearance of warfarin sodium from poly(vinyl chloride) infusion bags can be gained from figure 10.20. The marked effect of pH is seen. Losses can obviously be significant. When it is present at an initial concentration of 40 $\mu\ell$ ml^{-1} in 100 ml normal saline, a 76 per cent loss of medazepam[43] occurs in 8 h at 22 °C (table 10.8).

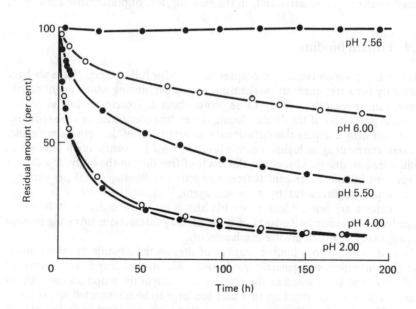

figure 10.20 Disappearance of warfarin sodium from aqueous buffered solutions stored in 100 ml PVC infusion bags at room temperature. From reference 43 with permission

table 10.8   **Losses of drugs from normal saline solutions in poly(vinyl chloride) (PVC) bags stored at 22 °C for 8 h***

Drug	Initial solute conc. (μg mℓ^{-1})	Per cent loss PVC (100 mℓ)	Per cent loss PVC (500 mℓ)
Diazepam	40	60	31
	120	58	31
Medazepam	40	76	–
Oxazepam	40	22	12
Nitrazepam	40	15	10
Warfarin sodium**	20	49	–
Warfarin sodium	20	29	–
Nitroglycerin	200	54	–
Thiopental sodium†	30	25	–
Pentobarbital sodium	30	0	0
Hydrocortisone acetate	20	0	0

**At pH 2 and 4.
†At pH 4.0 and 7.2
*From reference 43

Preservative interactions with membranes and closures

Preservatives such as the methyl and propyl parabens present in formulations can be sorbed into rubber and plastic membranes and closures, thus leading to decreased levels of preservative and, in the extreme, loss of preservative activity[44].

10.9   Protein binding

This is a topic which requires a chapter to itself for full coverage. Here we have space only for a treatment of mechanisms of protein binding which might enable recognition of molecules likely to be protein-bound. Protein binding alters the biological properties of the drug molecule as free drug concentrations are reduced. The bound drug assumes the diffusional characteristics of the protein molecule. In cases where drug is highly protein-bound ($\sim$ 90 per cent), small changes in binding lead to drastic changes in the levels of free drug in the body. If a drug is 95 per cent bound, 5 per cent is free. Reduction of binding to 90 per cent by, for example, displacement by a second agent, doubles the level of free drug. Such changes are not evident when binding is of a low order. Part B of the appendix to this chapter consists of a list of drug interactions involving protein binding, many of which involve displacements.

Most drugs bind to a limited number of sites on the albumin molecule. Binding to plasma albumin is generally easily reversible, so that drug molecules bound to albumin will be released as the level of free drug in the blood declines. Drugs bound to protein are attached to a unit too large to be transported across membranes. They are thus prevented from reacting with receptors or from entering the sites of drug metabolism or drug elimination.

What is the nature of plasma albumin? It consists of a single polypeptide chain of molecular weight 67 000 ± 2000. Human albumin contains between 569 and 613 amino acid residues. It is a globular molecule with a diameter of 5.6 nm (calculated assuming the molecule to be an anhydrous sphere). With an isoelectric point of 4.9 albumin has a net negative charge at pH 7.4, but it is amphoteric and capable of binding acidic and basic drugs. Subtle structural changes can occur on binding small molecules. Fatty acid binding produces a volume increase and a decrease in the axial ratio. This is due primarily to non-polar interaction between the hydrocarbon tail of the fatty acid molecule and the binding site, and reflects the adaptability of the albumin molecule. When a hydrophobic chain penetrates into the interior of the globular albumin molecule the helices of the protein separate, producing a small change in the tertiary structure of the protein.

What are the binding sites? Organic anions are thought to bind to a site containing the amino acid sequence Lys-Ala-Try-Ala-Val-Ala-Arg. The five non-polar side chains of these residues form a hydrophobic 'pool'. A cationic group is present at each end. It has been proposed that the ε-amino group of lysine is the point of attachment for anionic drugs. Albumin is, however, capable of binding molecules, such as cortisone or chloramphenicol, which are not ionised at all. Acidic drugs such as phenylbutazone, indomethacin and the salicylates may well establish primary contact with the albumin by electrostatic interactions but the bond may be strengthened by hydrophobic interactions. The dual interaction would therefore limit the number of possible binding sites for anionic organic drugs. In spite of electrostatic interactions a number of drugs bind to albumin according to their degree of lipophilicity. The importance of the hydrophobic interaction has received further support from studies of the binding of phenol derivatives, which depends mainly on the hydrophobic character of the substituent, the phenolic hydroxy group playing no significant role in the binding process. Similarly, penicillin derivatives and phenothiazines bind in a manner which depends on the hydrophobic characteristics of side-chains. So it has been concluded that binding to albumin is a process comparable to partitioning of drug molecules from a water phase to a non-polar phase. The hydrophobic sites are not necessarily 'preformed'. Fatty acids and warfarin are both capable of inducing conformational changes which result in the formation of hydrophobic 'pools' in the protein.

Plasma proteins other than albumin may also be involved in binding; examples of such interactions are shown in figure 10.21. Blood plasma normally contains on average about 6.72 g of protein per 100 mℓ, the protein comprising 4.0 g of albumin, 2.3 g of globulins and 0.24 g of fibrinogen. Drugs are normally bound to albumin by non-specific, reversible processes. Albumin is the main binding protein although dicoumarol is also bound to β- and γ-globulins, and a number of steroid hormones are specifically and preferentially bound to particular globulin fractions.

Protein binding can be considered to be an adsorption process obeying the law of mass action. If D represents drug and P the protein we can write

$$D + P \rightleftharpoons (DP)$$ (10.8)
(protein–drug complex)

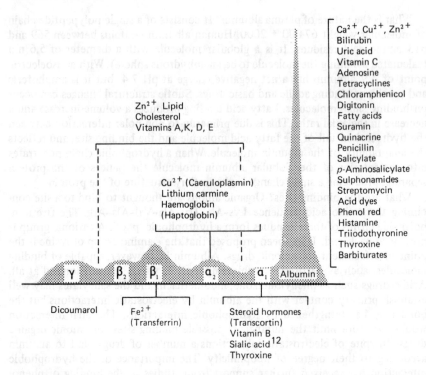

figure 10.21 Interaction of drugs and chemicals with plasma proteins. Plasma proteins are depicted according to their relative amounts. Adapted from F. W. Putnam. *The Proteins*, vol. 3 (ed. H. Neurath), 2nd edition, Academic Press, New York, 1965

At equilibrium

$$D_f + (P_t - D_b) \rightleftharpoons D_b \qquad (10.9)$$

where D_f is the molar concentration of unbound drug, P_t is the total molar concentration of protein, and D_b is the molar concentration of bound drug (= molar concentration of complex).

If one assumes one binding site per molecule the equilibrium constant is given by

$$K = \frac{k_1}{k_{-1}} = \frac{\text{rate constant for association}}{\text{rate constant for dissociation}} \qquad (10.10)$$

From equation 10.9

$$K = \frac{D_b}{D_f(P_t - D_b)} \qquad (10.11)$$

It is obvious that $k_1 > k_{-1}$. The rate constant for dissociation, k_{-1}, is the rate-limiting step in the exchange of drug between free and bound forms. From equation 10.11 we obtain

$$KD_f P_t - KD_f D_b = D_b \qquad (10.12)$$

That is
$$KD_f P_t = D_b + KD_f D_b = D_b(1 + KD_f) \qquad (10.13)$$

Therefore
$$\frac{KD_f P_t}{1 + KD_f} = D_b \qquad (10.14)$$

or
$$\frac{D_b}{P_t} = r = \frac{KD_f}{1 + KD_f} \qquad (10.15)$$

where r is the number of moles of drug bound to total protein in the system. If there are not one, but n, binding sites per protein molecule

$$r = \frac{nKD_f}{1 + KD_f} \qquad (10.16)$$

or
$$\boxed{\frac{1}{r} = \frac{1}{n} + \frac{1}{nKD_f}} \qquad (10.17)$$

Protein binding results are often quoted as the fraction of drug bound (β). This fraction varies generally with the concentration of both drug and protein as is shown in the equation relating β with n, K and concentration

$$\beta = \frac{1}{1 + D_f/(nP_t) + 1/(nKP_t)} \qquad (10.18)$$

Estimation of the thermodynamic parameters of binding allows one to interpret mechanisms of interaction. Table 10.9 gives the thermodynamic parameters of binding of a number of agents to bovine serum albumin. The negative sign of ΔG implies that binding is spontaneous. ΔH is negative, signifying an exothermic process and a reduction in the strength of the association as temperature increases. ΔS is positive, most likely signifying loss of structured water on binding. A diagrammatic representation of this latter process is shown in figure 10.22. This diagram shows the change in the ordered water in the hydrophobic cavity of the albumin and around the non-polar portion of the drug. The loss of the structured water gives rise to the positive entropy change, which contributes to a neagive free-energy change.

table 10.9 **Thermodynamic parameters of binding of drugs to bovine serum albumin***

Drug	Percentage bound	ΔG (kJ mol^{-1}) at 22 °C	ΔH (kJ mol^{-1})	ΔS (J mol^{-1} K^{-1})
Aminopyrine	24.7	−17.7	−5.19	+42.2
Antipyrine	21.4	−17.1	−4.98	+41.0
4-Aminoantipyrine	22.7	−17.4	−5.27	+41.0
Phenylbutazone	84.6	−28.9	−6.1	+77.4
Oxyphenylbutazone	80.3	−27.4	−5.9	+73.2

*From S. Ozeki and K. Tejima. *Chem. pharm. Bull.*, 22, 1297 (1974)

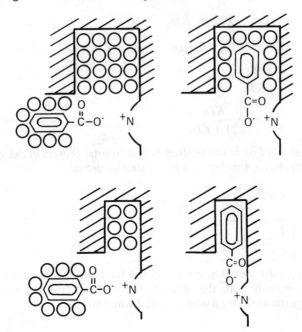

figure 10.22 Electrostatic contact and hydrophobic binding of a molecular model with anionic and hydrophobic properties. From A. Hasseblatt. *Proc. Eur. Soc. Study Drug Toxicity*, 13, 89 (1972)

Lipophilicity and protein binding

The extent of protein binding of many drugs has been shown to be a linear function of partition coefficient P or of log P. The binding of penicillins has been widely studied and a linear equation of the form log (percentage bound/percentage free) = 0.5 log P − 0.665 has been found to be applicable to serum binding of penicillins. Although there may be an electrostatic component to the interaction, the binding increases with the degree of lipophilicity suggesting, as is often the case, that more than one binding interaction is in force. Binding to protein outside of the plasma may determine the characteristics of drug action or transport. Muscle protein may bind drugs such as digoxin and so act as a depot. Concentrations of 1.2 ± 0.8, 11.3 ± 4.9 and 77.7 ± 43.3 ng mℓ^{-1} have been reported for digoxin in plasma, skeletal and cardiac muscle, respectively. Differences in the bioavailability of two antibiotics following intramuscular administration have been ascribed to differences in protein binding. Dicloxacillin, 95 per cent bound to protein, is absorbed more slowly from muscle than ampicillin, which is bound only to the extent of 20 per cent.

Drug binding often changes with drug concentration and with protein concentration. On increasing the drug/protein ratio, saturation of some sites can occur and there may be a decrease in binding (as indicated by total percentage). Hence the importance of determining binding at realistic albumin concentrations (about 40 g ℓ^{-1}). Figure 10.23 shows binding of phenylbutazone at different

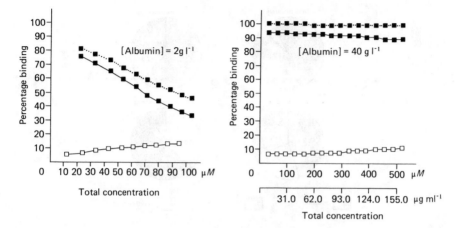

figure 10.23 Binding of phenylbutazone to human albumin at two concentrations, 2 g ℓ^{-1} and 40 g ℓ^{-1}. (■——■) Binding to type I sites; (□——□), binding to type II sites; (■. . .■), total binding. From J. P. Tillement, R. Zini, P. d'Athis and G. Vassent. *Eur. J. clin. Pharm.*, 7, 307 (1974)

sites on the albumin molecule (termed type I and type II sites) at concentrations of 2 and 40 g ℓ^{-1}. Total binding percentages are also shown.

In determining the pharmacological importance of protein binding, several factors have to be considered. If drug molecules not bound to plasma protein are freely distributed throughout the body on leaving the blood they enter a volume thirteen times as large as the plasma volume. Entry into the cerebrospinal fluid (CSF) depends on the concentration of free, diffusible drug in the plasma. Sulphanilamide enters CSF faster than sulphamethoxypyridazine does because less is bound to serum albumin. Such binding factors may override the intrinsic lipophilicities, which may suggest a different order of penetration to that observed in some cases.

When binding occurs with high affinity and the total amount of drug in the body is low, drug will be present almost exclusively in the plasma. Drugs with lower association constants ($K \sim 10^6$ or 10^7) will be distributed more in the body water spaces. When the number of available binding sites is reduced by a second drug, it will appear as if there has been an increase in over-all drug concentration. Although the greater proportion of most drugs is bound to albumin, the amount taken up by erythrocytes must not be neglected. The differences are shown diagrammatically in figure 10.24.

More directly, the effect of protein binding on antibiotic action is worth considering. Penicillins and cephalosporins bind reversibly to albumin. Only the free antibiotic has antibacterial activity. Oxacillin in serum at a concentration of 100 μg mℓ^{-1} exhibits an antibacterial effect similar to that of 10 μg mℓ^{-1} of the drug in water. A high degree of serum protein binding may nullify the apparent advantage of higher serum levels of some agents (table 10.10). An increase in the lipophilicity of penicillins results in decreased activity although one normally expects that this should enhance penetration of bacterial cell walls and enhance

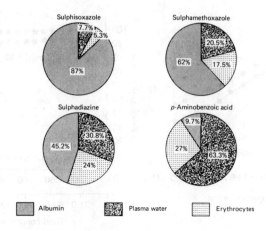

figure 10.24 Distribution of three sulphonamides and *p*-aminobenzoic acid between albumin, plasma water and erythrocytes in man. Log *P* values for octanol/water are sulphisoxazole 1.15, sulphamethoxazole 0.88, and sulphadiazine 0.13. Binding to albumin correlates with lipophilicity. From K. Berneis and W. Boguth. *Chemotherapy*, 22, 390 (1976)

table 10.10 **Protein binding and other characteristics of some penicillins and cephalosporins***

Antibiotic	log P**	Serum protein binding (per cent)	Serum concentration (μg mℓ^{-1}) during continuous infusion (500 mg h^{-1})		Peak serum levels after 500 mg kg^{-1}	
			Total	Free	Total	Free
Dicloxacillin	3.24	96–98	25	1	15	0.3
Cloxacillin	2.49	94–96	15	0.9	—	—
Oxacillin	2.38	92–94	10	0.8	15	0.8
Nafcillin	—	89–90	9	0.9	6	0.6
Penicillin G	1.72	59–65	16	5.6	4.5	1.6
Methicillin	1.06	37–60			16	10
Ampicillin	—	20–29	29	23.2	12	9
Carbenicillin	—	50	73	36.5		
Cefazolin†		89				
Cephalothin		65	18	6.3	7.6	2.6
Cephaloridine		20	24.7	19.7	18.5	14.8
Cephalexin		15	27	23		
Cephradine†		14				

*From reference 45
†Additional values from S. M. Singhri *et al. Chemotherapy*, 24, 121 (1978)
**log *P* values from A. Ryrfeldt. *J. Pharm. Pharmacol.*, 23, 463 (1971)

adsorption. However, the hydrophobic binding of penicillins to serum proteins reduces their potency *in vivo* by decreasing their effective concentration. Thus, comparisons between antibiotics are best made with activity-time plots and not serum concentration profiles, as the free levels often differ from the total antibiotic levels (see, for example, cloxacillin and penicillin G in table 10.10).

The second important consequence of protein binding is related to the fact that only free drug is able to cross the pores of the capillary endothelium. At equilibrium, levels of free drug on both sides of the capillary wall are equal. Albumin levels in most sites are considerably less than those in serum so there is little bound drug in extravascular regions. There is, however, no correlation between the degree of protein binding and peak serum levels of the penicillins and cephalosporins of table 10.10; other factors such as rates of elimination also determine the peak levels. However, protein binding will affect transport into other tissues as the gradient that determines movement is the gradient caused by free drug. Both ampicillin (50 mg kg^{-1} every 2 h) and oxacillin (50 mg kg^{-1} h^{-1}) produced similar peak levels in the serum given as repeated intravenous boluses. Levels of free drug were markedly different, however, as oxacillin is 75 per cent bound and ampicillin is 17.5 per cent bound in rabbit serum.

Penetration of specialised sites

Muscle, bone, synovial and interstitial fluid are readily accessible to intravascular antibiotics by way of aqueous pores in the capillary supply. The CSF, brain, eye, intestines and prostate lack such pores and entry into these areas is by way of a lipoidal barrier. There is evidence that extensive protein binding may prevent the access of antibiotics into the eye and the inflamed meninges.

It has been demonstrated that the free drug in serum is important in determining the amount of drug which reaches tissue spaces[46]. The total concentration of a drug in tissue fluid can be predicted from the serum concentration, the extent of serum protein binding, and the protein binding in the tissue fluid. The following formula has been shown to be highly predictive under equilibrium conditions[46]

$$C_t = \frac{C_s f_s}{f_t} \tag{10.19}$$

where C_t is the tissue fluid drug concentration, C_s the serum drug concentration, and f_s and f_t the free fractions of drug in serum and tissue fluid, respectively. If protein binding in the serum and tissue fluid is identical then, provided the same proteins are present in both 'phases', the tissue fluid drug concentration will equal the serum concentration; this is true for both high and low protein-containing extravascular fluid[46].

Where the specialised compartment in the body is separated from plasma by a lipoprotein membrane it is vital when attempting to medicate that zone to achieve therapeutic concentrations of drug, but not to achieve toxic levels. An example is in the treatment of bacterial prostatitis; most currently available antibiotics do not readily pass across the prostatic epithelium. The pH of the prostatic fluid is lower than that of plasma, being around 6.6. Milk has a pH of 6.8; hence, in passage of drugs into both prostatic fluid and into milk, the degree of dissociation of the drug appears to be the most important factor determining the degree of penetration[45]. The partition coefficient plays a secondary role. When the prostatic fluid/plasma ratios of sulphonamides were studied it was found[45] that the results could be predicted more closely from the ratio of undissociated/ dissociated drug rather than from log P values. It is probable that this is because lipophilic drugs will be highly protein bound and thus unable to penetrate the boundary membranes. The same conclusion has been reached over diffusion of drugs into milk of nursing mothers. The log P_0 (see chapter 9) for maximum diffusion into milk is lower than that for maximum gastro-intestinal, buccal or percutaneous absorption, because the drugs with high log P values, although liposoluble, are bound to protein. Therefore the plot of milk levels versus log P peaks at a lower log P value – in fact 0.53 (see table 9.1).

Appendix – Drug interactions based on physical mechanisms

The following information has been taken from a comprehensive listing of drug interactions published in New Zealand (R. Ferguson. *Drug Interactions of Clinical Significance*, I.M.S., New Zealand, 1977). The compiler, Dr Ruth I. Ferguson MSc, MPS, was formerly at the University of Strathclyde. **This selection is intended for illustrative purposes only and must *not* be used as a definitive source.**

A: Interactions affecting absorption of drugs

Drugs	Effect	Mechanism/Note	Recommendation
Atropine + Magnesium trisilicate	With concurrent administration up to 50 per cent of atropine is un-available for absorption	Surface adsorption effect	Space doses by 2–3 hours
Belladonna alkaloids + Magnesium oxide, Bismuth subnitrate	Loss of up to 90 per cent activity of belladonna	Surface adsorption with prevention of absorption into body	Space doses by 2–3 hours or use alternative ant-acids
Cloxacillin, Erythromycin, Penicillin, Tetracycline + Food	Reduced blood levels of oral antibiotics. May be below minimum effective concentration	Reduced absorption	Give antibiotic dose one hour before food
Calcium therapy + Aluminium hydroxide	Chronic aluminium hydroxide antacid intake (600 mℓ weekly) can cause osteomalacia	Calcium absorption is prevented	Use alternative antacid. Especial care should be taken with people with calcium deficiency

Drugs	Effect	Mechanism/Note	Recommendation
Chlorpromazine + Aluminium hydroxide, Magnesium oxide, Magnesium trisilicate	Reduced blood levels of chlor-promazine, with possible inhibition of therapeutic effect	Possibly due to alteration of gastro-intestinal pH by antacid	Doses should be spaced by 2–3 hours
Dicoumarol, Warfarin + Cholestyramine	Anticoagulant absorption is reduced and slowed (25 per cent lower blood levels)	Binding effect	Space doses by 2–3 hours
Dicoumarol, Warfarin + Laxatives	Chronic laxative administration may cause loss of anticoagulant control	Two possible effects: lubricant oil action may cause decreased absorption of anticoagulant; decreased absorption of vitamin K, particularly with oils and emulsions	Periodic long-term monitoring of pro-thrombin times in all suspect patients
Dicoumarol + Magnesium hydroxide, Magnesium oxide	Opposite effects have been reported: (i) Decreased absorption and lower blood levels (up to 75 per cent) of dicoumarol (ii) Increased absorption	Possibly adsorption phenomenon Chelation (see section 10.5)	Space doses by 2–3 hours
Digitalis glycosides + Dioctyl sodium sulphosuccinate	Increased absorption of digitalis to a possibly toxic level	Surface active properties of DOSS increase solubility and absorption of poorly absorbed digitalis	Space doses by 2–3 hours
Digoxin, Digitoxin + Cholestyramine	Greatly reduced absorption of digitalis alkaloids with concurrent administration	Binding effect (anionic resin)	Space doses by about 8 hours
Digoxin + Disodium edetate	Disodium edetate chelates calcium ions so antagonis-ing the action of digoxin	Chelation of digoxin in the plasma	May be necessary to monitor blood calcium levels
Digoxin + Magnesium trisilicate	Absorption of digoxin reduced by up to 90 per cent	Surface adsorption effect	Space doses 2–3 hours. Use alter-native antacid
Griseofulvin + Food	The absorption of griseofulvin is determined by the fat content in the food	Griseofulvin is a fat-soluble antibiotic absorbed with oil	Combination useful

Drugs	Effect	Mechanism/Note	Recommendation
Lincomycin + Kaolin	Reduced gastro-intestinal absorption of lincomycin by up to 90 per cent	Adsorption on to kaolin	Space doses by 2–3 hours
Oxyphenonium bromide + Magnesium trisilicate	Reduced blood levels of oxyphenonium with reduced effect	Oxyphenonium adsorbed onto the surface of magnesium trisilicate and so not available for absorption into the body	Space doses by at least 2–3 hours
p-Aminosalicylic acid (granule formulation) + Rifampicin	Reduced blood levels of rifampicin and hence impaired effect	Impaired gastro-intestinal absorption of rifampicin due to binding to bentonite	Space doses by 8–12 hours
Penicillin + Antacids, Food	Reduced blood levels of penicillin. May be below minimum effective concentration	Alteration of gastric pH and dilution effect	Space doses. Give penicillin before food
Propantheline bromide + Magnesium trisilicate	Reduced blood levels (with reduced action) of pro-pantheline	Adsorption of propantheline onto magnesium trisilicate, not available for absorption into the body	Space doses by 2–3 hours
Salicylates + Cholestyramine	Greatly decreased absorption of salicylates	Adsorption onto ionic exchange resin	Space doses by 4–5 hours
Tetracycline + Iron preparations	Lower blood levels of tetracycline, possibly below minimal active concentration	Chelation effect	Either space doses at least 2–3 hours or use parenteral or alternative antibiotic therapy
Tetracycline + Milk products, Aluminium, Bismuth, Calcium and Magnesium ions	Gastro-intestinal absorption of tetracycline and blood levels of tetracycline greatly reduced	Chelation effect	Space doses by 2–3 hours; possible exception mino-cycline
Tetracycline + Sodium bicarbonate	Impaired absorption of tetracycline	pH increase in gastro-intestinal tract prevents dissolution of complete dose	Space doses by 2–3 hours. (Sodium bicarbonate pre-sent in many antacid and effervescent preparations)

Drugs	Effect	Mechanism/Note	Recommendation
Thiazide diuretics + Cholestyramine	Decreased absorption and effect of diuretic	Cholestyramine can bind	Space doses 2–3 hours
Thyroid hormones + Cholestyramine	Absorption of thyroid hormone prevented	Thyroid hormones become bound to anionic resin	Space doses 2–3 hours

B. Interactions involving protein binding

Drugs	Effect	Mechanism/Note	Recommendation
Chlorpropamide, Tolbutamide + Sulphonamides	Increased blood level of hypo-glycaemic drug. Hypoglycaemic shock reported	Displacement from protein binding and possible metabolic inhibition, competition for excretion	Not reported with sulphadimethoxine, sulphafurazole, sulphamethoxazole. Monitor patient
Chlorpropamide, Tolbutamide + Salicylates	Introduction of oral salicylate therapy can result in a hypoglycaemic reaction	Displacement from protein binding, salicylates increase tissue glucose uptake	Warn patients against high dose or chronic salicylate therapy. With long-term salicylate therapy patients should be monitored and restabilised if necessary
Chlorpropamide + Dicoumarol	Prolongation of chlorpropamide half-life. Hypo-glycaemia. Also possible enhanced anticoagulant effect	Possible displacement from protein binding	Patients should be monitored — especially anti-coagulant response with initiation and cessation of concurrent therapy
Dicoumarol + Chloral hydrate	Immediate transient rise in warfarin blood level. Long-term decrease in warfarin effect	Transient – displacement of warfarin from protein binding by metabolite trichloroacetic acid. Long-term – metabolic induction	Owing to complexity of interaction, combination best avoided unless patient well monitored
Dicoumarol, Warfarin + Indomethacin	Possible elevated blood levels of anti-coagulant with risk of haemorrhage. Risk of gastro-intestinal bleeding	Displacement from protein binding	This combination best avoided

Drugs	Effect	Mechanism/Note	Recommendation
Dicoumarol, Warfarin + Methandrostenolone, Norethandrolone, Oxymetholone	Increased hypoprothrombinaemia with several reported cases of haemorrhage	Possibly due to displacement from protein binding	If combination is definitely indicated (on C_{17} alkylated anabolic steroids) patients should be carefully monitored
Dicoumarol, Warfarin + p-Aminosalicylic acid	Increased anti-coagulant serum levels with risk of haemorrhage	Displacement from protein binding	Monitor anti-coagulant effect especially with initiation or cessation of therapy
Dicoumarol, Warfarin + Paracetamol	With especially large doses increase in anti-coagulant effect	Possibly displacement from protein binding. Depression of clotting factor synthesis	Use low doses of paracetamol. Probably the safest mild analgesic to use with oral anticoagulants
Dicoumarol, Warfarin + Sulphadimethoxine, Sulphamethoxypyridazine, Sulphaphenazole	Increased anti-coagulant blood levels with risk of haemorrhage	Displacement from protein binding by strongly bound sulphonamide	Use alternative chemotherapeutic therapy. Monitor patient with initiation and cessation of concurrent therapy
Dicoumarol + Tolbutamide	Dependent on which drug is introduced to the other. Tolbutamide can displace dicoumarol from binding causing hypoprothrombinaemia. Dicoumarol can displace tolbutamide causing hypoglycaemia	Displacement from protein binding, and inhibition of metabolism	This type of combination requires careful monitoring and control
Methotrexate + Salicylates	Potentiation of methotrexate toxicity	Displacement of methotrexate from protein binding sites	Owing to severity of methotrexate toxicity, salicylates should not be given to patients receiving methotrexate
Methotrexate + Sulphonamides	Potentiation of methotrexate with high risk of toxicity	Displacement from protein binding (sulphafurazole reduces methotrexate binding from 70 to 28 per cent)	Owing to severity of methotrexate toxicity, sulphonamides that are protein bound should never be administered

Drugs	Effect	Mechanism/Note	Recommendation
Penicillin + Probenecid	Three-fold increase in serum penicillin level with increase also in spinal fluid levels but decreased brain level	Displacement from protein binding, decreased renal tubule secretion and decreased biliary excretion	Combination can be used when high serum and spinal fluid levels required
Quinine + Pyrimethamine	Quinine may reach toxic levels especially in sensitive patients	Displacement from protein binding sites	When commencing concurrent administration, a dose reduction of quinine should be considered
Thyroid hormones, Thyroxine + Phenytoin	Rise in thyroxine blood levels with reported tachycardia	Displacement of thyroxine by phenytoin from protein binding	Great care should be taken in administering parenteral phenytoin to people on thyroid therapy
Warfarin + Aspirin	Significant increase in anticoagulant effect with possible severe gastro-intestinal blood loss	Displacement from protein binding	Combination should be avoided if possible. Use alternative analgesic such as paracetamol or pentazocine
Warfarin + Chloral hydrate	Elevated warfarin serum levels with potentiation of hypoprothrombin-aemia with risk of haemorrhage	(proposed) Displacement of active metabolite, trichloroacetic acid, from protein binding	Substitute chloral hydrate with another hypnotic such as nitrazepam. Carefully watch patient and monitor
Warfarin + Clofibrate	Higher blood levels of warfarin with risk of haemorrhage	Displacement from protein binding	Warfarin dose may require a reduction of one-third to one-half according to coagulation test
Warfarin + Diazoxide	Rise in anti-coagulant blood levels with resultant haemorrhage risk	Displacement from protein binding	Monitor patient
Warfarin + Ethacrynic acid	Increased blood level of warfarin with haemorrhage risk	Displacement from protein binding	Monitor patient with initiation and cessation of therapy
Warfarin + Ethchlorvynol	Enhanced anti-coagulant effects. Haemorrhage reported	Metabolised to trichloroacetic acid which can displace warfarin from protein binding	Patients should be monitored

Drugs	Effect	Mechanism/Note	Recommendation
Warfarin + Mefenamic acid	Enhanced anti-coagulant blood levels with risk of haemorrhage and enhanced risk of gastro-intestinal bleeding	Displacement from protein binding	Concurrent administration is best avoided. If necessary coagulation times should be monitored
Warfarin + Nalidixic acid	Possible hypo-thrombinaemia with associated haemorrhage risk	Displacement from protein binding (N.B. *in vitro* evidence)	Combination administration should be used with care and patient coagulation monitored
Oral anticoagulants + Naproxen	Increased anti-coagulant effect and risk of gastro-intestinal bleeding	Naproxen has ulcerogenic potential. Displacement from protein binding	Best to avoid concurrent administration. Monitor patient
Warfarin + Phenylbutazone, Oxyphenbutazone	Potentiation of hypothrombinaemia with gastro-intestinal bleeding and haemorrhage. Reported fatalities	Displacement from protein binding	This combination should be avoided
Warfarin + Triclofos	Increased blood level of warfarin with potentiation of prothrombinaemia	Possibly due to displacement of warfarin from protein binding by active metabolite trichloroethanol	Patients should be monitored on initiation and cessation of concurrent administration

Reminder: Do *not* use this appendix as an authoritative source of drug interaction data. It is included here to illustrate the range of problems that may arise from interactions *in vivo* which can be interpreted as physicochemical. The references from the original list have been omitted

References

1. M. Edward. *Am. J. hosp. Pharm.*, 24, 440 (1967)
2. C. L. J. Coles and K. A. Lees. *Pharm. J.*, 206, 153 (1971)
3. B. Lynn. *ibid.*, 206, 154 (1971)
4. K. A. Fitzgerald and M. W. MacKay. *Am. J. hosp. Pharm.*, 43, 88 (1986)
5. S. J. Meldrum, *et al. Br. med. J.*, 2, 104 (1972)
6. A. J. Fairfax, J. Adam and F. S. Pagan. *Br. med. J.*, 1, 820 (1978)
7. G. Levy and B. A. Hayes. *New. Engl. J. Med.*, 262, 1053 (1960)
8. J. H. Lipton. *New Engl. J. Med.*, 268, 92 (1963)
9. W. Jusko, M. Gretch and R. Gassett. *J. Am. med. Ass.*, 225, 176 (1973)
10. D. E. Langdon, *et al. J. Am. med. Ass.*, 223, 184 (1973)
11. S. H. Yalkowsky and S. G. Valvani. *Drug Intell. clin. Pharm.*, 11, 417 (1977)

12. J. B. Crapper. *Br. med. J.*, 1, 33 (1975)
13. E. Tomlinson. *Pharm. Int.*, 1, 156 (1980)
14. G. M. Irvin, H. B. Kostenbauder, *et al. J. pharm. Sci.*, 58, 313 (1969)
15. T. R. Marshall, J. T. Ling, G. Follis and M. Russell. *Radiology*, 84, 536 (1965)
16. M A. El-Nakeeb. *Acta pharm. suec.*, 5, 1 (1968)
17. A. C. Barbara, C. Clemente and E. Wagman. *New Engl. J. Med.*, 274, 1316 (1966)
18. C. F. Bell. *Metal Chelation: Principles and Applications*, Calendar Press, Oxford, 1977
19. P. R. Clink and J. L. Colaizzi. *J. pharm. Sci.*, 62, 97 (1973)
20. J. T. Doluisio and A. N. Martin. *J. med. Chem.*, 6, 16 (1963)
21. A. E. Mahgoub, E. M. Khairy and A. Kasem. *J. pharm. Sci.*, 63, 1451 (1974)
22. W. Kohn. *Anal. Chem.*, 33, 862 (1962)
23. A. E. Martell and M. Calvin. *Chemistry of the Metal Chelate Compound*, Prentice-Hall, New York, 1952, p. 54
24. J. J. Ambre and L. T. Fischer. *Clin. Pharm. Ther.*, 14, 231 (1973)
25. M. A. Zoglio. *J. pharm. Sci.*, 58, 222 (1969)
26. K. J. Ikeda. *Chem. pharm. Bull.*, 23, 201 (1975)
27. K. J. Ikeda. *J. pharm. Sci.*, 63, 1168 (1974)
28. D. E. Tutt and M. A. Schwartz. *J. Am. chem. Soc.*, 93, 767 (1971)
29. L. Gross and M. Brotman. *Ann. intern. Med.*, 72, 95 (1970)
30. A. Coronato and G. R. J. Glass. *Proc. exp. Biol. Med.*, 142, 1341 (1973)
31. S. A. H. Khalil and M. A. Moustafa. *Pharmazie*, 28, 116 (1972)
32. G. Boman, P. Lundgren and G. Stjernstrom. *Eur. J. clin. Pharm.*, 8, 293 (1975)
33. T. J. McCarthy. *J. mond. Pharm.*, 12, 321 (1969)
34. E. G. Beveridge and I. A. Hope. *Pharm. J.*, 198, 457 (1967)
35. S. A. H. Khalil and R. N. Nasipuri. *J. Pharm. Pharmacol.*, 25, 138 (1973)
36. S. El-Masry and S. A. H. Khalil. *ibid.*, 26, 243 (1974)
37. J. Dony and J. Conter. *J. Pharm. belg.*, 11, 338 (1956)
38. C. Trolle-Lassen. *Arch. pharm. Chem.*, 67, 504 (1960)
39. I. Moriguchi and N. Kaneniwa. *Chem. pharm. Bull.*, 17, 394 (1969)
40. S. T. Anik and J.-Y. Hwang. *Int. J. Pharmaceut.*, 16, 181 (1983)
41. P. F. D'Arcy. *Drug Intell. clin. Pharm.*, 17, 726 (1983)
42. A. W. Malik, A. H. Amman, D. M. Baske and R. G. Stohl. *J. pharm. Sci.*, 70, 798 (1981)
43. L. Illum and H. Bundgaard. *Int. J. Pharmaceut.*, 10, 339 (1982)
44. N. R. Anderson and J. J. Motzo. *J. parenteral Sci. Tech.*, 36, 161 (1982)
45. E. J. Lien, J. Kuwahara and R. T. Koda. *Drug Intell. clin. Pharm.*, 8, 470 (1974)
46. L. R. Peterson and D. N. Gerding. *Lancet*, ii, 376 (1978)

11 Assessment of Dosage Forms
in vitro

Previous chapters have considered the underlying theory of pharmaceutical systems. In this chapter we discuss the measurement of drug solution rate from dosage forms, some aspects of rheological properties, and adhesivity in the assessment of dosage forms. The objective of most *in vitro* tests is to ensure consistency of quality and performance of product (e.g. batch-to-batch) and, while not necessarily mimicking *in vivo* conditions, each test should be designed so as to have some relevance to *in vivo* situations. This is best illustrated by thinking of the dissolution testing of dosage forms in which the rate of drug release is determined in an appropriate medium. There will not necessarily be any correlation with *in vivo* results, but a well designed test with a drug whose dissolution *in vivo* is the rate-limiting step in the absorption process will produce data which can be correlated on a rank-order or linear fashion with absorption *in vivo*.

11.1 Dissolution testing of dosage forms

Equations cannot readily be applied to deal with the situation of tablet and capsule dissolution because of the constantly changing surface area of the disintegrated tablet components and because of the presence of additives which may alter solubility and wettability. However, the rate of solution of a drug substance in solid form from a granule or a tablet is dependent to a large extent on its solubility in the solvent phase and its concentration in that phase. The rationale of testing solid dosage forms *in vitro* will not be considered here. Suffice it to say that the test conditions should provide a reasonable challenge to the dosage form in terms of degree of agitation, temperature, volume and pH of the dissolution medium. *In vitro* tests provide the opportunity to make precise and reproducible release measurements to distinguish between different formulations of the same drug or the same formulation after ageing or processing changes. They do not replace the need for clinical work but the *in vitro* test can pinpoint formulation factors which are of importance in determining drug release.

In vitro methods may be divided into two types involving either natural convection or forced convection. Most practical methods fall into the latter category as there is a degree of agitation *in vivo*. Natural convection methods, for example, in which a pellet of material is suspended from a balance arm in the dissolution medium, create unnatural conditions. In forced convection methods a degree of agitation is introduced. These can, in turn, be divided into those that employ non-sink and those that achieve sink conditions in the dissolution medium. Figure 11.1 illustrates diagrammatically methods involving forced convection and non-sink conditions, non-sink because there is no mechanism for replenishing the solvent: for most drugs the concentration in solution will increase rapidly to approach C_∞. Most methods rely on the assay of the dissolution medium for drug content; the Coulter Counter method directly measures the change in particle size as dissolution (and disintegration of granules) occurs.

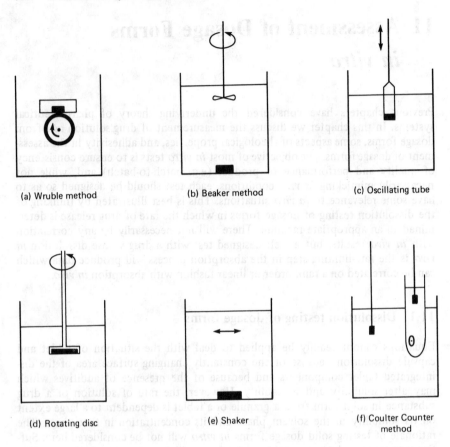

(a) Wruble method   (b) Beaker method   (c) Oscillating tube

(d) Rotating disc   (e) Shaker   (f) Coulter Counter method

figure 11.1 Diagrammatic representation of forced convection methods of following the dissolution of solid dosage forms. In forced convection agitation is achieved by the methods shown. Method (d) is useful only for non-disintegrating discs. Method (f) follows the changing particle size distribution following first on disintegration of the dosage form and then on dissolution of the particles

The USP and United States National Formulary method now widely used and adopted by the BP is a variant of the rotating basket method (figure 11.2). Tablets or capsules are placed in a basket of wire mesh, the mesh being small enough to retain broken pieces of tablet but large enough to allow entry of solvent without wetting problems. The basket may be rotated at any suitable speed but most USP monographs specify 50, 100 or 150 rpm. With drugs of very low solubility it is sometimes necessary to consider the use of *in vitro* tests which allow sink conditions to be maintained. This generally involves the use of a lipid phase into which the drug can partition, or it may involve dialysis or physical replacement of the solvent phase. In all the methods the appropriate pH for the dissolution medium must be chosen and a reasonable degree of agitation in those techniques which allow it. The considerable effect of rotation speed on the dissolution

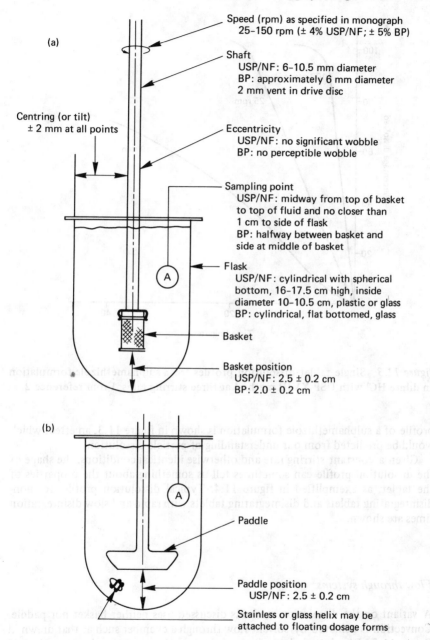

(a)

Speed (rpm) as specified in monograph
25–150 rpm (± 4% USP/NF; ± 5% BP)

Shaft
USP/NF: 6–10.5 mm diameter
BP: approximately 6 mm diameter
2 mm vent in drive disc

Centring (or tilt)
± 2 mm at all points

Eccentricity
USP/NF: no significant wobble
BP: no perceptible wobble

Sampling point
USP/NF: midway from top of basket
to top of fluid and no closer than
1 cm to side of flask
BP: halfway between basket and
side at middle of basket

Flask
USP/NF: cylindrical with spherical
bottom, 16–17.5 cm high, inside
diameter 10–10.5 cm, plastic or glass
BP: cylindrical, flat bottomed, glass

Basket

Basket position
USP/NF: 2.5 ± 0.2 cm
BP: 2.0 ± 0.2 cm

(b)

Paddle

Paddle position
USP/NF: 2.5 ± 0.2 cm

Stainless or glass helix may be
attached to floating dosage forms

figure 11.2 The rotating basket and rotating paddle versions of the official method for dissolution testing of solid oral dosage forms. (a) The rotating basket – method 1, USP/NF. This method is official for USP/NF and BP. Current specifications describing geometry and positions for each compendium are shown. (b) The rotating paddle – method 2, USP/NF. This method is official for USP/NF and is likely to be accepted in the European Pharmacopoeia. It is not official for BP. From reference 1 with permission

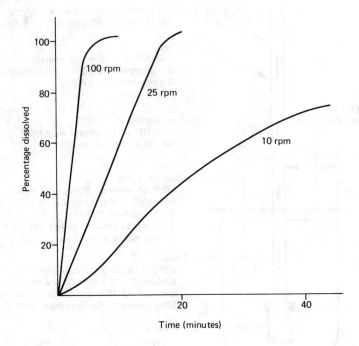

figure 11.3 Single tablet dissolution profiles for a sulphamethizole formulation in dilute HCl with USP XVIII method at three stirring rates. From reference 2

profile of a sulphamethizole formulation is shown in figure 11.3, an effect which would be predicted from our understanding of particle dissolution.

Given a constant stirring rate and otherwise identical conditions, the shape of the dissolution profile can sometimes tell us something about the properties of the tablet, as exemplified in figure 11.4. Typical dissolution profiles for non-disintegrating tablets and disintegrating tablets with rapid and slow disintegration times are shown.

Flow-through systems

A variant on the dissolution methods discussed uses neither basket nor paddle. Convection is achieved by solvent flow through a chamber such as that drawn in figure 11.5. Dissolution data obtained from such a system where continuous monitoring of drug concentration is achieved must be interpreted with care as the concentration–time profile will be dependent on the volume of solvent, its flow rate and the distance of the detection device from the flow cell, or rather the void volume of solvent. This emphasises that there is no absolute method of dissolution testing and that, with whatever form of test adopted, results are

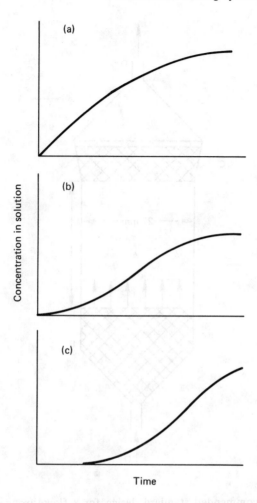

figure 11.4 Typical dissolution curve profiles observed with various tablet types. (a) *Non-disintegrating tablet:* surface area gradually reduced during test; dissolution rate determined by process of dissolution and diffusion. (b) *Disintegrating tablet:* rapid disintegration and deaggregation time; slow process of dissolution and diffusion; dissolution is rate-limiting. (c) *Disintegrating tablet:* slow disintegration and deaggregation time; rapid process of dissolution and diffusion; disintegration is rate-limiting. After reference 1

only useful on a comparative basis – batch vs batch, brand vs brand or formulation vs formulation.

The release of drugs from other dosage forms can be usefully examined. These include topical ointment and cream preparations, suppositories and transdermal devices.

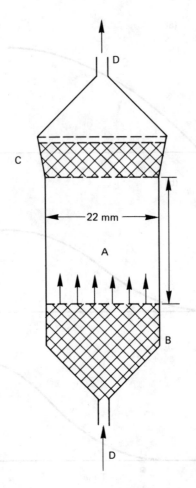

figure 11.5 A recommended standard design for a flow-through cell. The cell is cylindrical in shape and constructed of glass or other suitable material. A, an internal volume not exceeding 20 mℓ between barrier and filter; B, a bottom barrier of either a porous glass plate or a bed of 1 mm diameter glass beads designed to disperse flow and provide uniform distribution over the dosage chamber A; C, a suitable filter of approximately 25 mm diameter; D, fluid flow from bottom to top. From reference 1 with permission

11.2 *In vitro* evaluation of suppository formulations

Suppositories are probably more difficult to study *in vitro* than many other dosage forms because it is not easy to simulate the conditions in the rectal ampulla. The system shown in figure 11.6 has been used by Ritschel[3]. The suppository is placed in pH 7.8 buffer in a dialysis bag. This bag is placed in a

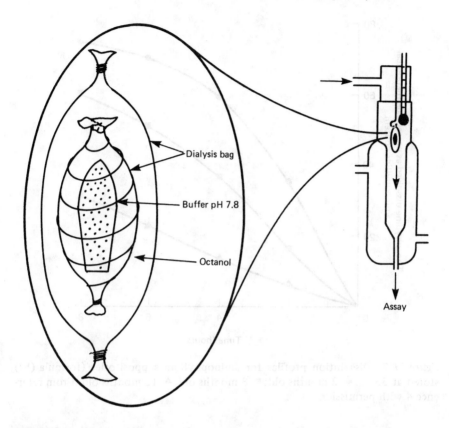

figure 11.6 An *in vitro* system for studying drug release from suppositories. After W. Ritschel, reference 3

second dialysis bag filled with octanol and the whole suspended in a flow system at 37 °C, as shown. Drug released into the outer liquid is monitored. The results from this set-up must be interpreted with care as it is possible for substances which complex with the drug to reduce transport across the dialysis membrane – something which may or may not happen with a biological membrane.

A variant on this arrangement was employed by Taylor and Simpkins[4] to obtain the data shown in figure 11.7 on the reduction in aminophylline release from Witepsol H15 bases on ageing.

11.3   *In vitro* release from transdermal systems

Using the NF XIII rotating bottle apparatus, release of nitroglycerin from Deponit transdermal patches gives the results shown in figure 11.8. The influence of temperature and pH on release was examined using the same technique, the results confirming that release is essentially determined by the diffusion of drug

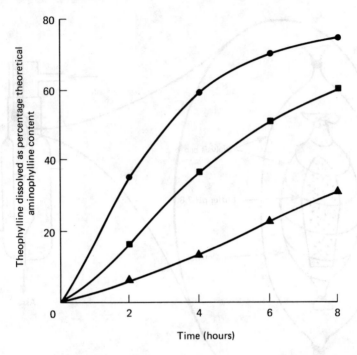

figure 11.7 Dissolution profiles for aminophylline suppositories (formula (b)) stored at 35 °C. ●, 2 months old; ■, 8 months old; ▲, 12 months old. From reference 4 with permission

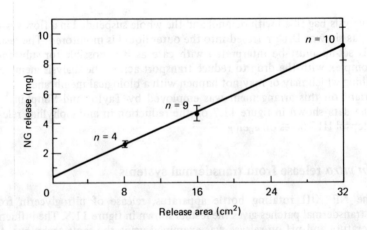

figure 11.8 *In vitro* nitroglycerin release (24 hours) into 80 mℓ isotonic NaCl versus release area of Deponit TTS. From reference 5 with permission

in the adhesive layer of the product (see chapter 8), but clearly being proportional to surface area, as expected.

Release increases as temperature increases – between 32 and 37 °C an increase in release of between 1 and 2 μg cm^{-2} h^{-1} is detected. Shah and colleagues[6] have carried out a comparative dissolution study of three brands of transdermal nitroglycerin using the paddle method (USP apparatus 2).

11.4 Rheological characteristics of products

The rheological behaviour of liquids and semisolids can be described by the terminology discussed in section 8.2.1. Viscosity can be used as a quality control procedure, but some very practical rheological tests may be carried out. Some are discussed here.

The injectability of non-aqueous injections, which are often viscous and difficult to inject, can be assessed by a test for syringeability[7]. Sesame oil has a viscosity of 56 cP, but added drugs and adjuvants can increase the viscosity of such vehicles. Figure 11.9 shows a syringeability testing apparatus comprising a

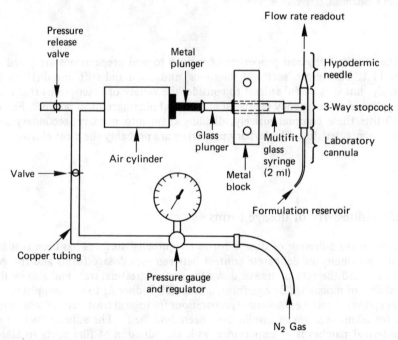

figure 11.9 A 'syringeability' testing apparatus for investigating the rheological properties of injectable formulations which influence their ease of injection. A glass syringe (2 mℓ) is fitted with a three-way stopcock connected to a hypodermic needle on one outlet and via a cannula to a reservoir containing formulation. A controlled pressure of nitrogen (10–50 kPa) is applied via a series of pressure regulators and a metal plunger to the glass plunger. The time (in seconds) to deliver 1.6 mℓ of a formulation through a given size of hypodermic needle is recorded. Flow rates in mℓ/s could thus be calculated. From reference 7

glass syringe in which can be applied a controlled pressure range from 10 to 15 kPa directly into the liquid formulation. The time taken to deliver 1.6 mℓ of the formulation is measured.

table 11.1   **Properties of some topical preparations***

Class	Nature	Required physical properties
1	Ophthalmic ointments	Softest type of ointment
2	Boric acid ointment, and other commonly used ointments	Soft and unctuous, but stiff enough to remain in place when applied
3	Protective ointments, e.g. zinc oxide paste	Hard and stiff, so that remains in place when applied to moist ulcerated areas

*From P. Sherman, reference 8

The required physical properties of various topical preparations are listed in table 11.1. The terms 'soft and unctuous' and 'hard and stiff' are difficult to quantify, but it is useful at least to consider the variety of descriptions that may be applied in a consistency profile of an external pharmaceutical product. Figure 11.10 lists these comprehensively, dividing them into primary, secondary and tertiary characteristics. The tertiary properties are probably the most elusive.

11.5   Adhesivity of dosage forms

Interest in the adhesion of dosage forms to epithelial surfaces has been aroused by the possibility of deliberate contact between oral dosage forms and the gut wall to retard the rate of transit down the gastro-intestinal tract and also by the possibility of moistened dosage forms accidentally adhering to the oesophagus or other epithelial surfaces. Adhesive preparations for topical treatment of stomatitis and for administration of insulin have been described[9]. The adhesive nature of transdermal patches is of importance, as is the adhesion of film coats to tablet surfaces. Adhesion of erythrocytes and bacterial cells to polymer surfaces is of increasing importance in the understanding, respectively, of blood compatibility of polymers and bacterial infection mediated by catheters.

Some examples of solid/solid adhesion processes are shown diagrammatically in figure 11.11.

Testing dosage forms for adhesivity has been carried out by several groups[10,11]. One method[10] involved the pulling up of a dosage form through an oesophagus

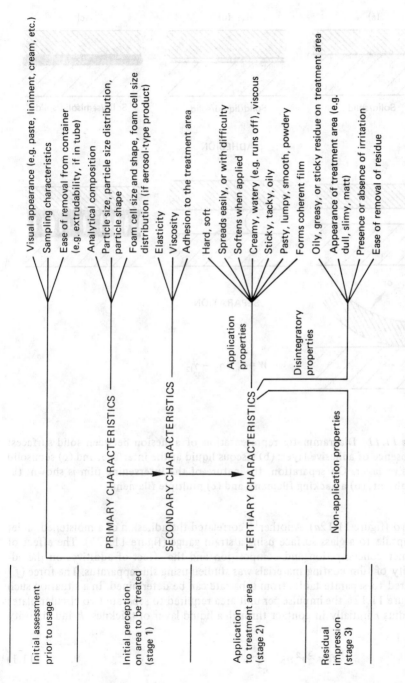

figure 11.10 Consistency profile for pharmaceuticals employed externally. From reference 8

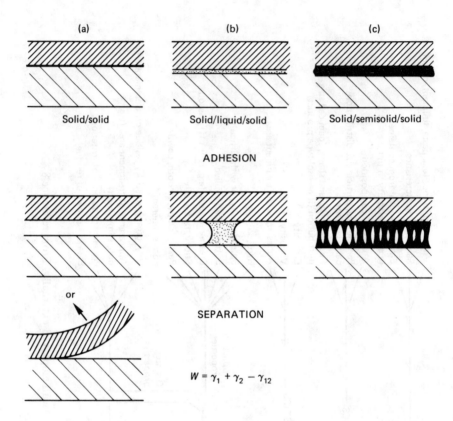

figure 11.11 Diagrammatic representation of adhesion between solid surfaces: (a) absence of adhesive layer; (b) viscous liquid at the interface; and (c) semisolid adhesive layer. On separation, the nature of the intervening film is shown: (a) film absent; (b) a necking filament; and (c) multiple filaments

ex vivo (figure 11.12*a*). Another[11] correlated the adhesion of a moistened tablet or capsule to a glass surface using a strain gauge (figure 11.12*b*). The effect of polymer concentration and composition and the effect of additives on the adhesivity of film coating materials was studied using this apparatus. The force (f) required to separate tablet from substrate can be determined. In a situation such as figure 11.12*b* the impulse per unit area required to separate two circular plates of radius r initially in contact through a liquid layer of thickness h and viscosity η is

$$ft = \tfrac{3}{4}r^2 \eta k' \, \frac{1}{h^2} \tag{11.1}$$

where k' is a constant. This equation applies only when the surfaces are pulled

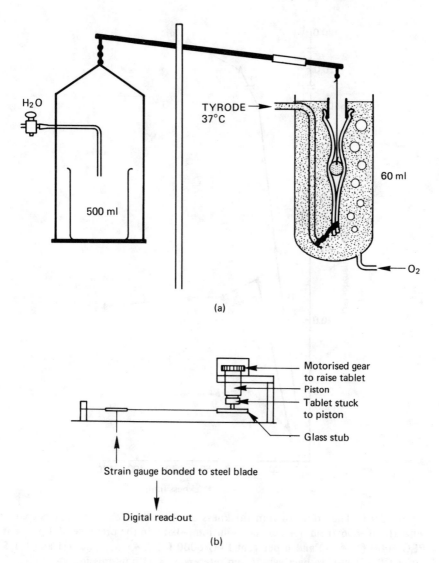

(a)

(b)

figure 11.12 (a) A system for measurement of the force required to detach a solid dosage form from the isolated porcine oesophagus (from reference 12). (b) Apparatus for measuring tablet adhesiveness in moist state to glass (or other surfaces). Digital readout calibrated against weights applied to a glass slide. Detachment force recorded when tablet separates from glass stub (from reference 11)

apart slowly. The tackiness of a system is not, however, simply related to viscosity. High molecular weight materials must be present in aqueous solutions at least to provide an elastic element to the viscous flow. Rubbery polymers which

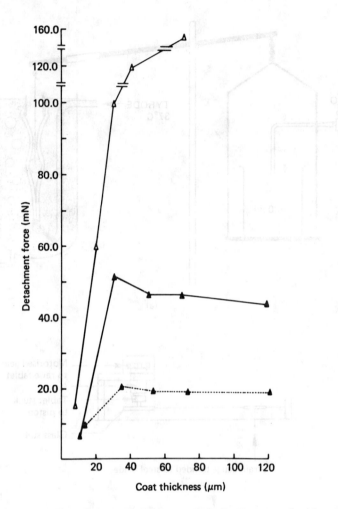

figure 11.13 The effect of film thickness on the adhesiveness of tablets coated
with HPMC-606 from a 4 per cent solution (△) and in the presence of 4 per cent
PEG 6000 (——▲——) and 6 per cent PEG 6000 (. . . .▲. . . .), i.e. 1:1 and 1:1.5
ratios HPMC: PEG respectively. From reference 13 with permission

have partly liquid and partly elastic characteristics are employed as adhesives:
surgical dressings, adhesive tapes, etc. What factors influence the adhesivity of
high molecular weight soluble polymers such as hydroxypropylmethylcellulose
is still unclear. HPMC is a component of film coating materials. Figure 11.13
shows the detachment force required to separate tablets from a glass surface
when coated with HPMC-606 from a 4 per cent solution, shown here as a function
of the thickness of the film coat. The effect of the presence of high concentrations of polyoxyethylene glycol 6000 is to reduce adhesivity, presumably
because the glycol solution is less tacky.

References

1. W. A. Hanson. *Handbook of Dissolution Testing*, Pharmaceutical Technology Publications, Springfield, 1982
2. G. Mattock and I. J. McGilveray. *J. pharm. Sci.*, 61, 746 (1972)
3. W. Ritschel. *Arzneim. Forsch.*, 23, 1031 (1973)
4. J. D. Taylor and D. E. Simpkins. *Pharm. J.*, 227, 601 (1981)
5. M. Wolff, G. Cordes and V. Luckow. *Pharm. Res.*, 1, 23 (1985)
6. V. P. Shah, N. W. Tymes, L. A. Yamamoto and J. P. Skelly. *Int. J. Pharmaceut.*, 32, 243 (1986)
7. Y. W. Chien, P. Przybyszewski and E. G. Shami. *J. parenteral Drug Ass.*, 35, 281 (1981)
8. P. Sherman. *Rheol. Acta*, 10, 121 (1971)
9. T. Nagai. *Medicinal Res. Rev.*, 6, 227 (1986)
10. M. Marvola, K. Vahervoo, A. Sothman, *et al. J. pharm. Sci.*, 71, 975 (1982)
11. H. Al-Dujaili, A. T. Florence and E. G. Salole. *Int. J. Pharmaceut.*, 34, 67 (1986)
12. M. Marvola *et al. J. pharm. Sci.*, 72, 1034 (1983)
13. H. Al-Dujaili, A. T. Florence and E. G. Salole. *Int. J. Pharmaceut.*, 34, 75 (1986)

Index*

Absorption 335 *et seq.*
  of acidic and basic drugs 340, 341t
  and antacids 411
  buccal 354–357
  from emulsions 251
  from the eye 380
  from the gastrointestinal tract
    341, 348 *et seq.*
  from injections 357 *et seq.*
  intranasal 397 *et seq.*
  $\log P$ and 338
  percutaneous 370
  rectal 399 *et seq.*
  vaginal 386
Absorption bases 370
Acetanilide
  absorption from emulsions 253
  derivatives, water solubility of 135
Acetylsalicylic acid
  adsorption on to charcoal 198
  binding of 431
  decomposition in solid state 102,
    117, 118
  decomposition in suppositories 119
  hydrolysis in suspensions 90
  ionisation in solution 64
  transacetylation of 117
Acid
  absorption of 341t
  conjugate 65
  defined 64
  polyprotic 72, 73
  weak, dissociation of 64, 65
Acid–base catalysis
  general 81, 110
  specific 81, 107–109, 114
Acid–base pairs 64
Acidic drugs, solubility of 141, 142
Acidity constant, *see* Dissociation
  constant
Activated charcoal 198
Activation, energy of 105, 367
Activity (thermodynamic)
  and biological activity 162
  and concentration 48
  defined 48

  of ionised drugs 49
  mean ionic 49, 51
  and osmotic pressure 58
  of solvents 52, 53
  in topical preparations 372, 373
  of water 52, 53
Activity coefficient
  defined 48
  mean ionic 50, 51
  solvent and 373
Adhesion
  of particles 266
  between solids 466
  and wetting 268
Adhesivity, of dosage forms 464 *et seq.*
Adsorption
  by antacids 197, 433, 434
  by clays 196, 434
  of drugs 432 *et seq.*, 446 *et seq.*
  factors affecting 194–197
  of insulin 298
  isotherms 190–195
  of macromolecules 297
  medical and pharmaceutical appli-
    cations of 197
  physical and chemical 190
  on plastics 199
  of preservatives 169
  of proteins and peptides 435
  at solid–liquid interface 189 *et seq.*
Aerosil, *see under* Silica
Aerosols 229, 230
  deposition in respiratory tract
    388 *et seq.*
  formulation of 10 *et seq.*
  generation of 395
  particle size distribution of 396,
    397
  propellants 3, 13t
  in vitro assessment of 396, 397
Albumin 439
  adsorption of 289
  binding to 438, 439
Alginates 304, 305
Aluminium hydroxide, adsorption by
  197, 434

*A 't' following a number refers to a table.